Orthogonal Transforms for Digital Signal Processing

N. Ahmed · K. R. Rao

Orthogonal Transforms for Digital Signal Processing

Springer-Verlag
Berlin · Heidelberg · New York 1975

Nasir Ahmed
Associate Professor
Department of Electrical Engineering,
Kansas State University, Manhattan, Kansas

Kamisetty Ramamohan Rao
Professor
Department of Electrical Engineering,
University of Texas at Arlington, Arlington, Texas

With 129 Figures

ISBN 3-540-06556-3 Springer-Verlag Berlin · Heidelberg · New York
ISBN 0-387-06556-3 Springer-Verlag New York · Heidelberg · Berlin

Printed in GDR.

Library of Congress Catalog Card Number 73-18912.

To the memory of
my mother and grandmother

N. Ahmed

Preface

This book is intended for those wishing to acquire a working knowledge of orthogonal transforms in the area of digital signal processing. The authors hope that their introduction will enhance the opportunities for interdisciplinary work in this field.

The book consists of ten chapters. The first seven chapters are devoted to the study of the background, motivation and development of orthogonal transforms, the prerequisites for which are a basic knowledge of Fourier series transform (e.g., via a course in differential equations) and matrix algebra. The last three chapters are relatively specialized in that they are directed toward certain applications of orthogonal transforms in digital signal processing. As such, a knowlegde of discrete probability theory is an essential additional prerequisite. A basic knowledge of communication theory would be helpful, although not essential.

Much of the material presented here has evolved from graduate level courses offered by the Departments of Electrical Engineering at Kansas State University and the University of Texas at Arlington, during the past five years. With advanced graduate students, all the material was covered in one semester. In the case of first year graduate students, the material in the first seven chapters was covered in one semester. This was followed by a problems project-oriented course directed toward specific applications, using the material in the last three chapters as a basis.

N. Ahmed · K. R. Rao

Acknowledgements

Since this book has evolved from course material, the authors are grateful for the interest and cooperation of their graduate students, and in particular to Mr. T. Natarajan, Department of Electrical Engineering, Kansas State University. The authors thank the Departments of Electrical Engineering at Kansas State University and the University of Texas at Arlington for their encouragement. In this connection, the authors are grateful to Dr. W. W. Koepsel, Chairman of the Department of Electrical Engineering, Kansas State University. Special thanks are due to Dr. A. E. Salis, Dean, and Dr. R. L. Tucker, Associate Dean, College of Engineering, and Dr. F. L. Cash, Chairman, Department of Electrical Engineering, University of Texas at Arlington, for providing financial, technical, and secretarial support throughout the development of this book. Thanks are also due to Ms. Dorothy Bridges, Ann Woolf, Marsha Pierce, Linda Dusch, Eugenie Joe, Eva Hooper, Dana Kays, Kay Morrison, and Sharon Malden for typing various portions of the manuscript.

Finally, personal notes of gratitude go to our wives, Esther and Karuna, without whose encouragement, patience, and understanding this book could not have been written.

N. Ahmed · K. R. Rao

Contents

Chapter One

Introduction

Chapter Two

Fourier Representation of Signals

Chapter Three

Fourier Representation of Sequences

Chapter Four

Fast Fourier Transform

Chapter Five

A Class of Orthogonal Functions

Chapter Six

Walsh-Hadamard Transform

Chapter Seven

Miscellaneous Orthogonal Transforms

Chapter Eight

Generalized Wiener Filtering

Chapter Nine

Data Compression

Chapter Ten

Feature Selection in Pattern Recognition

Chapter One

Introduction

1.1 General Remarks

In recent years there has been a growing interest regarding the study of orthogonal transforms in the area of digital signal processing [1–14, 40]. This is primarily due to the impact of high speed digital computers and the rapid advances in digital technology and consequent development of special purpose digital processors [4, 15–17]. Research efforts and related applications of such transforms include image processing [18–27], speech processing [23, 28, 29], feature selection in pattern recognition [23, 30–32], analysis and design of communication systems [23, 33, 34], generalized Wiener filtering [35, 36], and spectroscopy [23, 37]. As such, orthogonal transforms have been used to process various types of data including seismic, sonar, biological, and biomedical. The scope of interdisciplinary work in this area of study is thus apparent.

This book is directed at one who wishes to acquire a working knowledge of orthogonal transforms in the area of digital signal processing. To this end, the first seven chapters are devoted to the study of the background and development of orthogonal transforms. Because of the limitation of the scope of this book, not all topics presented will be treated in depth. Rather, a balance between rigor and clarity will be sought. The last three chapters deal with three applications of orthogonal transforms: (1) generalized Wiener filtering, (2) data compression, and (3) feature selection in pattern recognition. The motivation for selecting these particular applications is that they best provide the setting in which the reader can grasp some theoretical foundations and techniques. Subsequently he can apply or extend these to other areas.

1.2 Terminology

The terminology which will be used is best introduced by referring to Fig. 1.1, in which $x(t)$ represents a *signal* (or *analog signal*) that is a continuous function of time, t. If $x(t)$ is input to an ideal sampler that samples at a specified rate of N samples per second, then output of the sampler is a *discrete* (or *sampled*) signal $x^*(t)$, given by

$$x^*(t) \simeq \Delta t \sum_{m=0}^{N-1} x(m\Delta t)\, \delta(t - m\Delta t) \tag{1.2-1}$$

In Eq. (1), Δt is the sampling interval, and $\delta(t)$ denotes the delta or Dirac function.

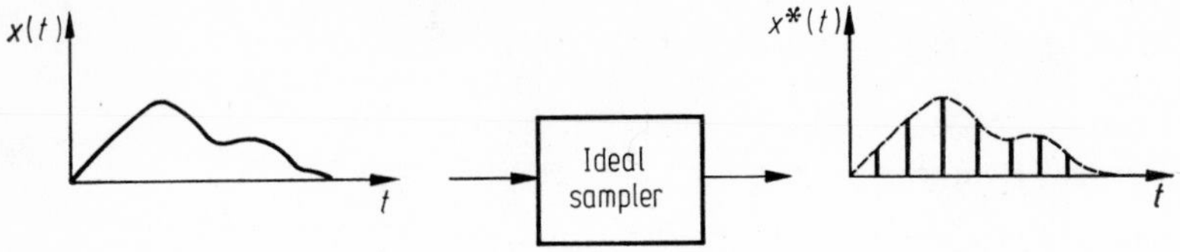

Fig. 1.1 Pertaining to terminology

The above sampling results in the *sequence* (or *data sequence*) $\{X(m)\}$, $m = 0, 1, \ldots, N-1$, where $X(m) = x(m\Delta t)$. The term digital sequence implies that each $X(m)$ has been quantized and coded in digital form. Correspondingly the term digital signal $x^*(t)$ implies that each sampled value $x(m\Delta t)$ in Eq. (1) has been quantized and coded in digital form [38].

1.3 Signal Representation Using Orthogonal Functions [39]

A set of real-valued[1] continuous functions $\{u_n(t)\} = \{u_0(t), u_1(t), \ldots\}$ is said to be orthogonal on the interval $(t_0, t_0 + T)$ if

$$\int_T u_m(t)\, u_n(t)\, dt = \begin{cases} c, & \text{if } m = n \\ 0, & \text{if } m \neq n \end{cases} \tag{1.3-1}$$

where the notation $\int_T$ denotes $\int_{t_0}^{t_0+T}$. When $c = 1$, $\{u_n(t)\}$ is said to be an *orthonormal set.*

Suppose $x(t)$ is a real-valued signal defined on the interval $(t_0, t_0 + T)$, which is represented by the expansion

$$x(t) = \sum_{n=0}^{\infty} a_n u_n(t) \tag{1.3-2}$$

where a_n denotes the n-th coefficient of the expansion. To evaluate the a_n, we multiply both sides of Eq. (2) by $u_m(t)$ and integrate over $(t_0, t_0 + T)$ to obtain

$$\int_T x(t)\, u_m(t)\, dt = \int_T \sum_{n=0}^{\infty} a_n u_n(t)\, u_m(t)\, dt \tag{1.3-3}$$

Application of Eq. (1) to Eq. (3) leads to

$$a_m = \frac{1}{c} \int_T x(t)\, u_m(t)\, dt, \qquad m = 0, 1, \ldots \tag{1.3-4}$$

[1] The development which follows can be extended to complex-valued functions, some aspects of which are discussed in Prob. 1-4.

Now, the orthogonal set $\{u_n(t)\}$ with

$$\int_T u_n^2(t)\, dt < \infty \tag{1.3-5}$$

is said to be *complete* or *closed* if either of the following statements is true:

(1) There exists no signal $x(t)$ with

$$\int_T x^2(t)\, dt < \infty \tag{1.3-6}$$

such that

$$\int_T x(t)\, u_n(t)\, dt = 0, \qquad n = 0, 1, \ldots \tag{1.3-7}$$

(2) For any piecewise continuous signal $x(t)$ with

$$\int_T x^2(t)\, dt < \infty$$

and an $\varepsilon > 0$, however small, there exists an N and a finite expansion

$$\hat{x}(t) = \sum_{n=0}^{N-1} a_n u_n(t) \tag{1.3-8}$$

such that

$$\int_T |x(t) - \hat{x}(t)|^2\, dt < \varepsilon \tag{1.3-9}$$

From the above development it is apparent that the orthogonal function expansion in Eq. (2) enables a representation of $x(t)$ by the infinite but countable[1] set $\{a_0, a_1, a_2, \ldots\}$. Further, when $\{u_n(t)\}$ is complete, such a representation is possible by the finite set $\{a_0, a_1, \ldots, a_{N-1}\}$.

Physical significance. Squaring both sides of the representation in Eq. (2), we obtain

$$x^2(t) = \sum_{n=0}^{\infty} a_n^2 u_n^2(t) + \sum_{p=0}^{\infty} \sum_{\substack{q=0 \\ p \neq q}}^{\infty} a_p a_q u_p(t)\, u_q(t) \tag{1.3-10}$$

Integration of both sides of Eq. (10) leads to

$$\int_T x^2(t)\, dt = \sum_{n=0}^{\infty} a_n^2 \int_T u_n^2(t)\, dt + \sum_{p=0}^{\infty} \sum_{\substack{q=0 \\ p \neq q}}^{\infty} a_p a_q \int_T u_p(t)\, u_q(t)\, dt \tag{1.3-11}$$

Application of the orthogonal property in Eq. (1) to Eq. (11) results in

$$\frac{1}{T} \int_T x^2(t)\, dt = \frac{c}{T} \sum_{n=0}^{\infty} a_n^2 \tag{1.3-12}$$

The result in Eq. (12) is known as *Parseval's Theorem.*

[1] Countable set implies that there is a one-to-one correspondence between the elements of the set and the positive integers.

Now, if $x(t)$ is a voltage or current signal which is connected across the terminals of a one ohm pure resistor, then the left hand side of Eq. (12) represents the average power dissipated by the resistor. Thus the set of numbers

$$\left\{\frac{c}{T}\, a_n^2\right\}$$

represents the distribution of the power in $x(t)$.

In conclusion, we remark that the above signal representation using orthogonal functions can be divided into two main categories: (1) $\{u_n(t)\}$ consists of sinusoidal functions, and (2) $\{u_n(t)\}$ consists of nonsinusoidal functions. We shall study both these categories during the course of this book.

1.4 Book Outline

The book consists of ten chapters, the first seven of which are devoted to the background, development, and a study of the properties of discrete orthogonal transforms. The last three deal with specific applications.

Chapter 2 presents a review of Fourier methods of signal representation. It also enables a systematic transition from the Fourier representation of signals to that of digital signals.

In Chapter 3, Fourier representation of discrete and digital signals via the discrete Fourier transform is introduced. In this connection, properties of the discrete Fourier transform which parallel those of the Fourier series/transform are studied. Again a recursive technique to compute Fourier spectra is developed.

The fourth chapter is concerned with a detailed development of the fast Fourier transform, which is an algorithm that enables efficient computation of the discrete Fourier transform. Various applications of the fast Fourier transform are illustrated by means of numerical examples.

Chapter 5 introduces a class of nonsinusoidal orthogonal functions along with an appropriate notation to represent them. The concept of sequency as a generalized frequency is explained. The material in this chapter is used in Chapters 6 and 7 in connection with the development of various nonsinusoidal orthogonal transforms.

The sixth chapter presents a detailed study of the Walsh-Hadamard transform. Algorithms to compute the Walsh-Hadamard transform are developed. Power and phase spectra are defined and compared with corresponding discrete Fourier spectra. Throughout the chapter, the analogy between the Walsh-Hadamard and discrete Fourier transforms and their properties is explored.

In Chapter 7 we study a set of miscellaneous transforms which consists of the generalized transform, Haar transform, slant transform and discrete cosine transform. Fast algorithms to compute these transforms are developed. It is shown that the generalized transform enables a systematic transition from the Walsh-Hadamard transform to the discrete Fourier transform. The motivation for studying the Haar, slant, and discrete cosine transforms is that they are considered for certain applications in the chapters that follow.

Chapter 8 concerns the application of orthogonal transforms with respect to the classical signal processing technique known as Wiener filtering. It is shown that orthogonal transforms can be used to extend Wiener filtering to the processing of digital signals with an emphasis on reduction of computational requirements.

As a second application, data compression via orthogonal transforms is discussed in Chapter 9. In this connection, an optimum transform called the Karhunen-Loeve transform is developed. Applications of data compression in the areas of image processing and electrocardiographic data processing are illustrated by examples.

Chapter 10 is concerned with the application of orthogonal transforms with respect to feature selection in pattern recognition. The main objective of this chapter is to illustrate how orthogonal transforms can be used to achieve a substantial reduction in the number of features required, with a relatively small increase in classification error. To this end, some simple classification algorithms and related implementations are also presented.

References

1. Andrews, H. C., and Caspari, K. L.: A Generalized Technique for Spectral Analysis. *IEEE Trans. Computers* C-19 (1970) 16–25.
2. Andrews, H. C., and Kane, J.: Kronecker Matrices, Computer Implementation, and Generalized Spectra, *J. of the ACM* 17 (1970) 260–268.
3. Glassman, J. A.: A Generalization of the Fast Fourier Transform. *IEEE Trans. Computers* C-19 (1970) 105–116.
4. Special Issues on Fast Fourier Transform, *IEEE Trans. Audio and Electroacoustics.* AU-15 and AU-17, 1967 and 1969.
5. Ahmed, N., Rao, K. R., and Schultz, R. B.: A Generalized Discrete Transform. *Proc. IEEE* 59 (1971) 1360–1362.
6. Rao, K. R., Mrig, L. C., and Ahmed, N.: A Modified Generalized Discrete Transform. *Proc. IEEE* 61 (1973) 668–669.
7. Special Issue on Digital Pattern Recognition. *Proc. IEEE* 60, October, 1972.
8. Special Issue on Digital Picture Processing. *Proc. IEEE* 60, July, 1972.
9. Special Issue on Two-Dimensional Digital Signal Processing. *IEEE Trans. Computers* C-21, July, 1972.
10. Special Issue on Digital Signal Processing. *IEEE Trans. Audio and Electroacoustics* AU-18, December, 1970.
11. Special Issue on Feature Extraction and Pattern Recognition. *IEEE Trans. Computers* C-20, September, 1971.

12. Special Issue on Signal Processing for Digital Communications. *IEEE Trans. Communications Technology* COM-19, December, 1971.
13. Special Issue on 1972 Conference on Speech Communication and Processing. *IEEE Trans. Audio and Electroacoustics* AU-21, June, 1973.
14. Special Issue on Two Dimensional Digital Filtering and Image Processing. *IEEE Trans. Circuit Theory* CT-21, November, 1974.
15. Ristenbatt, M. P.: Alternatives in Digital Communications. *Proc. IEEE* 61 (1973) 703–721.
16. Carl, J. W., and Swartwood, R. V.: A Hybrid Walsh Transform Computer. *IEEE Trans. Computers* C-22 (1973) 669–672.
17. Wishner, H. D.: Designing a Special-Purpose Digital Image Processor. *Computer Design* 11 (1972) 71–76.
18. Pratt, W. K., and Andrews, H. C.: Two-Dimensional Transform Coding of Images. *Internat. Symp. Information Theory*, 1969.
19. Pratt, W. K., Kane, J., and Andrews, H. C.: Hadamard Transform Image Coding. *Proc. IEEE* 57 (1969) 58–68.
20. Andrews, H. C.: *Computer Techniques in Image Processing*. New York, London: Academic Press, 1970, 73–179.
21. Wintz, P. A.: Transform Picture Coding. *Proc. IEEE* 60 (1972) 809–820. (Special issue on Digital Picture Processing.)
22. Habibi, A., and Wintz, P. A.: Image Coding By Linear Transformations and Block Quantization. *IEEE Trans. Communication Technology* COM-19 (1971) 50–62.
23. Several papers in the Proc. 1970–1974 Symp. *Applications of Walsh Functions*, Washington, D. C.
24. Andrews, H. C., Tescher, H. G., and Kruger, R. P.: Image Processing by Digital Computer. *IEEE Spectrum* 9, 1972, 20–32.
25. Huang, T. S., Schreiber, W. F., and Tretiak, O. J.: Image Processing, *Proc. IEEE* 59 (1971) 1586–1609.
26. Pratt, W. K.: Spatial Transform Coding of Color Images. *IEEE Trans. Communication Technology* COM-19 (1971) 980–992.
27. Fukinuki, T., Miyata, M.: Intraframe Image Coding by Cascaded Hadamard Transforms. *IEEE Trans. Communications* COM-21 (1973) 175–180.
28. Campanella, S. J., and Robinson, G. S.: Comparison of Orthogonal Transformations for Digital Speech Processing. *IEEE Trans. Communication Technology* COM-19 (1971) 1045–1050.
29. Shum, F. Y. Y., Elliot, A. R., and Brown, W. O.: Speech Processing with Walsh-Hadamard Transforms, *IEEE Trans. Audio and Electroacoustics* AU-21 (1973) 174–179.
30. Welchel, J. E., and Guinn, E. F.: The Fast Fourier-Hadamard Transform and its use in Signal Representation and Classification. *Eascon '68 Record*, Electronic and Aerospace Systems Convention, Washington, D. C., Sept. 9–11, 1968, published by IEEE Group on Aerospace and Electronic Systems.
31. Andrews, H. C.: Multidimensional Rotations in Feature Selection. *IEEE Trans. Computers* C-20 (1971) 1045–1051.
32. Andrews, H. C.: *Introduction to Mathematical Techniques in Pattern Recognition*. New York, London: Wiley-Interscience, 1972, 24–32, and 211–234.
33. Harmuth, H. F.: *Transmission of Information by Orthogonal Functions*. New York, Heidelberg, Berlin: Springer 1972.
34. Pearl, J., Andrews, H. C., and Pratt, W. K.: Performance Measures for Transform Data Coding. *IEEE Trans. Communications* COM-20 (1972) 411–415.

35. Pearl, J.: Walsh Processing of Random Signals. *IEEE Trans. Electromagnetic Compatability* EMC-13 (1971) 137–141.
36. Pratt, W. K.: Generalized Wiener Filtering Computation Techniques. *IEEE Trans. Computers* C-21 (1972) 636–641.
37. Gibbs, J. E., and Gobbie, H. A.: Application of Walsh Functions to Transform Spectroscopy. *Nature* 224 (1969) 1012–1013.
38. Rabiner, L. R. et al.: Terminology in Digital Signal Processing. *IEEE Trans. Audio and Electroacoustics*, AU-20 (1972) 703–721.
39. Lee, Y. W.: *Statistical Theory of Communication.* New York: John Wiley, 1961.
40. Rabiner, L. R., and Rader, C. M. (Editors): *Digital Signal Processing.* New York: IEEE Press, 1972.

Problems

1-1 If $Z_1 = x_1 + iy_1$ and $Z_2 = x_2 + iy_2$, where $i = \sqrt{-1}$, show that

$$\overline{Z_1 + Z_2} = \bar{Z}_1 + \bar{Z}_2$$

and

$$\overline{Z_1 Z_2} = \bar{Z}_1 \bar{Z}_2$$

the bar indicating complex conjugate.

1-2 Given a set of real-valued functions $\{u_n(t)\}$ defined on $(0, T)$ such that

$$\int_T u_m(t)\, u_n(t)\, dt = \begin{cases} 0, & m \neq n \\ T/4, & m = n \end{cases}$$

Consider the expansion

$$x(t) = \sum_{n=0}^{\infty} a_n u_n(t)$$

(a) What formula would you use to compute the coefficients a_n?

Answer:
$$a_n = \frac{4}{T} \int_T x(t)\, u_n(t)\, dt.$$

(b) Show that

$$\frac{1}{T} \int_T x^2(t)\, dt = \frac{1}{4} \sum_{n=0}^{\infty} a_n^2$$

1-3 If $Z_n = x_n + iy_n$, $n = 1, 2, 3$, show that:

(a)
$$\overline{Z_1 Z_2 Z_3} = \bar{Z}_1 \bar{Z}_2 \bar{Z}_3$$

(b)
$$(\overline{Z_n^4}) = (\bar{Z}_n)^4$$

(c)
$$\overline{\left(\frac{Z_1}{Z_2 Z_3}\right)} = \frac{\bar{Z}_1}{\bar{Z}_2 \bar{Z}_3}$$

and

$$\left|\frac{Z_1}{Z_2 Z_3}\right| = \frac{|Z_1|}{|Z_2|\,|Z_3|}$$

1-4 A set of complex-valued orthonormal functions $\{u_n(t)\}$ on the interval $(0, T)$ is defined as

$$\int_T u_m(t)\,\bar{u}_n(t)\,dt = \begin{cases} 1, & m = n \\ 0, & m \neq n \end{cases}$$

where $\bar{u}_n(t)$ denotes the complex conjugate of $u_n(t)$. Consider the expansion

$$x(t) = \sum_{n=0}^{\infty} a_n u_n(t)$$

where $x(t)$ is a real or complex-valued signal.
(a) What formula would you use to compute the coefficients a_n?

Answer: $$a_n = \int_T x(t)\,\bar{u}_n(t)\,dt, \qquad n = 0, 1, 2, \ldots$$

(b) Show that the Parseval's theorem for the above representation is given by

$$\frac{1}{T}\int_T |x(t)|^2\,dt = \frac{1}{T}\sum_{n=0}^{\infty} |a_n|^2$$

1-5 The Legendre polynomials $p_n(x)$ are defined by the recursive formula

$$(n + 1)\,p_{n+1}(x) = (2n + 1)\,x p_n(x) - n p_{n-1}(x), \qquad n = 1, 2, 3, \ldots$$

where

$$p_0(x) = 1, \quad \text{and} \quad p_1(x) = x.$$

These polynomials are orthogonal over the interval $|x| < 1$. That is

$$\int_{-1}^{1} p_m(x)\,p_n(x)\,dx = \begin{cases} 0, & m \neq n \\ \dfrac{2}{2n + 1}, & m = n \end{cases} \tag{P1-5-1}$$

(a) Find $p_2(x)$, $p_3(x)$ and $p_4(x)$.
(b) Verify that Eq. (P1-5-1) is valid for the set $\{p_0(x), p_1(x), p_2(x)$, and $p_3(x)\}$.

Answer: $p_2(x) = \frac{1}{2}(3x^2 - 1)$, $p_3(x) = \frac{1}{2}(5x^3 - 3x)$, and $p_4(x) = \frac{1}{8}(35x^4 - 30x^2 + 3)$

1-6 Let $f(x) = 1 - |x|$, $|x| < 1$
Consider the approximation

$$f(x) \simeq \sum_{n=0}^{4} a_n p_n(x)$$

where $p_n(x)$ are the Legendre polynomials defined in Prob. 1-5. Compute the coefficients a_0, a_1, a_2, a_3 and a_4.

Answer: $$a_0 = 1,\ a_1 = 0,\ a_2 = -5/8,\ a_3 = 0,\ a_4 = 3/16.$$

Chapter Two

Fourier Representation of Signals

The purpose of this chapter is twofold. First, it presents a review of the Fourier methods of representing signals. Second, it provides the foundation for a systematic transition from the Fourier representation of analog signals to that of digital signals.

2.1 Fourier Representation

As an example of the general treatment of orthogonal representations in Section 1.3, we consider the case when the set of functions $\{u_n(t)\}$ are the Fourier sinusoidal functions $\{1, \cos n\omega_0 t, \sin n\omega_0 t\}$. Then, the series expansion corresponding to Eq. (1.3-2) is given by

$$x(t) = a_0 + \sum_{n=1}^{\infty} a_n \cos n\omega_0 t + \sum_{n=1}^{\infty} b_n \sin n\omega_0 t \tag{2.1-1}$$

where, ω_0 (radians per second) is the fundamental angular frequency which is related to the period T (seconds) of the function by the formula $T = 2\pi/\omega_0$. The fundamental angular frequency is equal to 2π times the fundamental frequency f_0 (cycles per second or Hertz). The frequencies $n\omega_0$ or nf_0 are called *harmonics* since they are integral multiples of the fundamental frequencies ω_0 and f_0 respectively.

It is assumed that $x(t)$ satisfies the following conditions in addition to the condition in Eq. (1.3-6):

(1) $x(t)$ has at most a finite number of discontinuities in one period.

(2) $x(t)$ has at most a finite number of maxima and minima in one period.

With the above assumptions it can be shown that the set of coefficients $\{a_n\}$ and $\{b_n\}$ are uniformly bounded. Alternately, we say that the series in Eq. (1) converges uniformly in the interval $(0, T)$ and hence term by term integration is permissible. At points of discontinuity in $x(t)$, there is convergence in the mean. The coefficients $\{a_0, a_n, b_n\}$ in Eq. (1) can be computed by using the fact that the set of functions $\{\cos n\omega_0 t, \sin n\omega_0 t\}$

have orthogonal properties over the period T. That is

$$\int_T \cos n\omega_0 t \cos m\omega_0 t \, dt = \begin{cases} T/2, & m = n \\ 0, & m \neq n \end{cases}$$

$$\int_T \cos n\omega_0 t \sin m\omega_0 t \, dt = 0, \text{ for all } m \text{ and } n \tag{2.1-2}$$

$$\int_T \sin n\omega_0 t \sin m\omega_0 t \, dt = \begin{cases} T/2, & m = n \\ 0, & m \neq n \end{cases}$$

Using Eqs. (1) and (2) it can be shown that (Prob. 2-1)

$$a_0 = \frac{1}{T} \int_T x(t) \, dt$$

$$a_n = \frac{2}{T} \int_T x(t) \cos n\omega_0 t \, dt$$

and

$$b_n = \frac{2}{T} \int_T x(t) \sin n\omega_0 t \, dt \tag{2.1-3}$$

From the above discussion we conclude that the signal $x(t)$ can be represented by the set of real numbers $\{a_0, a_n, b_n\}$. To relate these coefficients to the distribution of power in $x(t)$, we need the relation corresponding to Eq. (1.3-12) which is Parseval's theorem. It can be shown that (Prob. 2-2) Parseval's theorem for the representation in Eq. (1) is

$$\frac{1}{T} \int_T x^2(t) \, dt = a_0^2 + \frac{1}{2} \sum_{n=1}^{\infty} (a_n^2 + b_n^2) \tag{2.1-4}$$

From Eq. (4) it is clear that the distribution of power in $x(t)$ is given by the set of real numbers $\left\{a_0^2, \frac{1}{2}(a_n^2 + b_n^2)\right\}$.

Complex Fourier series. As a second example, we consider the case when the set of functions $\{u_n(t)\}$ in Eq. (1.3-2) are complex. Starting with the ordinary form of Fourier series discussed above, it is straightforward to obtain the complex Fourier series using standard trigonometric manipulations.

Using the formulas

$$\cos n\omega_0 t = \frac{1}{2} (e^{i n\omega_0 t} + e^{-i n\omega_0 t}) \tag{2.1-5}$$

and

$$\sin n\omega_0 t = \frac{1}{2i} (e^{i n\omega_0 t} - e^{-i n\omega_0 t}) \tag{2.1-6}$$

$i = \sqrt{-1}$, we can write Eq. (1) in the form

$$x(t) = a_0 + \frac{1}{2} \sum_{n=1}^{\infty} \{a_n (e^{i\,n\omega_0 t} + e^{-i\,n\omega_0 t}) - ib_n (e^{i\,n\omega_0 t} - e^{-i\,n\omega_0 t})\}$$

$$= a_0 + \frac{1}{2} \sum_{n=1}^{\infty} \{(a_n - ib_n)\, e^{i\,n\omega_0 t} + (a_n + ib_n)\, e^{-i\,n\omega_0 t}\} \tag{2.1-7}$$

We define

$$c_n = \frac{1}{2}(a_n - ib_n) \tag{2.1-8}$$

From Eqs. (3) and (8) it follows that

$$c_n = \frac{1}{T} \int_T x(t) \left[\cos n\omega_0 t - i \sin n\omega_0 t\right] dt$$

That is

$$c_n = \frac{1}{T} \int_T x(t)\, e^{-i\,n\omega_0 t}\, dt \tag{2.1-9}$$

Also

$$c_{-n} = \bar{c}_n = \frac{1}{2}(a_n + ib_n) \tag{2.1-10}$$

Substituting Eqs. (8) and (10) in Eq. (7) we obtain

$$x(t) = a_0 + \sum_{n=1}^{\infty} [c_n e^{i\,n\omega_0 t} + c_{-n} e^{-i\,n\omega_0 t}] = a_0 + \sum_{\substack{n=-\infty \\ n \neq 0}}^{\infty} c_n e^{i\,n\omega_0 t} \tag{2.1-11}$$

Again, from Eqs. (3) and (9), we have

$$c_0 = \frac{1}{T} \int_T x(t)\, dt = a_0 \tag{2.1-12}$$

Thus Eqs. (11) and (12) yield

$$x(t) = \sum_{n=-\infty}^{\infty} c_n e^{i\,n\omega_0 t} \tag{2.1-13}$$

Equation (13) is the complex Fourier series representation of $x(t)$ – that is, $x(t)$ is represented by a set of complex numbers $\{c_n\}$.

Remark: From Eqs. (9) and (13) it is apparent that the factor $1/T$ could be included either as a part of the integration, or as a part of the summation, or $\sqrt{1/T}$ associated with each one of them in the interest of symmetry.

Using the relation

$$\int_T e^{i(m-n)\omega_0 t}\, dt = \begin{cases} T, & m = n \\ 0, & m \neq n \end{cases} \tag{2.1-14}$$

and Eq. (13), it can be shown that (Prob. 2-3)

$$\frac{1}{T}\int_T x^2(t)\,dt = c_0^2 + 2\sum_{n=1}^{\infty}|c_n|^2 \qquad (2.1\text{-}15)$$

Equation (15) is Parseval's theorem corresponding to Eq. (13). We note that Eqs. (15) and (4) are identical with $c_0 = a_0$, and $c_n = \frac{1}{2}(a_n - ib_n)$. From Eq. (15) it follows that the distribution of power in the signal $x(t)$ is represented by the set of real numbers $\{c^0, 2|c_n|^2\}$.

Figure 2.1 shows a geometric interpretation of a complex number $c_n = \frac{1}{2}(a_n - ib_n)$. That is, $c_n = |c_n|e^{i\phi_n}$, where

$$\phi_n = \begin{cases} \tan^{-1}\left(\dfrac{-b_n}{a_n}\right) & n = 1, 2, 3, \ldots \\ 0, & n = 0 \end{cases} \qquad (2.1\text{-}16)$$

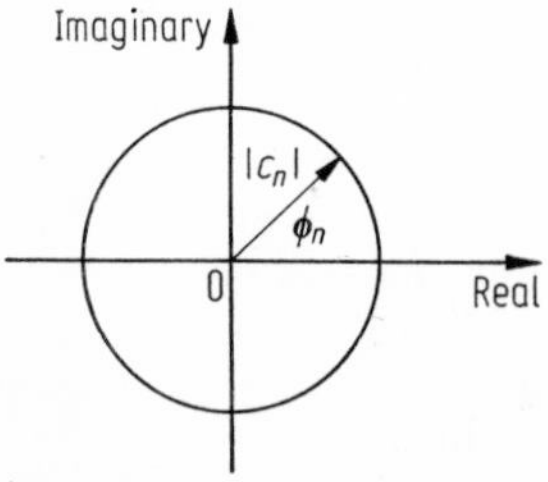

Fig. 2.1 Geometric interpretation of a complex number

In engineering applications, the angle ϕ_n in Eq. (16) is usually referred to as a "phase" angle. However, in more general terms it represents the *orientation* of c_n with respect to a reference, which happens to be the real axis in Fig. 2.1.

2.2 Power, Amplitude and Phase Spectra

Power and amplitude spectra. Figure 2.2 shows a periodic extension $x_p(t)$ of a typical signal $x(t)$. Now, suppose we shift $x_p(t)$ to the right by an amount τ. Then, the resulting signal $x_p(t - \tau)$ is as shown in Fig. 2.2.

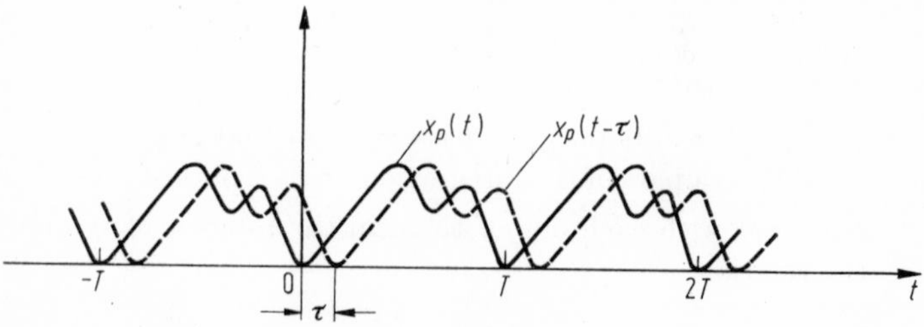

Fig. 2.2 Periodic extension of $x(t)$ and $x(t - \tau)$

From Eq. (2.1-13) we have

$$x_p(t) = \sum_{n=-\infty}^{\infty} c_n e^{i n\omega_0 t} \tag{2.2-1}$$

which yields

$$x_p(t-\tau) = \sum_{n=-\infty}^{\infty} c_n e^{i n\omega_0 (t-\tau)} = \sum_{n=-\infty}^{\infty} (c_n e^{-i n\omega_0 \tau})\, e^{i n\omega_0 t}$$

That is

$$x_p(t-\tau) = \sum_{n=-\infty}^{\infty} c_{n,\tau} e^{i n\omega_0 t} \tag{2.2-2}$$

where

$$c_{n,\tau} = c_n\, e^{-i n\omega_0 \tau} \tag{2.2-3}$$

From Eq. (3) it is clear that

$$|c_{n,\tau}|^2 = |c_n|^2 = |c_{-n}|^2, \text{ for all } \tau \in [0, T) \tag{2.2-4}$$

We now define the Fourier power spectrum as

$$P_n = |c_n|^2, \qquad n = 0, \pm 1, \pm 2, \ldots \tag{2.2-5}$$

where P_n denotes the n-th power spectral point. From the above discussion, it is apparent that this power spectrum has the following properties as evident from Eqs. (4) and (5):

(i) P_n is *invariant* to the amount of time shift, τ.
(ii) P_n is nonnegative.
(iii) P_n is an even function of n.

Corresponding to the power spectrum in Eq. (5) the Fourier amplitude spectrum is defined as

$$p_n = \sqrt{P_n}, \qquad n = 0, \pm 1, \pm 2, \ldots$$

where, p_n is the nonnegative square root of the power spectral point P_n.

Phase spectrum. The Fourier phase spectrum of a periodic signal $x_p(t)$ is defined as

$$\Psi_n = \begin{cases} \tan^{-1}\left(\dfrac{\mathrm{Im}\,[c_n]}{\mathrm{Re}\,[c_n]}\right), & n = \pm 1, \pm 2, \ldots \\ 0, & n = 0 \end{cases} \tag{2.2-6}$$

where the symbols Im[ ] and Re[ ] denote the imaginary and real parts of the terms enclosed respectively. If $x_p(t)$ is multiplied by a real constant K, then its Fourier series representation is

$$K x_p(t) = \sum_{n=-\infty}^{\infty} (K c_n)\, e^{i n\omega_0 t} \tag{2.2-7}$$

From Eqs. (3) and (7) it follows that the Fourier phase spectrum has the following properties:

(i) Ψ_n is a function of τ. That is, Ψ_n changes when the signal is subjected to a shift, unlike the power spectrum which is independent of τ.
(ii) Ψ_n is independent of K. That is, Ψ_n is invariant with respect to amplification or attenuation of the signal. In contrast, the power spectrum is a function K.
(iii) $\Psi_{-l} = -\Psi_l$, $l = 1, 2, 3, \ldots$ (see Prob. 2-4). That is, Ψ_n is an odd function of n.

Remarks: From the geometrical interpretation of c_n in Fig. 2.1 and the above discussion, it follows that c_n can be expressed in terms of the Fourier power and phase spectra as

$$c_n = \sqrt{P_n}\, e^{i\Psi_n}, \qquad n = 0, \pm 1, \pm 2, \ldots \tag{2.2-8}$$

or

$$c_n = p_n e^{i\Psi_n}, \qquad n = 0, \pm 1, \pm 2, \ldots \tag{2.2-9}$$

where

$$p_n = |c_n| \quad \text{and} \quad \Psi_n = \phi_n$$

Now, since

$$x(t) = \sum_{n=-\infty}^{\infty} c_n e^{in\omega_0 t},$$

Eqs. (8) and (9) imply that if the amplitude (or power) and phase spectra are known, then $x(t)$ can be reconstructed *uniquely*. In what follows we consider a simple example.

Example 2.2-1

Consider a signal $x(t)$ defined as

$$x(t) = \begin{cases} A, & \dfrac{-\tau}{2} \leq t \leq \dfrac{\tau}{2} \\ 0, & \text{elsewhere} \end{cases}$$

Let $x_p(t)$ be a periodic extension of $x(t)$ with period T as shown in Fig. 2.3.

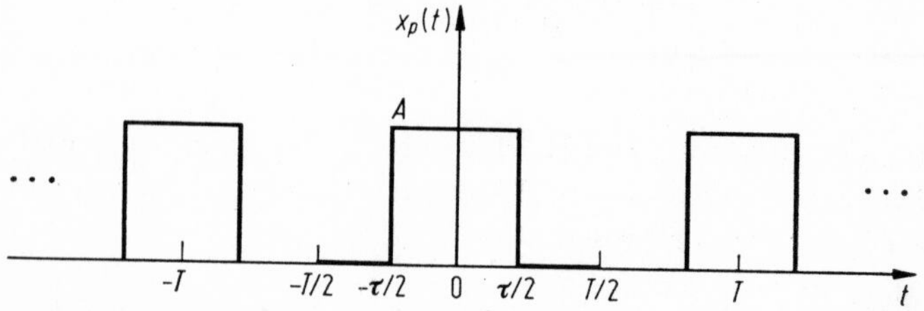

Fig. 2.3 $x_p(t)$, the periodic extension of $x(t)$

(a) Obtain a complex Fourier series representation of $x_p(t)$ as follows:

$$x_p(t) = \frac{1}{T} \sum_{n=-\infty}^{\infty} c_n e^{in\omega_0 t} \tag{2.2-10}$$

where

$$c_n = \int_{-T/2}^{T/2} x(t)\, e^{-in\omega_0 t}\, dt \tag{2.2-11}$$

(b) Sketch the Fourier amplitude and phase spectra.
Solution: From Eq. (11) we have

$$c_n = \int_{-\tau/2}^{\tau/2} A e^{-in\omega_0 t}\, dt = \frac{2A}{n\omega_0} \left[\frac{e^{in\omega_0 \tau/2} - e^{-in\omega_0 \tau/2}}{2i} \right]$$

$$= A\tau \left[\frac{\sin (n\omega_0 \tau/2)}{(n\omega_0 \tau/2)} \right], \qquad n = 0, \pm 1, \pm 2, \ldots \tag{2.2-12}$$

(b) Plots of the corresponding amplitude and phase spectra which are obtained using Eq. (9) are shown in Figs. 2.4a and 2.4b respectively.

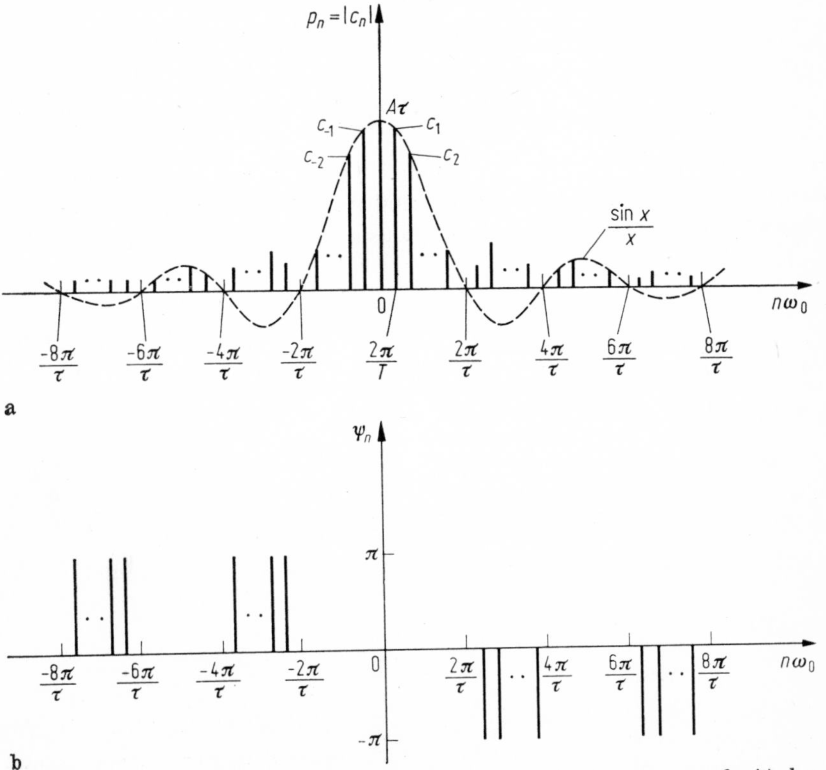

Fig. 2.4 Fourier amplitude and phase spectra. (a) Amplitude spectrum of $x_p(t)$ shown in Fig. 2.3, (b) Phase spectrum of $x_p(t)$ shown in Fig. 2.3

With respect to the above plots, we make the following observations:

(1) The spacing between successive $p_n = |c_n|$ in Fig. 2.4a equals $2\pi/T$, which is the fundamental angular frequency ω_0.

(2) The phase spectrum is either 0, π or $-\pi$. This follows from the relation in Eq. (9):

$$c_n = p_n e^{i\Psi_n}, \qquad n = 0, \pm 1, \pm 2, \ldots$$

2.3 The Fourier Transform

The transition from the Fourier series to the Fourier transform is best illustrated by means of Example 2.2-1 described in the previous section. Suppose we let T tend to infinity. Then, from Fig. 2.3 it is clear that the train of pulses reduces to a *single* pulse which is an aperiodic (i.e., non-periodic) function $x(t)$. Again, in Fig. 2.4a we observe that as T increases, the spectral lines crowd in and ultimately approach the continuous function denoted by the $(\sin x)/x$ function. Next, we examine the behavior of the complex Fourier coefficient c_n, where

$$c_n = \int_{-T/2}^{T/2} x(t)\, e^{-in\omega_0 t}\, dt, \quad \omega_0 = \frac{2\pi}{T} \tag{2.3-1}$$

Now, as T tends to infinity, the fundamental angular frequency ω_0 becomes a differential angular frequency $d\omega$, and $n\omega_0$, which is the n-th harmonic angular frequency, becomes a continuous angular frequency ω.

Thus, Eq. (1) yields

$$\lim_{T\to\infty} c_n = \int_{-\infty}^{\infty} x(t)\, e^{-i\omega t}\, dt \tag{2.3-2}$$

The right hand side of Eq. (2) is defined as the *Fourier transform* of $x(t)$ which we denote by $F_x(\omega)$; that is,

$$F_x(\omega) = \int_{-\infty}^{\infty} x(t)\, e^{-i\omega t}\, dt \tag{2.3-3}$$

Let us now consider the effect of T tending to infinity on Eq. (2.2-10), which can be written as

$$x_p(t) = \frac{1}{2\pi} \sum_{n=-\infty}^{\infty} c_n e^{in\omega_0 t}\, (\omega_0) \tag{2.3-4}$$

As T tends to infinity, the summation over the harmonics in Eq. (4) becomes an integration over the entire continuous range $(-\infty, \infty)$. Again, from the limiting arguments pertaining to the transition from Eq. (1) to Eq. (2), it is known that ω_0 becomes $d\omega$ and $n\omega_0$ becomes ω, while c_n becomes $F_x(\omega)$. Hence the limiting form of Eq. (4) is given by

$$\lim_{T\to\infty} x_p(t) = x(t) = \frac{1}{2\pi}\int_{-\infty}^{\infty} F_x(\omega)\, e^{i\omega t}\, d\omega \tag{2.3-5}$$

which is defined as the *inverse Fourier transform* of $F_x(\omega)$.

Equations (3) and (5) can be grouped as follows:

$$F_x(\omega) = \int_{-\infty}^{\infty} x(t)\, e^{-i\omega t}\, dt \qquad \text{Fourier transform} \tag{2.3-3}$$

$$x(t) = \frac{1}{2\pi}\int_{-\infty}^{\infty} F_x(\omega)\, e^{i\omega t}\, d\omega \qquad \text{inverse Fourier transform} \tag{2.3-5}$$

Table 2.3-1 Summary of Fourier Series and Fourier Transform pairs

Form Number[1]	Fourier Series Pair	Fourier Transform Pair
I	$x(t) = \frac{1}{T}\sum_{n=-\infty}^{\infty} c_n e^{in\omega_0 t}$ $c_n = \int_T x(t)\, e^{-in\omega_0 t} dt$	$x(t) = \frac{1}{2\pi}\int_{-\infty}^{\infty} F_x(\omega)\, e^{i\omega t} d\omega$ $F_x(\omega) = \int_{-\infty}^{\infty} x(t) e^{-i\omega t} dt$
II	$x(t) = \frac{1}{\sqrt{T}}\sum_{n=-\infty}^{\infty} c_n e^{in\omega_0 t}$ $c_n = \frac{1}{\sqrt{T}}\int_T x(t)\, e^{-in\omega_0 t} dt$	$x(t) = \frac{1}{\sqrt{2\pi}}\int_{-\infty}^{\infty} F_x(\omega)\, e^{i\omega t} d\omega$ $F_x(\omega) = \frac{1}{\sqrt{2\pi}}\int_{-\infty}^{\infty} x(t)\, e^{-i\omega t} dt$
III	$x(t) = \sum_{n=-\infty}^{\infty} c_n e^{in\omega_0 t}$ $c_n = \frac{1}{T}\int_T x(t)\, e^{-in\omega_0 t} dt$	$x(t) = \int_{-\infty}^{\infty} F_x(\omega)\, e^{i\omega t} d\omega$ $F_x(\omega) = \frac{1}{2\pi}\int_{-\infty}^{\infty} x(t)\, e^{-i\omega t} dt$

[1] Three more forms can be obtained by replacing i by $-i$ in each of the series and transform pairs.

These two equations are called the *Fourier transform pair*. A sufficient condition for the existence of the Fourier transform of an aperiodic function $x(t)$ is that $x(t)$ be absolutely integrable in the interval $(-\infty, \infty)$; that is

$$\int_{-\infty}^{\infty} |x(t)|\, dt < \infty$$

From Eq. (3) it is clear that $F_x(\omega)$ is a continuous function of ω which can be expressed as

$$F_x(\omega) = A(\omega) + iB(\omega) \tag{2.3-6}$$

where $A(\omega)$ and $B(\omega)$ are respectively the real and imaginary parts of $F_x(\omega)$. Thus the corresponding amplitude, power and phase spectra of $x(t)$ are given by

$$p(\omega) = |F_x(\omega)| \tag{2.3-7}$$

$$P(\omega) = |F_x(\omega)|^2 \tag{2.3-8}$$

and

$$\Psi(\omega) = \tan^{-1}\left[\frac{B(\omega)}{A(\omega)}\right], \qquad |\omega| < \infty \tag{2.3-9}$$

respectively.

In conclusion, we summarize the various forms of the "Fourier series pairs" along with their corresponding Fourier transform pairs in Table 2.3-1 (p. 17).

2.4 Relation Between the Fourier Series and the Fourier Transform

Consider a signal $x(t)$ defined over a finite interval, say L, as shown in Fig. 2.5.

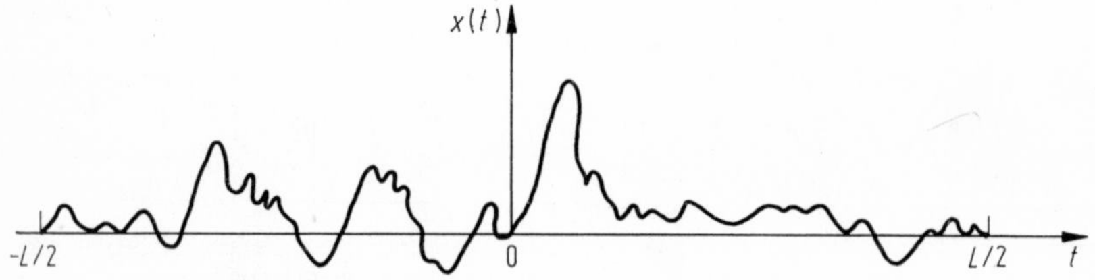

Fig. 2.5 Signal defined over a finite interval, L

Using Eq. (2.3-3), the Fourier transform of $x(t)$ in Fig. 2.5 is obtained as

$$F_x(\omega) = \int_{-L/2}^{L/2} x(t)\, e^{-i\omega t}\, dt \tag{2.4-1}$$

Substitution of $n\omega_0$ for ω in Eq. (1) results in

$$F_x(n\omega_0) = \int_{-L/2}^{L/2} x(t)\, e^{-in\omega_0 t}\, dt, \qquad n = 0, \pm 1, \pm 2, \ldots$$

$$= \int_L x(t)\, e^{-in\omega_0 t}\, dt \tag{2.4-2}$$

where we define $\omega_0 = 2\pi/L$.

Again, the Fourier series representation of a L-periodic extension of $x(t)$ is given by Eqs. (2.2-10) and (2.2-11) as

$$x(t) = \frac{1}{L} \sum_{n=-\infty}^{\infty} c_n e^{in\omega_0 t}$$

where

$$c_n = \int_L x(t)\, e^{-in\omega_0 t}\, dt, \qquad n = 0, \pm 1, \pm 2, \ldots \tag{2.4-3}$$

A comparison of Eq. (3) with Eq. (2) leads to the following fundamental remark:

If a signal is defined over a *finite* interval L, then its Fourier transform $F_x(\omega)$ is exactly specified by the Fourier series at a set of equally spaced points on the ω axis. The distance between these points of specification is $2\pi/L$ radians.

2.5 Crosscorrelation, Autocorrelation and Convolution

Crosscorrelation. If $x_p(t)$ and $y_p(t)$ respectively denote the T-periodic extensions of two signals $x(t)$ and $y(t)$, then their *crosscorrelation* function is defined as

$$\hat{Z}_{xy}(\tau) = \frac{1}{T} \int_T x_p(t)\, y_p(t + \tau)\, dt \tag{2.5-1}$$

where τ is a continuous time displacement in the range $(-\infty, \infty)$, independent of t. In the general theory of harmonic analysis, the crosscorrelation function $\hat{Z}_{xy}(\tau)$ is of considerable interest.

Before proceeding further, it is worthwhile considering a simple example which illustrates the graphical implications of Eq. (1).

If we consider,

$$x(t) = \begin{cases} 1, & 0 < t \le t_1 \\ 0, & \text{elsewhere} \end{cases} \tag{2.5-2}$$

and

$$y(t) = \begin{cases} 2, & 0 < t \le t_1 \\ 0, & \text{elsewhere} \end{cases} \tag{2.5-3}$$

then, the corresponding $x_p(t)$ and $y_p(t + \tau)$ are as shown in Fig. 2.6.

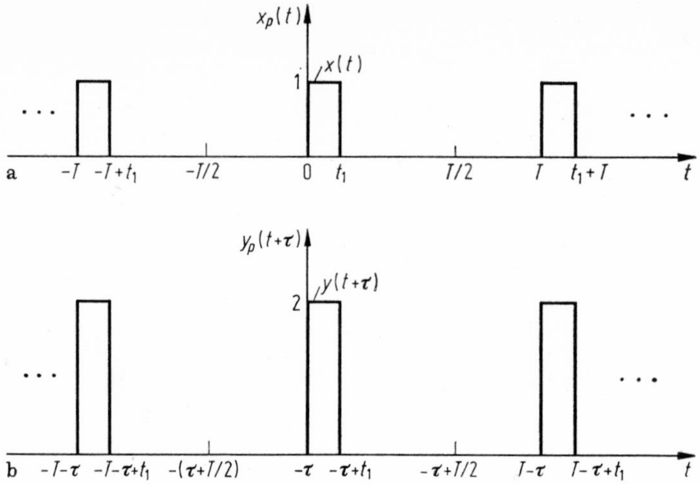

Fig. 2.6 Periodic extensions of $x(t)$ and $y(t + \tau)$

From Eq. (1) it is clear that $\hat{Z}_{xy}(\tau)$ is also a periodic function. Thus it suffices to compute it over one period. Now, consider the product of $x_p(t)$ and $y_p(t + \tau)$ as shown in Fig. 2.7, where $y_p(t + \tau)$ is moved across $x_p(t)$ from the right to the left.

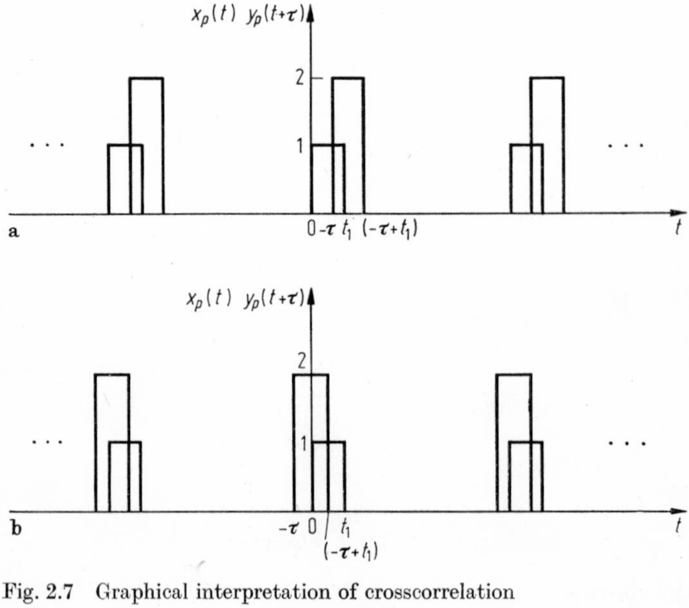

Fig. 2.7 Graphical interpretation of crosscorrelation

From Figs. 2.7a and 2.7b respectively, we obtain

$$\hat{Z}_{xy}(\tau) = \frac{1}{T}\int_{-\tau}^{t_1}(2)(1)\,dt = \frac{2}{T}(t_1+\tau), \qquad -t_1 \le \tau < 0$$

and

$$\hat{Z}_{xy}(\tau) = \frac{1}{T}\int_{0}^{-\tau+t_1}(2)(1)\,dt = \frac{2}{T}(-\tau+t_1), \qquad 0 \le \tau < t_1 \tag{2.5-4}$$

Using Eq. (4) we sketch $\hat{Z}_{xy}(\tau)$ as shown in Fig. 2.8.

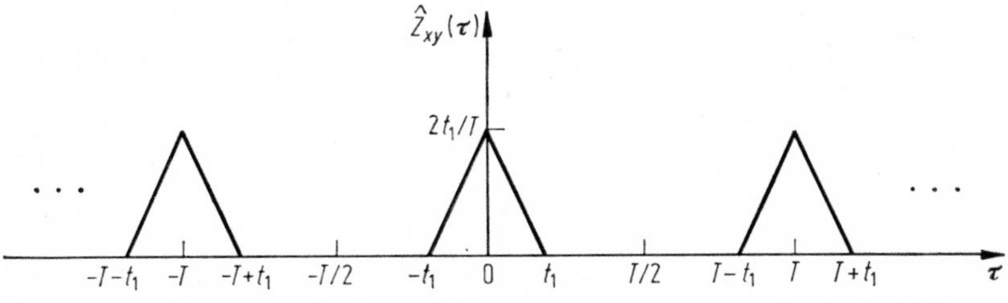

Fig. 2.8 Crosscorrelation of $x_p(t)$ and $y_p(t)$

From the above graphical considerations it follows that the process of crosscorrelating two T-periodic signals reduces to shifting one signal with respect to the other, and subsequently averaging over one period T.

Correlation Theorem. If $(c_n)_x$ and $(c_n)_y$ are respectively the Fourier series coefficients of $x_p(t)$ and $y_p(t)$, then,

$$(c_n)_{\hat{z}} = \overline{(c_n)}_x(c_y), \qquad n = 0, \pm 1, \pm 2, \ldots \tag{2.5-5}$$

where $(c_n)_{\hat{z}}$ is the n-th Fourier series coefficient of $\hat{Z}_{xy}(\tau)$ and $\overline{(c_n)}_x$ is the complex conjugate of $(c_n)_x$.

The proof is straightforward. From Eq. (2.1-13) we obtain

$$x_p(t) = \sum_{n=-\infty}^{\infty} (c_n)_x\, e^{in\omega_0 t} \tag{2.5-6}$$

$$y_p(t) = \sum_{n=-\infty}^{\infty} (c_n)_y\, e^{in\omega_0 t} \tag{2.5-7}$$

and

$$\hat{Z}_{xy}(\tau) = \sum_{n=-\infty}^{\infty} (c_n)_{\hat{z}}\, e^{in\omega_0 \tau} \tag{2.5-8}$$

where

$$(c_n)_x = \frac{1}{T} \int_T x_p(t)\, e^{-i n \omega_0 t}\, dt \tag{2.5-9}$$

$$(c_n)_y = \frac{1}{T} \int_T y_p(t)\, e^{-i n \omega_0 t}\, dt \tag{2.5-10}$$

and

$$(c_n)_{\hat{z}} = \frac{1}{T} \int_T \hat{Z}_{xy}(\tau)\, e^{-i n \omega_0 \tau}\, d\tau \tag{2.5-11}$$

Substitution of Eq. (7) in Eq. (1) results in

$$\hat{Z}_{xy}(\tau) = \frac{1}{T} \int_T x_p(t) \left[\sum_{n=-\infty}^{\infty} (c_n)_y\, e^{i n \omega_0 (t+\tau)}\, dt \right]$$

$$= \sum_{n=-\infty}^{\infty} \left\{ (c_n)_y\, e^{i n \omega_0 \tau} \left[\frac{1}{T} \int_T x_p(t)\, e^{i n \omega_0 t}\, dt \right] \right\} \tag{2.5-12}$$

From Eq. (9) it follows that the term enclosed by the square brackets in Eq. (12) equals $\overline{(c_n)}_x$. Thus Eq. (12) yields

$$\hat{Z}_{xy}(\tau) = \sum_{n=-\infty}^{\infty} \overline{(c_n)}_x\, (c_n)_y\, e^{i n \omega_0 \tau} \tag{2.5-13}$$

Comparison of Eq. (13) with Eq. (8) yields Eq. (5).

Obviously, an alternate form of Eq. (5) would be

$$(c_n)_{\hat{z}} = (c_n)_x\, \overline{(c_n)}_y, \qquad n = 0, \pm 1, \pm 2, \ldots$$

Autocorrelation. In Eq. (1), if $x_p(t)$ and $y_p(t)$ are one and the same functions, then

$$\hat{Z}_{xx}(\tau) = \frac{1}{T} \int_T x_p(t)\, x_p(t+\tau)\, dt \tag{2.5-14}$$

where, $\hat{Z}_{xx}(\tau)$ is defined as the *autocorrelation* function. Again, denoting $(c_n)_x$ and $(c_n)_y$ by c_n in Eq. (5), we obtain

$$(c_n)_{\hat{z}} = |c_n|^2, \qquad n = 0, \pm 1, \pm 2, \ldots \tag{2.5-15}$$

From Eqs. (14) and (15) it follows that Eq. (8) may be written in the form

$$\frac{1}{T} \int_T x_p(t)\, x_p(t+\tau)\, dt = \sum_{n=-\infty}^{\infty} |c_n|^2\, e^{i n \omega_0 \tau} \tag{2.5-16}$$

If we further consider the special case when $\tau = 0$, then Eq. (16) reduces to

$$\frac{1}{T} \int_T x_p^2(t)\, dt = \sum_{n=-\infty}^{\infty} |c_n|^2$$

That is

$$\frac{1}{T} \int_T x_p^2(t)\, dt = c_0^2 + 2 \sum_{n=1}^{\infty} |c_n|^2 \tag{2.5-17}$$

Clearly, Eq. (17) is Parseval's theorem which was derived earlier [see Eq. (2.1-15)].

Convolution. If $x_p(t)$ and $y_p(t)$ denote the T-periodic extensions of two signals $x(t)$ and $y(t)$ respectively, then

$$Z_{xy}(\tau) = \frac{1}{T} \int_T x_p(t)\, y_p(\tau - t)\, dt \tag{2.5-18}$$

is defined as the *convolution* function. We note the similarity between $\hat{Z}_{xy}(\tau)$ and $Z_{xy}(\tau)$ in Eqs. (1) and (18) respectively.

As in the case of crosscorrelation, it is instructive to study the graphical interpretation of Eq. (18). To do so, we consider the case where $x(t)$ and $y(t)$ are as defined in Eqs. (2) and (3) respectively. Then, $x_p(t)$ and $y_p(\tau - t)$ are as shown in Fig. 2.9. We observe that $y_p(\tau - t)$ is a "mirror image" of $y_p(t)$. Again, since $Z_{xy}(\tau)$ in Eq. (18) is also periodic, it is sufficient to compute it over one period.

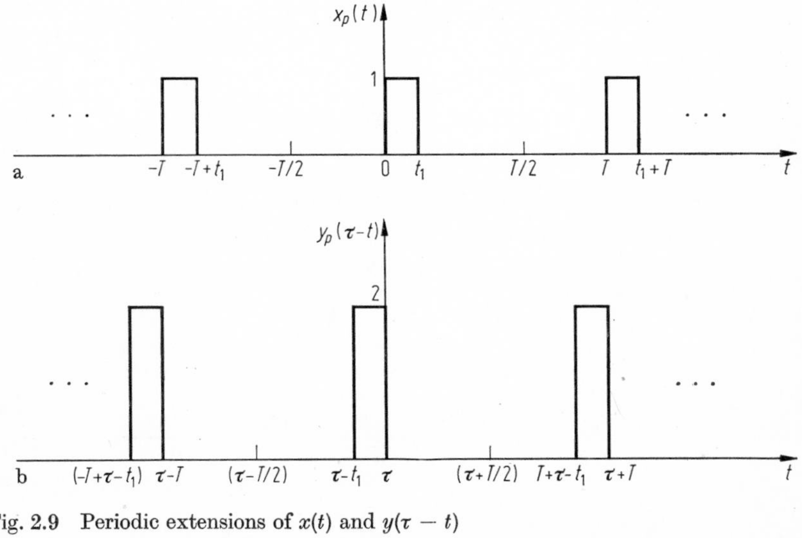

Fig. 2.9 Periodic extensions of $x(t)$ and $y(\tau - t)$

Figures 2.10 a and 2.10 b show the product of $x_p(t)$ and $y_p(\tau - t)$, where $y_p(\tau - t)$ is moved across $x_p(t)$ from the left to the right.

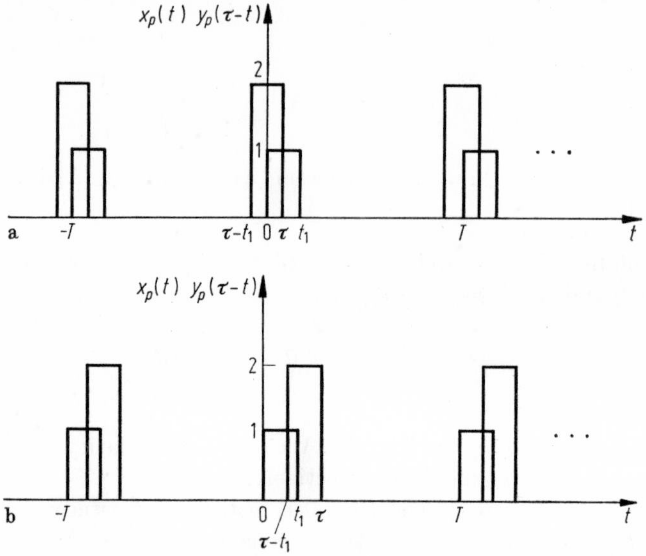

Fig. 2.10 Graphical interpretation of convolution

From Figs. 2.10 a and 2.10 b respectively it follows that

$$Z_{xy}(\tau) = \frac{1}{T} \int_0^{\tau} (1)\,(2)\,dt = 2\frac{\tau}{T}, \qquad 0 < \tau \leq t_1 \tag{2.5-19}$$

and

$$Z_{xy}(\tau) = \frac{1}{T} \int_{\tau - t_1}^{t_1} (1)\,(2)\,dt = \frac{2}{T}\,(2t_1 - \tau), \qquad t_1 < \tau \leq 2t_1 \tag{2.5-20}$$

A sketch of $Z_{xy}(\tau)$ which results from Eqs. (19) and (20) is shown in Fig. 2.11.

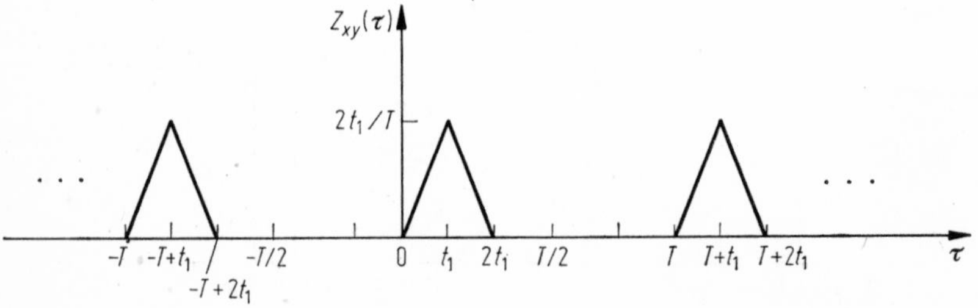

Fig. 2.11 Convolution of $x_p(t)$ and $y_p(t)$

The above graphical considerations imply that the process of convolution is identical to that of correlation, except that the "mirror image" of $x_p(t)$ or $y_p(t)$ is crosscorrelated with the other.

Convolution Theorem: If $(c_n)_x$ and $(c_n)_y$ are the Fourier series coefficients of $x_p(t)$ and $y_p(t)$ as defined in Eqs. (6) and (7) respectively, then,

$$(c_n)_z = (c_n)_x \, (c_n)_y, \qquad n = 0, \pm 1, \pm 2, \ldots \tag{2.5-21}$$

where the $(c_n)_z$ are related to $Z_{xy}(\tau)$ by the following Fourier series pair:

$$Z_{xy}(\tau) = \sum_{n=-\infty}^{\infty} (c_n)_z \, e^{in\omega_0\tau}$$

$$(c_n)_z = \frac{1}{T} \int_T Z_{xy}(\tau) \, e^{-in\omega_0\tau} \, d\tau$$

Since the proof of the above theorem is similar to that pertaining to the correlation theorem in Eq. (5), it is left as an exercise (Prob. 2-5).

2.6 Sampling Theorem

If $x(t)$ is a signal of duration L, then its L-periodic extension has the Fourier series representation

$$x_p(t) = \sum_{n=-\infty}^{\infty} c_n \, e^{in\omega_0 t} \tag{2.6-1}$$

where

$$c_n = \frac{1}{L} \int_L x_p(t) \, e^{-in\omega_0 t} \, dt, \qquad \omega_0 = \frac{2\pi}{L}$$

From Eq. (1) it is clear that an infinite number of coefficients c_n are required to describe $x_p(t)$ exactly. Therefore, in the strict mathematical sense, there is no signal $x(t)$ with a finite duration L, such that,

$$c_n = 0, \qquad |n| > \hat{N} \tag{2.6-2}$$

where $\hat{N}$ is some finite positive integer. However, it is well-known that for all practical purposes, the transmission characteristics of any physical system vanish at "very large" frequencies. Examples of such systems include the human vocal, audio and visual mechanisms, and various kinds of communication systems. Therefore, it is reasonable to assume that for sufficiently large $\hat{N}$, the condition in Eq. (2) is satisfied. Signals of this type are said to be bandlimited, and have a bandwidth of $\hat{N}\omega_0$ (radians per second).

In essence, the sampling theorem states that if a signal $x(t)$ is band-limited with bandwidth B Hz, then it can be *uniquely* determined by its sampled representation obtained by sampling it at f_s samples per second, where $f_s \geq 2B$. Thus this theorem enables us to work with the digital signal $x^*(t)$ corresponding to $x(t)$. In what follows, a proof for the sampling theorem is presented.

Theorem. If $x(t)$ is a signal of duration L, such that the Fourier series representation of its periodic extension has no harmonics above the $N/2$ harmonic[1], then $x(t)$ is completely determined by the set of values

$$x\left(\frac{kL}{N+1}\right), \qquad k = 0, 1, 2, \ldots, N \tag{2.6-3}$$

Furthermore, it may be obtained from these values as follows:

$$x(t) = \sum_{k=0}^{N} x\left(\frac{kL}{N+1}\right) \frac{\sin\left\{\frac{(N+1)\pi}{L}\left(t - \frac{kL}{N+1}\right)\right\}}{(N+1)\sin\left\{\frac{\pi}{L}\left(t - \frac{kL}{N+1}\right)\right\}}, \qquad 0 \leq t < L \tag{2.6-4}$$

Proof: The Fourier series representation of $x(t)$ is given by

$$x(t) = \sum_{n=-N/2}^{N/2} c_n e^{\frac{i n 2\pi t}{L}}, \qquad 0 \leq t < L \tag{2.6-5}$$

Substituting $t = kL/(N+1)$ in Eq. (5) we obtain

$$x\left(\frac{kL}{N+1}\right) = \sum_{n=-N}^{N} c_n e^{\frac{i 2\pi nk}{N+1}}$$

$$= \sum_{n=-N}^{N} c_n U^{nk} \tag{2.6-6}$$

where

$$U = e^{\frac{i 2\pi}{N+1}}$$

Multiplying both sides of Eq. (6) by U^{-mk}, and summing over $k = 0$ to N, results in

$$\sum_{k=0}^{N} x\left(\frac{kL}{N+1}\right) U^{-mk}, = \sum_{k=0}^{N} \sum_{n=-N}^{N} c_n U^{(n-m)k} \tag{2.6-7}$$

Now, it can be shown that the set of functions $\{U^{mk}\}$ are orthogonal. That is,

$$\sum_{k=0}^{N} U^{(n-m)k} = \begin{cases} N+1, & n = m \\ 0, & n \neq m \end{cases} \tag{2.6-8}$$

[1] $N/2$ is assumed to be an integer, without any lass of generality.

female intuition

Thus, from Eqs. (7) and (8) it follows that

$$\sum_{k=0}^{N} x\left(\frac{kL}{N+1}\right) U^{-mk} = (N+1)\, c_m$$

That is

$$c_n = \frac{1}{N+1} \sum_{k=0}^{N} x\left(\frac{kL}{N+1}\right) U^{-nk}, \qquad n = 0, \pm 1, \pm 2, \ldots, \pm N/2 \qquad (2.6\text{-}9)$$

From Eq. (9) it is clear that the c_n are uniquely determined by the N sampled values of $x(t)$ which are defined in Eq. (3). Substituting Eq. (9) in Eq. (5) we obtain

$$x(t) = \frac{1}{N+1} \sum_{n=-N/2}^{N/2} \left\{ \sum_{k=0}^{N} x\left(\frac{kL}{N+1}\right) U^{-nk} \right\} e^{\frac{i 2\pi n t}{L}} \qquad (2.6\text{-}10)$$

Since $U = e^{i 2\pi/(N+1)}$, Eq. (10) yields

$$x(t) = \sum_{k=0}^{N} x\left(\frac{kL}{N+1}\right) \left\{ \frac{1}{(N+1)} \sum_{n=-N/2}^{N/2} e^{\frac{i 2\pi n}{L}\left(t - \frac{kL}{N+1}\right)} \right\} \qquad (2.6\text{-}11\,\text{a})$$

Let $z = \left(t - \dfrac{kL}{N+1}\right)$. Then Eq. (11a) can be written as

$$x(t) = \sum_{k=0}^{N} x\left(\frac{kL}{N+1}\right) \left\{ \frac{1}{(N+1)} \sum_{n=-N/2}^{N/2} e^{i \omega_0 n z} \right\} \qquad (2.6\text{-}11\,\text{b})$$

In Eq. (11b) we observe that

$$\begin{aligned} \sum_{n=-N/2}^{N/2} e^{i \omega_0 n z} &= e^{-i\omega_0 \frac{N}{2} z} + e^{-i\omega_0 \left(\frac{N}{2} - 1\right) z} \\ &\quad + \cdots + 1 + \cdots + e^{i\omega_0 \left(\frac{N}{2} - 1\right) z} + e^{i\omega_0 \frac{N}{2} z} \\ &= 2\left[\frac{1}{2} + \cos\omega_0 z + \cos 2\omega_0 z + \cdots + \cos\left(\frac{N}{2}\omega_0 z\right)\right] \end{aligned} \qquad (2.6\text{-}12)$$

We now use the trigonometric identity[1]

$$\frac{1}{2} + \cos\omega_0 z + \cos 2\omega_0 z + \cdots + \cos\frac{N}{2}\omega_0 z = \frac{\sin\left\{(N+1)\,\omega_0 \frac{z}{2}\right\}}{2\sin\left(\omega_0 \frac{z}{2}\right)} \qquad (2.6\text{-}13)$$

[1] Guillemin, E. A.: *Mathematics of Circuit Analysis.* New York: Wiley, 1949, p. 437.

Substitution of Eqs. (12) and (13) in Eq. (11b) results in

$$x(t) = \sum_{k=0}^{N} x\left(\frac{kL}{N+1}\right) \frac{\sin\left\{(N+1)\,\omega_0 \frac{z}{2}\right\}}{(N+1)\sin\left(\omega_0 \frac{z}{2}\right)} \tag{2.6-14}$$

Substituting $z = \left(t - \dfrac{kL}{N+1}\right)$ and $\omega_0 = 2\pi/L$ in Eq. (14) we obtain

$$x(t) = \sum_{k=0}^{N} x\left(\frac{kL}{N+1}\right) \frac{\sin\left\{\frac{(N+1)\pi}{L}\left(t - \frac{kL}{N+1}\right)\right\}}{(N+1)\sin\left\{\frac{\pi}{L}\left(t - \frac{kL}{N+1}\right)\right\}}, \qquad 0 \le t < L$$

which is the desired result.

With respect to the above discussion, we make the following comments:

1. The sampling theorem establishes a *minimum rate* of sampling for preserving all the information in the signal $x(t)$; any higher rate of sampling also preserves the information. Again, since the minimum number of samples is $(N+1)$ in time L, the time interval Δt between successive samples must satisfy the condition

$$\Delta t \le \frac{L}{N+1} \tag{2.6-15}$$

2. Since $B = Nf_0/2$ Hz is the bandwidth of the signal where $f_0 = 1/L$ is the fundamental frequency, the inequality in Eq. (15) becomes

$$\Delta t \le \frac{N}{2B(N+1)} \tag{2.6-16}$$

3. Equation (16) implies that the sampling frequency $f_s = 1/\Delta t$ must satisfy the condition

$$f_s \ge 2B\left(1 + \frac{1}{N}\right)$$

which yields

$$f_s \ge 2B \text{ samples per second, for } N \gg 1 \tag{2.6-17}$$

2.7 Summary

In this chapter we started with the Fourier series, which is the most familiar type of signal representation. For a T-periodic signal, it was shown that as T approaches infinity, the Fourier series yields the Fourier transform. Again it was established that the Fourier transform $F_x(\omega)$ of a signal $x(t)$, of finite duration L, is exactly specified by the Fourier series at a set of equally spaced points on the ω axis. The spacing between these points is $2\pi/L$ radians. Finally, some aspects of convolution and correlation were discussed, and a systematic transition from the representation of signals to that of digital signals was facilitated by the sampling theorem.

References

1. Hancock, J. C.: *An Introduction to the Principles of Communication Theory.* New York: McGraw-Hill, 1961, Chap. 1.
2. Harman, W. N.: *Principles of Statistical Theory of Communication.* New York: McGraw-Hill, 1963, Chap. 1.
3. Guillemin, E. A.: *The Mathematics of Circuit Analysis.* New York: Wiley, 1949, Chap. VII.
4. Goldman, S.: *Information Theory.* New York: Prentice-Hall, 1953, Chap. 2.
5. Lee, Y. W.: *Statistical Theory of Communication.* New York: Wiley, 1961, Chap. 1.
6. Panter, P. F.: *Modulation, Noise and Spectral Analysis.* New York: McGraw-Hill, 1965, Chaps. 2 and 3.
7. Reza, F. M.: *An Introduction to Information Theory.* New York: McGraw-Hill, 1961, Chap. 9.
8. Shannon, C. E.: A Mathematical Theory of Communication. *Bell System Tech. J.* 27 (1948) 379–423.
9. Robbins, W. P., and Fawcett, R. L.: A Classroom Demonstration of Correlation, Convolution and The Superposition Integral. *IEEE Trans. Education* E-16 (1973) 18–23.

Problems

2-1 Using Eqs. (2.1-1) and (2.1-2), verify Eq. (2.1-3).
2-2 Using Eqs. (2.1-1) and (2.1-2), derive Eq. (2.1-4).
2-3 Using Eqs. (2.1-13) and (2.1-14), derive Eq. (2.1-15).
2-4 The Fourier phase spectrum is defined as

$$\Psi_n = \begin{cases} \tan^{-1}\left(\dfrac{\mathrm{Im}[c_n]}{\mathrm{Re}[c_n]}\right), & n = \pm 1, \pm 2, \ldots \\ 0, & n = 0. \end{cases}$$

Using this definition and Eq. (2.1-3), show that

$$\Psi_{-l} = -\Psi_l, \qquad l = 1, 2, 3, \ldots$$

2-5 If $Z_{xy}(\tau) = \dfrac{1}{T}\displaystyle\int_T x_p(t)\, y_p(\tau - t)\, dt,$

show that

$$(c_n)_z = (c_n)_x\,(c_n)_y, \qquad n = 0, \pm 1, \pm 2, \ldots$$

where

$$(c_n)_x = \frac{1}{T} \int_T x_p(t)\, e^{-in\omega_0 t}\, dt$$

$$(c_n)_y = \frac{1}{T} \int_T y_p(t)\, e^{-in\omega_0 t}\, dt$$

and

$$(c_n)_z = \frac{1}{T} \int_T Z_{xy}(\tau)\, e^{-in\omega_0 \tau}\, d\tau$$

2-6 Consider the signal $x(t)$ shown in Fig. P2-6-1(a)

(a) Show that $F_x(\omega) = (2/\omega^2\varepsilon)(1 - \cos \omega\varepsilon)$
(b) Evaluate $F_x(0)$. [Answer: $F_x(0) = \varepsilon$]
(c) Let $x_p(t)$ be the periodic extension of $x(t)$ as shown in Fig. P2-6-1(b). Compute the Fourier series coefficients c_n such that

$$x_p(t) = \frac{1}{T} \sum_{n=-\infty}^{\infty} c_n\, e^{in\omega_0 t}$$

where ω_0 is the fundamental radian frequency and T is the period of $x_p(t)$.

Answer:

$$c_n = \begin{cases} 0, & n \text{ even} \\ \dfrac{4\varepsilon}{n^2\pi^2}, & n \text{ odd} \end{cases}$$

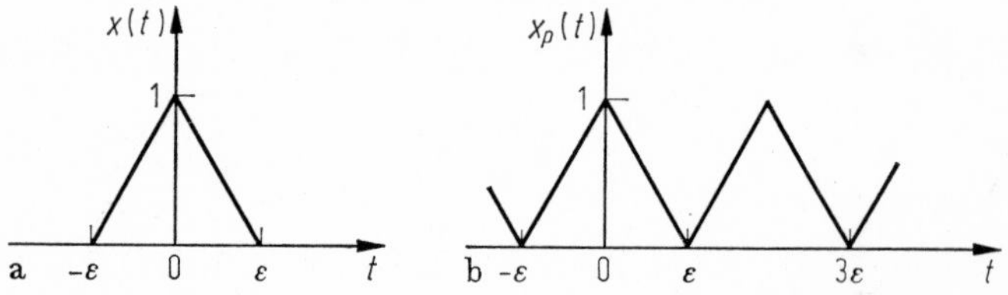

Fig. P2.6-1.

Chapter Three

Fourier Representation of Sequences

The Fourier representation of analog signals was discussed in the previous chapter. This representation is now extended to data sequences, and digital signals. To this end, the discrete Fourier transform (DFT) is defined and several of its properties are developed. Specifically, the convolution and correlation theorems are described and the spectral properties such as amplitude, phase, and power spectra are developed. By illustrating the 2-dimensional DFT, it is shown that the DFT can be extended to multiple dimensions. Finally, the concepts of time-varying Fourier power and phase spectra are introduced.

3.1 Definition of the Discrete Fourier Transform

If $\{X(m)\}$ denotes a sequence $X(m)$, $m = 0, 1, \ldots, N-1$ of N finite valued real or complex numbers, then its discrete [1] or finite [2] Fourier transform is defined as

$$C_x(k) = \frac{1}{N} \sum_{m=0}^{N-1} X(m) W^{km}, \qquad k = 0, 1, \ldots, N-1 \qquad (3.1\text{-}1)$$

where $W = e^{-i2\pi/N}$, $i = \sqrt{-1}$. The exponential functions W^{km} in Eq. (1) are orthogonal such that

$$\sum_{m=0}^{N-1} W^{km} W^{-lm} = \begin{cases} N, & \text{if } (k-l) \text{ is zero or an integer multiple of } N \\ 0, & \text{otherwise} \end{cases} \qquad (3.1\text{-}2)$$

Now, from Eq. (1) we have

$$\sum_{k=0}^{N-1} C_x(k)\, W^{-km}$$

$$= \frac{1}{N} \sum_{k=0}^{N-1} W^{-km} [X(0) + X(1) W^k + \cdots + X(m) W^{km} + \cdots + X(N-1) W^{k(N-1)}] \qquad (3.1\text{-}3)$$

Application of Eq. (2) to Eq. (3) results in the inverse discrete Fourier transform (IDFT) which is defined as

$$X(m) = \sum_{k=0}^{N-1} C_x(k)\, W^{-km}, \qquad m = 0, 1, \ldots, N-1 \qquad (3.1\text{-}4)$$

Since Eqs. (1) and (4) constitute a transform pair, it follows that the representation of the data sequence $\{X(m)\}$ in terms of the exponential functions W^{km} is unique.

We observe that functions W^{km} are N-periodic; that is,

$$W^{km} = W^{(k+N)m} = W^{k(m+N)}, \qquad k, m = 0, \pm 1, \pm 2, \ldots \tag{3.1-5}$$

Consequently the sequences $\{C_x(k)\}$ and $\{X(m)\}$ as defined by Eqs. (1) and (4) are also N-periodic. That is, the sequences $\{X(m)\}$ and $\{C_x(k)\}$ satisfy the following conditions:

$$\begin{aligned} X(\pm m) &= X(sN \pm m) \\ C_x(\pm k) &= C_x(sN \pm k), \end{aligned} \qquad s = 0, \pm 1, \pm 2, \ldots \tag{3.1-6}$$

Using Eqs. (5) and (6) it can be shown that

$$\sum_{m=p}^{q} X(m)\, W^{km} = \sum_{m=0}^{N-1} X(m)\, W^{km} \tag{3.1-7}$$

and

$$\sum_{k=p}^{q} C_x(k)\, W^{-km} = \sum_{k=0}^{N-1} C_x(k)\, W^{-km} \tag{3.1-8}$$

when p and q are such that $|p - q| = N - 1$.

It is generally convenient to adopt the convention [2]

$$X(m) \leftrightarrow C_x(k)$$

to represent the transform pair defined by Eqs. (1) and (4).

3.2 Properties of the DFT

A detailed discussion of the properties of the DFT is available in [2, 4]. In what follows we consider a few of these properties which are of interest to us.

1. Linearity Theorem. The DFT is a linear transform; i.e., if

$$X(m) \leftrightarrow C_x(k)$$

and

$$Z(m) = aX(m) + bY(m)$$

then

$$C_z(k) = aC_x(k) + bC_y(k) \tag{3.2-1}$$

2. Complex Conjugate Theorem. If $\{X(m)\} = \{X(0)\; X(1) \cdots X(N-1)\}$ is a real-valued sequence such that $N/2$ is an integer, and

$$X(m) \leftrightarrow C_x(k)$$

then

$$C_x(N/2 + l) = \bar{C}_x(N/2 - l), \qquad l = 0, 1, \ldots, N/2 \tag{3.2-2}$$

where $\bar{C}_x(k)$ denotes the complex conjugate of $C_x(k)$.

Proof: $X(m) \leftrightarrow C_x(k)$ implies that

$$C_x(k) = \frac{1}{N} \sum_{m=0}^{N-1} X(m)\, W^{km}$$

That is

$$C_x\left(\frac{N}{2} + l\right) = \frac{1}{N} \sum_{m=0}^{N-1} X(m)\, W^{(N/2+l)m}$$

$$= \frac{1}{N} \sum_{m=0}^{N-1} X(m)\, W^{-(N/2-l)m}\, W^{Nm}$$

$$= \bar{C}_x(N/2 - l)$$

since $W^{Nm} = 1$. Thus

$$C_x(N/2 + l) = \bar{C}_x(N/2 - l), \qquad l = 0, 1, \ldots, N/2$$

Note that this implies that $C_x(N/2)$ is always real.

3. Shift Theorem. If

$$X(m) \leftrightarrow C_x(k)$$

and

$$Z(m) = X(m + h), \qquad h = 0, 1, 2, \ldots, N - 1 \tag{3.2-3}$$

then

$$C_z(k) = W^{-kh}\, C_x(k) \tag{3.2-4}$$

Proof: $Z(m) \leftrightarrow C_z(k)$

That is

$$C_z(k) = \frac{1}{N} \sum_{m=0}^{N-1} Z(m)\, W^{km}, \qquad k = 0, 1, 2, \ldots, N - 1 \tag{3.2-5}$$

which yields

$$C_z(k) = \frac{1}{N} \sum_{m=0}^{N-1} X(m + h)\, W^{km} \tag{3.2-6}$$

Let $m + h = r$. Then Eq. (6) yields

$$C_z(k) = W^{-kh} \left\{ \frac{1}{N} \sum_{r=h}^{N+h-1} X(r)\, W^{kr} \right\} \tag{3.2-7}$$

From Eqs. (3.1-7) and (7) it follows that

$$C_z(k) = W^{-kh} C_x(k)$$

which is the desired result.

Similarly, if

$$Z(m) = X(m - h), \qquad h = 0, 1, \ldots, N - 1$$

then it can be shown that

$$C_z(k) = W^{kh} C_x(k) \tag{3.2-8}$$

4. Convolution Theorem. If $\{X(m)\}$ and $\{Y(m)\}$ are real-valued sequences such that

$$\begin{aligned} X(m) &\leftrightarrow C_x(k) \\ Y(m) &\leftrightarrow C_y(k) \end{aligned} \tag{3.2-9}$$

and their convolution is given by

$$Z(m) = \frac{1}{N} \sum_{h=0}^{N-1} X(h)\, Y(m - h), \qquad m = 0, 1, \ldots, N - 1 \tag{3.2-10}$$

then

$$C_z(k) = C_x(k)\, C_y(k) \tag{3.2-11}$$

Proof: Taking the DFT of $\{Z(m)\}$ we obtain

$$C_z(k) = \frac{1}{N} \sum_{m=0}^{N-1} Z(m)\, W^{km} \tag{3.2-12}$$

$$= \frac{1}{N^2} \sum_{m=0}^{N-1} \sum_{h=0}^{N-1} X(h)\, Y(m - h)\, W^{km}$$

$$= \frac{1}{N} \sum_{h=0}^{N-1} X(h) \left\{ \frac{1}{N} \sum_{m=0}^{N-1} Y(m - h)\, W^{km} \right\} \tag{3.2-13}$$

From the shift theorem, it follows that

$$\frac{1}{N} \sum_{m=0}^{N-1} Y(m - h)\, W^{km} = W^{kh} C_y(k) \tag{3.2-14}$$

Thus Eqs. (13) and (14) yield

$$C_z(k) = C_y(k) \frac{1}{N} \left\{ \sum_{h=0}^{N-1} X(h)\, W^{kh} \right\} = C_x(k)\, C_y(k)$$

Clearly, Eq. (11) is analogous to Eq. (2.5-21) which represents the convolution theorem for the Fourier series case. This theorem states that the convolution of data sequences is equivalent to multiplication of their DFT coefficients.

5. Correlation Theorem. If $\{X(m)\}$ and $\{Y(m)\}$ are real-valued sequences such that

$$X(m) \leftrightarrow C_x(k)$$

$$Y(m) \leftrightarrow C_y(k)$$

and their correlation is given by

$$\hat{Z}(m) = \frac{1}{N} \sum_{h=0}^{N-1} X(h)\, Y(m+h), \qquad m = 0, 1, \ldots, N-1 \tag{3.2-15}$$

then

$$C_{\hat{z}}(k) = \bar{C}_x(k)\, C_y(k) \tag{3.2-16}$$

Proof: By definition

$$C_{\hat{z}}(k) = \frac{1}{N} \sum_{m=0}^{N-1} \hat{Z}(m)\, W^{km} \tag{3.2-17}$$

Substituting Eq. (15) in Eq. (17) and subsequently interchanging the order of summation, we obtain

$$C_{\hat{z}}(k) = \frac{1}{N} \sum_{h=0}^{N-1} X(h) \left\{ \frac{1}{N} \sum_{m=0}^{N-1} Y(m+h)\, W^{km} \right\} \tag{3.2-18}$$

Applying the shift theorem [see Eq. (4)] to Eq. (18), we obtain

$$C_{\hat{z}}(k) = C_y(k) \left\{ \frac{1}{N} \sum_{h=0}^{N-1} X(h)\, W^{-kh} \right\} \tag{3.2-19}$$

Now,

$$C_x(k) = \frac{1}{N} \sum_{h=0}^{N-1} X(h)\, W^{kh}, \qquad k = 0, 1, \ldots, N-1$$

implies that

$$\bar{C}_x(k) = \frac{1}{N} \sum_{h=0}^{N-1} X(h)\, W^{-kh} \tag{3.2-20}$$

Hence Eq. (19) yields

$$C_{\hat{z}}(k) = \bar{C}_x(k)\, C_y(k), \qquad k = 0, 1, \ldots, N-1$$

which completes the proof of the theorem.

Comment: If the sequences $\{X(m)\}$ and $\{Y(m)\}$ are identical, then Eq. (16) reduces to

$$C_{\hat{z}}(k) = |C_x(k)|^2, \qquad k = 0, 1, \ldots, N-1 \tag{3.2-21}$$

Again, the IDFT of $\{C_{\hat{z}}(k)\}$ yields

$$\hat{Z}(m) = \sum_{k=0}^{N-1} C_{\hat{z}}(k)\, W^{-km} \tag{3.2-22}$$

Substituting Eqs. (15) and (21) in Eq. (22) we obtain

$$\frac{1}{N} \sum_{h=0}^{N-1} X(h) X(m+h) = \sum_{k=0}^{N-1} |C_x(k)|^2\, W^{-km} \tag{3.2-23}$$

For the special case when $m = 0$, this equation becomes

$$\frac{1}{N} \sum_{h=0}^{N-1} X^2(h) = \sum_{k=0}^{N-1} |C_x(k)|^2 \tag{3.2-24}$$

Comparing Eq. (24) with Eq. (2.5-17), it follows that Eq. (24) represents Parseval's theorem for the data sequence $\{X(m)\}$.

Example 3.2-1

Consider two 4-periodic sequences

$$\{X(m)\} = \{1 \;\; 2 \;\; -1 \;\; 3\}, \quad \text{and} \quad \{Y(m)\} = \{-1 \;\; 1 \;\; 4 \;\; 1\}.$$

Verify that

$$\frac{1}{4}\sum_{l=0}^{3} X(l)\,Y(2-l) = \sum_{k=0}^{3} C_x(k)\,C_y(k)\,W^{-2k}$$

Solution:

$$\begin{aligned}\frac{1}{4}\sum_{l=0}^{3} X(l)\,Y(2-l) &= \frac{1}{4}\{X(0)\,Y(2) + X(1)\,Y(1) + X(2)\,Y(0) + X(3)\,Y(-1)\}\\ &= \frac{1}{4}\{X(0)\,Y(2) + X(1)\,Y(1) + X(2)\,Y(0) + X(3)\,Y(3)\}\end{aligned} \tag{3.2-25}$$

since $Y(-l) = Y(-l+N)$, and $N = 4$.

Substituting numerical values for $X(m)$ and $Y(m)$, $m = 0, 1, 2, 3$ in Eq. (25), we obtain

$$\frac{1}{4}\sum_{l=0}^{3} X(l)\,Y(2-l) = 2.5 \tag{3.2-26}$$

Again,

$$C_x(k) = \frac{1}{4}\sum_{m=0}^{3} X(m)\,W^{km}, \qquad k = 0, 1, 2, 3 \tag{3.2-27}$$

where $W = e^{-i2\pi/4} = -i$. Evaluating Eq. (27) we get

$$C_x(0) = \frac{5}{4}, \quad C_x(1) = \frac{(2+i)}{4}, \quad C_x(2) = \frac{-5}{4}, \quad C_x(3) = \frac{(2-i)}{4}$$

Similarly,

$$C_y(0) = \frac{5}{4}, \quad C_y(1) = \frac{-5}{4}, \quad C_y(2) = \frac{1}{4}, \quad \text{and} \quad C_y(3) = \frac{-5}{4}$$

Thus

$$\sum_{k=0}^{3} C_x(k)\,C_y(k)\,W^{-2k} = \frac{1}{16}\{25 + 5(2+i) - 5 + 5(2-i)\} = 2.5$$

which agrees with Eq. (26).

3.3 Matrix Representation of Correlation and Convolution

In Section 3.2 the notions of correlation and convolution were introduced by means of the correlation and convolution theorems. If $\{X(m)\}$ and $\{Y(m)\}$ are two real N-periodic sequences, then the correlation and convolution operations are respectively defined as

$$\hat{Z}(m) = \frac{1}{N} \sum_{h=0}^{N-1} X(h)\, Y(m+h) \tag{3.3-1}$$

and

$$Z(m) = \frac{1}{N} \sum_{h=0}^{N-1} X(h)\, Y(m-h) \tag{3.3-2}$$

Correlation. In Eq. (1) we let $N = 4$ to obtain the following set of equations:

$$\begin{aligned}
4\hat{Z}(0) &= X(0)Y(0) + X(1)Y(1) + X(2)Y(2) + X(3)Y(3)\\
4\hat{Z}(1) &= X(0)Y(1) + X(1)Y(2) + X(2)Y(3) + X(3)Y(4)\\
4\hat{Z}(2) &= X(0)Y(2) + X(1)Y(3) + X(2)Y(4) + X(3)Y(5)\\
4\hat{Z}(3) &= X(0)Y(3) + X(1)Y(4) + X(2)Y(5) + X(3)Y(6)
\end{aligned} \tag{3.3-3}$$

Since $\{Y(m)\}$ is 4-periodic, Eq. (3) can be expressed in matrix form as

$$\begin{bmatrix} \hat{Z}(0) \\ \hat{Z}(1) \\ \hat{Z}(2) \\ \hat{Z}(3) \end{bmatrix} = \frac{1}{4} \begin{bmatrix} X(0) & X(1) & X(2) & X(3) \\ X(3) & X(0) & X(1) & X(2) \\ X(2) & X(3) & X(0) & X(1) \\ X(1) & X(2) & X(3) & X(0) \end{bmatrix} \begin{bmatrix} Y(0) \\ Y(1) \\ Y(2) \\ Y(3) \end{bmatrix} \tag{3.3-4}$$

Convolution. With $N = 4$, Eq. (2) yields

$$\begin{aligned}
4Z(0) &= X(0)Y(0) + X(1)Y(-1) + X(2)Y(-2) + X(3)Y(-3)\\
4Z(1) &= X(0)Y(1) + X(1)Y(0) \quad + X(2)Y(-1) + X(3)Y(-2)\\
4Z(2) &= X(0)Y(2) + X(1)Y(1) \quad + X(2)Y(0) \quad + X(3)Y(-1)\\
4Z(3) &= X(0)Y(3) + X(1)Y(2) \quad + X(2)Y(1) \quad + X(3)Y(0)
\end{aligned} \tag{3.3-5}$$

That is

$$\begin{bmatrix} Z(0) \\ Z(1) \\ Z(2) \\ Z(3) \end{bmatrix} = \frac{1}{4} \begin{bmatrix} X(0) & X(1) & X(2) & X(3) \\ X(1) & X(2) & X(3) & X(0) \\ X(2) & X(3) & X(0) & X(1) \\ X(3) & X(0) & X(1) & X(2) \end{bmatrix} \begin{bmatrix} Y(0) \\ Y(3) \\ Y(2) \\ Y(1) \end{bmatrix} \tag{3.3-6}$$

Equations (4) and (6) suggest the presence of simple rules to write down the matrix forms of correlation and convolution as indicated by the arrows. These rules are easily extended to the general case to obtain the following matrix equations:

Correlation:

$$\begin{bmatrix} \hat{Z}(0) \\ \hat{Z}(1) \\ \hat{Z}(2) \\ \cdot \\ \cdot \\ \cdot \\ \hat{Z}(N-2) \\ \hat{Z}(N-1) \end{bmatrix} = \frac{1}{N} \begin{bmatrix} X(0) & X(1) & X(2) \cdots & X(N-1) \\ X(N-1) & X(0) & X(1) \cdots & X(N-2) \\ X(N-2) & X(N-1) & X(0) \cdots & X(N-3) \\ \cdot & \cdot & \cdot & \cdot \\ \cdot & \cdot & \cdot & \cdot \\ \cdot & \cdot & \cdot & X(2) \\ X(2) & X(3) & X(4) \cdots & X(1) \\ X(1) & X(2) & X(3) \cdots & X(0) \end{bmatrix} \begin{bmatrix} Y(0) \\ Y(1) \\ Y(2) \\ \cdot \\ \cdot \\ \cdot \\ Y(N-2) \\ Y(N-1) \end{bmatrix} \tag{3.3-7}$$

and

Convolution:

$$\begin{bmatrix} Z(0) \\ Z(1) \\ Z(2) \\ \cdot \\ \cdot \\ \cdot \\ Z(N-2) \\ Z(N-1) \end{bmatrix} = \frac{1}{N} \begin{bmatrix} X(0) & X(1) & X(2) \cdots & X(N-1) \\ X(1) & X(2) & X(3) \cdots & X(0) \\ X(2) & X(3) & X(4) \cdots & X(1) \\ \cdot & \cdot & \cdot & \cdot \\ \cdot & \cdot & \cdot & \cdot \\ X(N-3) & \cdot & \cdot & \cdot \\ X(N-2) & X(N-1) & X(0) \cdots & X(N-3) \\ X(N-1) & X(0) & X(1) \cdots & X(N-2) \end{bmatrix} \begin{bmatrix} Y(0) \\ Y(N-1) \\ Y(N-2) \\ \cdot \\ \cdot \\ \cdot \\ \\ Y(1) \end{bmatrix} \tag{3.3-8}$$

In closing, we remark that if the sequences $\{X(m)\}$ and $\{Y(m)\}$ are identical, then Eq. (1) yields

$$\hat{Z}(m) = \frac{1}{N} \sum_{h=0}^{N-1} X(h)\,X(m+h), \qquad m = 0, 1, \ldots, N-1 \tag{3.3-9}$$

which is defined as the autocorrelation of $\{X(m)\}$.

3.4 Relation Between the DFT and the Fourier Transform/Series

Let $x(t)$ be a real-valued signal of duration L seconds, such that the Fourier series of its periodic extension has no harmonic above the $\tilde{N}/2$ harmonic[1]. Since $f_0 = 1/L$ Hz, the bandwidth B can be expressed as

$$B = \frac{\tilde{N}}{2L} \tag{3.4-1}$$

From Eq. (1) and the sampling theorem [see Eq. (2.6-17)], it follows that $x(t)$ can be represented by N equally spaced sampled values $X(m)$ such that

$$X(m) = x(m\Delta t), \qquad m = 0, 1, \ldots, N-1 \tag{3.4-2}$$

where $\Delta t = L/N$ is the sampling interval, and $N \geqslant \tilde{N}$.

We denote the sampled version of $x(t)$ by $x^*(t)$, which can be represented by a sequence of impulses as shown in Fig. 3.1. The impulse at $t = m\Delta t$ has the strength $\Delta t X(m)$. Thus $x^*(t)$ is expressed as

$$x^*(t) = \sum_{m=0}^{N-1} [\Delta t\, X(m)]\, \delta(t - m\Delta t) \tag{3.4-3}$$

where $\delta(t)$ is the Dirac or impulse function.

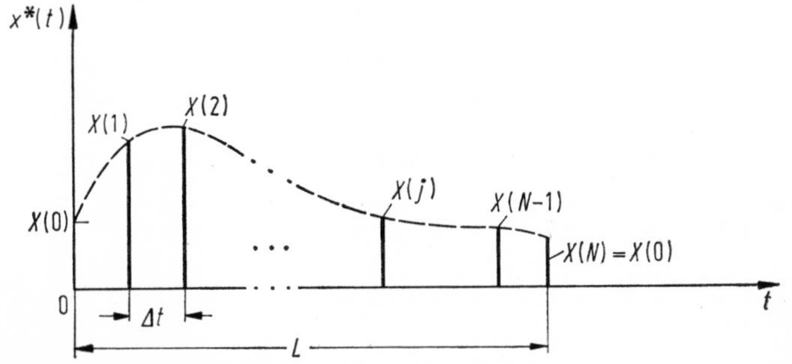

Fig. 3.1 Sampled representation of $x(t)$

Relation to the Fourier transform. Taking the Fourier transform of $x^*(t)$ in Eq. (3) we obtain

$$F_{x^*}(\omega) = \Delta t \int_{-\infty}^{\infty} \sum_{m=0}^{N-1} X(m)\, \delta(t - m\Delta t)\, e^{-i\omega t}\, dt \tag{3.4-4}$$

That is

$$F_{x^*}(\omega) = \Delta t \sum_{m=0}^{N-1} X(m)\, e^{-i\omega m \Delta t} \tag{3.4-5}$$

[1] For convenience, $\tilde{N}$ is considered to be an even number.

Now, Eq. (5) yields $F_{x^*}(\omega)$ for *all* values of ω. However, if we are only interested with values of $F_{x^*}(\omega)$ at a set of discrete points, Eq. (5) can be expressed as

$$F_{x^*}(k\omega_0) = \Delta t \sum_{m=0}^{N-1} X(m)\, e^{-ikm\omega_0 \Delta t}, \qquad k = 0, \pm 1, \pm 2, \ldots, \pm N/2 \tag{3.4-6}$$

where $\omega_0 = 2\pi/L$.

In Eq. (6) we note that

$$F_{x^*}(-r\omega_0) = \bar{F}_{x^*}(r\omega_0), \qquad r = 0, 1, \ldots, N/2$$

Thus without loss of generality, Eq. (6) can be written as

$$F_{x^*}(k\omega_0) = \Delta t \sum_{m=0}^{N-1} X(m)\, e^{-ikm\omega_0 \Delta t}, \qquad k = 0, 1, \ldots, N/2 \tag{3.4-7}$$

Recalling that $\Delta t = L/N$ and $\omega_0 L = 2\pi$, Eq. (7) yields

$$F_{x^*}(k\omega_0) = \frac{L}{N} \sum_{m=0}^{N-1} X(m)\, W^{km}, \qquad k = 0, 1, \ldots, N/2 \tag{3.4-8}$$

where $W = e^{-i2\pi/N}$.

Comparing Eq. (8) with Eq. (3.1-1) we obtain the desired relation

$$C_x(k) = \frac{1}{L}\{F_{x^*}(k\omega_0)\}, \qquad k = 0, 1, \ldots, N/2 \tag{3.4-9}$$

Relation to the Fourier series. The Fourier series representation of $x^*(t)$ (see Fig. 3.1) is given by

$$x^*(t) = \frac{1}{L} \sum_{k=-N/2}^{N/2} c_k\, e^{ik\omega_0 t} \tag{3.4-10}$$

where

$$c_k = \int_L x^*(t)\, e^{-ik\omega_0 t}\, dt, \qquad k = 0, \pm 1, \pm 2, \ldots, \pm N/2$$

and $\omega_0 = 2\pi/L$.

Substitution of Eq. (3) in the expression for c_k in Eq. (10) results in

$$c_k = \int_L \sum_{m=0}^{N-1} [\Delta t\, X(m)]\, \delta(t - m\Delta t)\, e^{-ik\omega_0 t}\, dt$$

$$= \Delta t \sum_{m=0}^{N-1} \int_L X(m)\, \delta(t - m\Delta t)\, e^{-ik\omega_0 t}\, dt$$

That is

$$c_k = \Delta t \sum_{m=0}^{N-1} X(m)\, e^{-ik\omega_0 m \Delta t} \tag{3.4-11}$$

Substituting $\Delta t = L/N$ and $\omega_0 = 2\pi/L$ in Eq. (11), we obtain

$$c_k = \frac{L}{N} \sum_{m=0}^{N-1} X(m)\, W^{km} \tag{3.4-12}$$

Comparing Eq. (12) with Eq. (3.1-1) we conclude that the $C_x(k)$ are directly related to the Fourier series coefficients as follows:

$$C_x(k) = \frac{1}{L}\{c_k\}$$

and

$$\bar{C}_x(k) = \frac{1}{L}\{c_{-k}\}, \qquad k = 0, 1, \ldots, N/2 \tag{3.4-13}$$

3.5 Power, Amplitude and Phase Spectra

The notion of amplitude, power and phase spectra follows naturally from the discussion in the last section, where the DFT coefficients were seen to be directly related to the Fourier transform/series.

Power spectrum. We recall Parseval's theorem which is expressed in Eq. (3.2-24) as

$$\frac{1}{N} \sum_{h=0}^{N-1} X^2(h) = \sum_{k=0}^{N-1} |C_x(k)|^2 \tag{3.5-1}$$

If $x(t)$ represents a voltage or current waveform, and a load of 1 ohm pure resistance is assumed, then the left hand side of Eq. (1) represents the average power dissipated in the 1 ohm resistor. Thus, each $|C_x(k)|^2$ in Eq. (1) represents the power contributed by the harmonic associated with the frequency number k. Thus the DFT power spectrum is defined as

$$P(k) = |C_x(k)|^2, \qquad k = 0, 1, \ldots, N-1. \tag{3.5-2}$$

An important observation results from Eq. (2):

There are only $(N/2 + 1)$ independent DFT spectral points when $\{X(m)\}$ is real by virtue of the complex conjugate property in Eq. (3.2-2). These points are

$$P(k) = |C_x(k)|^2, \qquad k = 0, 1, \ldots, N/2 \tag{3.5-3}$$

From the shift theorem in Eq. (3.2-4) it follows that the power spectrum defined in Eq. (2) is invariant with respect to shifts of the N-periodic data sequence $\{X(m)\}$.

The definition of an *amplitude spectrum* follows readily from that of a power spectrum. The amplitude spectrum is defined as

$$p(k) = |C_x(k)|, \qquad k = 0, 1, \ldots, N-1 \tag{3.5-4}$$

Again, the amplitude spectrum is also invariant with respect to shifts of the data sequence $\{X(m)\}$.

Phase spectrum. Given a data sequence $X(m)$, $m = 0, 1, \ldots, N - 1$, the phase spectrum is defined as

$$\psi_x(k) = \tan^{-1}\left\{\frac{I_x(k)}{R_x(k)}\right\}, \qquad k = 0, 1, \ldots, N - 1 \qquad (3.5\text{-}5)$$

where $R_x(k)$ and $I_x(k)$ are respectively the real and imaginary parts of $C_x(k)$. As in the case of the power spectrum, only $(N/2 + 1)$ of the DFT phase spectral points in Eq. (5) are independent when $\{X(m)\}$ is real. The independent spectral points are $\psi_x(k)$, $k = 0, 1, \ldots, N/2$.

From Eqs. (3.1-1) and (5) it follows that a fundamental property of the phase spectrum is that it is invariant with respect to multiplication of $\{X(m)\}$ by a constant. Again, from Eq. (5) it is clear that the phase spectral point $\psi_x(k)$ represents the orientation of $C_x(k)$ in a two-dimensional space as shown in Fig. 3.2.

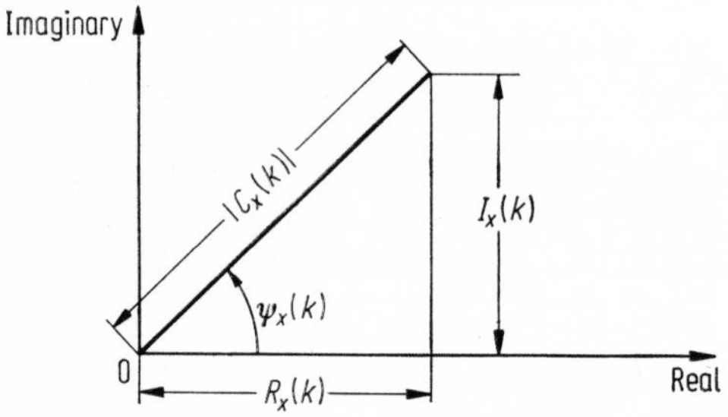

Fig. 3.2 Geometrical interpretation of the phase spectrum

In conclusion we note that by virtue of the complex conjugate property in Eq. (3.2-2), the DFT spectra have the following properties when $\{X(m)\}$ is real:

1. The power spectrum in Eq. (2) is an *even function* about the point $k = N/2$.

2. The phase spectrum in Eq. (5) is an *odd function* about the point $k = N/2$ [see Prob. 3-6].

3.6 2-dimensional DFT

The DFT can be extended to multiple dimensions of which the 2-dimensional case is the most useful one, since it is applicable in the area of image processing [10]. The 2-dimensional DFT is defined as

$$C_{xx}(k_1, k_2) = \frac{1}{N_1 N_2} \sum_{m_2=0}^{N_2-1} \sum_{m_1=0}^{N_1-1} X(m_1, m_2)\, W_1^{k_1 m_1} W_2^{k_2 m_2} \tag{3.6-1}$$

where

$$W_l = e^{\frac{-i2\pi}{N_l}}$$

and m_l, k_l vary from 0 through $(N_l - 1)$, $l = 1, 2$. The data is in the form of a $(N_1 \times N_2)$ matrix $[X(m_1, m_2)]$. That is,

$$[X(m_1, m_2)] = \begin{bmatrix} X(0,0) & X(0,1) & \cdots X(0, N_2-1) \\ X(1,0) & X(1,1) & \cdots X(1, N_2-1) \\ \cdots & \cdots & \cdots \\ X(N_1-1, 0) & X(N_1-1, 1) & \cdots X(N_1-1, N_2-1) \end{bmatrix} \tag{3.6-2}$$

In Eq. (1) we consider the inner summation which is given by

$$\frac{1}{N_1} \sum_{m_1=0}^{N_1-1} X(m_1, m_2)\, W_1^{k_1 m_1}$$
$$= \frac{1}{N_1} \{X(0, m_2) + X(1, m_2)\, W_1^{k_1} + \cdots + X(N_1-1, m_2) W_1^{k_1(N_1-1)}\} \tag{3.6-3}$$

From Eq. (3) it follows that its right-hand side represents the DFT of each *column* of the data matrix $[X(m_1, m_2)]$. Thus we introduce the notation

$$\frac{1}{N_1} \sum_{m_1=0}^{N_1-1} X(m_1, m_2) W_1^{k_1 m_1} = C_x(k_1, m_2) \tag{3.6-4}$$

The coefficients $C_x(k_1, m_2)$ in Eq. (4) can be written in the form of a $(N_1 \times N_2)$ matrix $[C_x(k_1, m_2)]$ as follows:

$$[C_x(k_1, m_2)] = \begin{bmatrix} C_x(0,0) & C_x(0,1) & \cdots C_x(0, N_2-1) \\ C_x(1,0) & C_x(1,1) & \cdots C_x(1, N_2-1) \\ \cdots & \cdots & \cdots \\ C_x(N_1-1, 0) & C_x(N_1-1, 1) & \cdots C_x(N_1-1, N_2-1) \end{bmatrix} \tag{3.6-5}$$

Substitution of Eq. (4) in Eq. (1) results in

$$C_{xx}(k_1, k_2) = \frac{1}{N_2} \sum_{m_2=0}^{N_2-1} C_x(k_1, m_2) W_2^{k_2 m_2} \tag{3.6-6}$$

That is

$$C_{xx}(k_1, k_2) = \frac{1}{N_2} \{C_x(k_1, 0) + C_x(k_1, 1) W_2^{k_2} + \cdots + C_x(k_1, N_2-1) W_2^{k_2(N_2-1)}\}$$

This implies that the coefficients $C_{xx}(k_1, k_2)$ are obtained by taking the DFT of each *row* of $[C_x(k_1, m_2)]$ in Eq. (5). This results in a set of $N_1 N_2$ coefficients which can be written in the form of a matrix to obtain

$$[C_{xx}(k_1, k_2)] = \begin{bmatrix} C_{xx}(0, 0) & C_{xx}(0, 1) & \cdots & C_{xx}(0, N_2 - 1) \\ C_{xx}(1, 0) & C_{xx}(1, 1) & \cdots & C_{xx}(1, N_2 - 1) \\ \cdots \cdots \cdots \cdots \cdots \cdots \cdots \cdots \cdots \cdots \\ C_{xx}(N_1 - 1, 0) & C_{xx}(N_1 - 1, 1) & \cdots & C_{xx}(N_1 - 1, N_2 - 1) \end{bmatrix}$$

$$(3.6\text{-}7)$$

From the above discussion it is observed that the 2-dimensional DFT in Eq. (1) can be viewed as using a 1-dimensional DFT a total of $N_1 N_2$ times, as follows: *then second appendice the FFT explain and N4 log2 N) formula.*

(i) With $N = N_1$, the DFT in Eq. (3.1-1) is used N_2 times to obtain $[C_x(k_1, m_2)]$ given by Eq. (5).

(ii) Next with $N = N_2$, the DFT in Eq. (3.1-1) is used N_1 times to obtain $[C_{xx}(k_1, k_2)]$ given by Eq. (7).

In conclusion, we note that as a consequence of (i) and (ii) above, Eq. (1) can equivalently be written in the form of the following matrix equation:

$$[C_{xx}(k_1, k_2)] = \frac{1}{N_1 N_2} \Lambda_1 [X(m_1, m_2)] \Lambda_2 \qquad (3.6\text{-}8)$$

where Λ_1 is a $(N_1 \times N_1)$ matrix whose elements are

$$\alpha_{rs} = W_1^{rs}; \qquad r, s = 0, 1, \ldots, N_1 - 1,$$

Λ_2 is a $(N_2 \times N_2)$ matrix whose elements are

$$\beta_{rs} = W_2^{rs}; \qquad r, s = 0, 1, \ldots, N_2 - 1,$$

$$W_1 = e^{\frac{-i 2\pi}{N_1}}, \quad \text{and} \quad W_2 = e^{\frac{-i 2\pi}{N_2}}$$

3.7 Time-varying Fourier Spectra

If $x(t)$ is a signal whose bandwidth is B Hz, then the corresponding digital signal $x^*(t)$ is given by

$$x^*(t) = \Delta t \sum_{m=0}^{N-1} X(m) \delta(t - m \Delta t)$$

where $X(m)$ is the m-th sample and $\Delta t \leq 1/2B$. The Fourier transform of $x^*(t)$ is given by

$$F_{x^*}(\omega) = \Delta t \sum_{m=0}^{N-1} X(m) e^{-im\omega \Delta t} \qquad (3.7\text{-}1)$$

We now show that the power and phase spectra corresponding to $F_{x^*}(\omega)$ can be computed *recursively* at times $t_s = s\Delta t$, $s = 0, 1, \ldots, N-1$. Consequently the spectra so obtained will be referred to as *time-varying Fourier spectra* [7–9], and it will be assumed that they are desired at a set of frequencies ω_k, $k = 1, 2, \ldots, M$.

Let $\hat{x}^*(t)$ denote the signal which is the "mirror image" of $x^*(t)$. That is,

$$\hat{x}^*(t) = \Delta t \sum_{m=0}^{N-1} X(N-1-m)\,\delta(t - m\Delta t) \tag{3.7-2}$$

The Fourier transform of $\hat{x}^*(t)$ at $\omega = \omega_k$ is given by

$$F_{\hat{x}^*}(\omega_k) = \Delta t \sum_{m=0}^{N-1} X(N-1-m)\,e^{-im\omega_k \Delta t}, \qquad k = 1, 2, \ldots, M \tag{3.7-3}$$

It can be shown that [see Appendix 3-1] the power and phase spectra of $x^*(t)$ and $\hat{x}^*(t)$ are related as follows:

$$|F_{x^*}(\omega_k)|^2 = |F_{\hat{x}^*}(\omega_k)|^2,$$

and

$$\psi_{x^*}(\omega_k) = -[\psi_{\hat{x}^*}(\omega_k) + (N-1)\,\omega_k \Delta t], \qquad k = 1, 2, \ldots, M \tag{3.7-4}$$

Recursive Computation of $|F_{\hat{x}^*}(\omega_k)|^2$ and $\psi_{\hat{x}^*}(\omega_k)$. From Eq. (3) we have

$$F_{\hat{x}^*}(\omega_k) = \Delta t\,\{R(\omega_k) - iI(\omega_k)\} \tag{3.7-5}$$

where

$$R(\omega_k) = \sum_{m=0}^{N-1} X(N-1-m)\cos(m\omega_k \Delta t)$$

and

$$I(\omega_k) = \sum_{m=0}^{N-1} X(N-1-m)\sin(m\omega_k \Delta t)$$

Expressing $F_{\hat{x}^*}(\omega_k)$ in terms of a (2×1) vector $\boldsymbol{F}_{\hat{x}^*}(\omega_k)$ one has

$$\boldsymbol{F}_{\hat{x}^*}(\omega_k) = \Delta t \begin{bmatrix} R(\omega_k) \\ I(\omega_k) \end{bmatrix} \tag{3.7-6}$$

Again, consider the (2×2) matrix

$$\mathbf{L}(\omega_k) = \begin{bmatrix} \cos(\omega_k \Delta t) & -\sin(\omega_k \Delta t) \\ \sin(\omega_k \Delta t) & \cos(\omega_k \Delta t) \end{bmatrix}$$

It can be shown that $\mathbf{L}(\omega_k)$ is orthogonal (see Prob. 3-9), and hence has the property

$$\mathbf{L}(\omega_k)^m = \begin{bmatrix} \cos(m\omega_k \Delta t) & -\sin(m\omega_k \Delta t) \\ \sin(m\omega_k \Delta t) & \cos(m\omega_k \Delta t) \end{bmatrix} \tag{3.7-7}$$

where $L(\omega_k)^m$ denotes the m-th power of $L(\omega_k)$.

Thus from Eqs. (6) and (7) it follows that

$$\boldsymbol{F}_{\hat{x}*}(\omega_k) = \Delta t \sum_{m=0}^{N-1} \mathbf{L}(\omega_k)^m \boldsymbol{b} X(N-1-m) \tag{3.7-8}$$

where

$$\boldsymbol{b} = \begin{bmatrix} 1 \\ 0 \end{bmatrix}$$

Now, we introduce the recurrence relation

$$\boldsymbol{Z}(\omega_k, s) = \mathbf{L}(\omega_k)\, \boldsymbol{Z}(\omega_k, s-1) + \boldsymbol{b} X(s), \qquad s = 0, 1, \ldots, N-1 \tag{3.7-9}$$

where s denotes time $t = s\Delta t$ and

$$\boldsymbol{Z}(\omega_k, -1) = \begin{bmatrix} 0 \\ 0 \end{bmatrix}$$

For example, with $s = 0, 1, 2$, Eq. (9) yields

$$\boldsymbol{Z}(\omega_k, 0) = \boldsymbol{b} X(0)$$

$$\boldsymbol{Z}(\omega_k, 1) = \mathbf{L}(\omega_k)\, \boldsymbol{b} X(0) + X(1)$$

and

$$\boldsymbol{Z}(\omega_k, 2) = \mathbf{L}(\omega_k)^2 \boldsymbol{b} X(0) + \mathbf{L}(\omega_k)\, \boldsymbol{b} X(1) + \boldsymbol{b} X(2)$$

Hence, using mathematical induction it can be shown that

$$\boldsymbol{Z}(\omega_k, N-1) = \sum_{m=0}^{N-1} \mathbf{L}(\omega_k)^m \boldsymbol{b} X(N-1-m) \tag{3.7-10}$$

That is

$$\mathbf{Z}(\omega_k, N-1) = \begin{bmatrix} \sum_{m=0}^{N-1} X(N-1-m) \cos(m\omega_k \Delta t) \\ \sum_{m=0}^{N-1} X(N-1-m) \sin(m\omega_k \Delta t) \end{bmatrix} = \begin{bmatrix} Z_1(\omega_k, N-1) \\ Z_2(\omega_k, N-1) \end{bmatrix} \tag{3.7-11}$$

A comparison of Eq. (5) with Eq. (11) leads to the fundamental result

$$F_{\hat{x}*}(\omega_k) = \Delta t \{Z_1(\omega_k, N-1) - iZ_2(\omega_k, N-1)\} \tag{3.7-12}$$

From Eq. (12) it follows that the power and phase spectra of $\hat{x}^*(t)$ are given by

$$|F_{\hat{x}*}(\omega_k)|^2 = (\Delta t)^2 \, ||\boldsymbol{Z}(\omega_k, N-1)||^2$$

and

$$\psi_{\hat{x}*}(\omega_k) = \tan^{-1}\left[\frac{-Z_2(\omega_k, N-1)}{Z_1(\omega_k, N-1)}\right] \tag{3.7-13}$$

where $||\boldsymbol{Z}(\omega_k, N-1)||$ denotes the norm of $\boldsymbol{Z}(\omega_k, N-1)$.

Time-varying spectra. Substituting Eq. (13) in Eq. (4) we obtain

$$|F_{x^*}(\omega_k)|^2 = (\Delta t)^2 \,||\mathbf{Z}(\omega_k, N-1)||^2$$

and

$$\psi_{x^*}(\omega_k) = -\left\{\tan^{-1}\left[\frac{-Z_2(\omega_k, N-1)}{Z_1(\omega_k, N-1)}\right] + (N-1)\,\omega_k \Delta t\right\} \qquad (3.7\text{-}14)$$

Equation (14) implies that the power and phase spectra of $x^*(t)$ are directly related to $\mathbf{Z}(\omega_k, N-1)$. However, since $\mathbf{Z}(\omega_k, s)$ is computed *recursively* using Eq. (9), we can define the following *time-varying* spectra for $x^*(t)$.

$$|F_{x^*}(\omega_k, s)|^2 = (\Delta t)^2 \,||\mathbf{Z}(\omega_k, s)||^2$$

and

$$\psi_{x^*}(\omega_k, s) = -\left\{\tan^{-1}\left[\frac{-Z_2(\omega_k, s)}{Z_1(\omega_k, s)}\right] + s\omega_k \Delta t\right\}, \qquad (3.7\text{-}15)$$

$$k = 1, 2, \ldots, M; \qquad s = 0, 1, \ldots, N-1$$

Clearly, the time-varying spectra have the property that when $s = (N-1)$, they are identical to the Fourier spectra defined in Eq. (14).

Computational considerations. $|F_{x^*}(\omega_k, s)|^2$ and $\psi_{x^*}(\omega_k, s)$ in Eq. (15) can be computed using a bank of M recurrence relations of the type given by Eq. (9). The steps pertaining to the computation of the time-varying spectra at $\omega = \omega_k$ are summarized in Fig. 3.3.

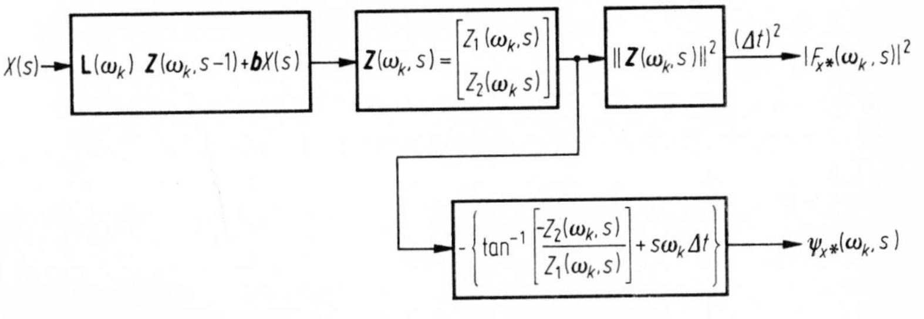

Fig. 3.3 Recursive computation of time-varying spectra

Special case. Consider the case when the frequencies ω_k are chosen such that $\omega_k = 2\pi k/(N\Delta t)$. Then Eq. (1) yields

$$F_{x^*}(\omega_k) = \Delta t \sum_{m=0}^{N-1} X(m)\, e^{\frac{-i\,2\pi km}{N}}$$

from which it follows that

$$C_x(k) = \left(\frac{1}{N\Delta t}\right) F_{x^*}(\omega_k), \qquad k = 0, 1, \ldots, N/2 \qquad (3.7\text{-}16)$$

where $C_x(k)$ is the k-th DFT coefficient. Combining Eqs. (15) and (16) we obtain the following *time-varying DFT* power and phase spectra:

$$|C_x(k, s)|^2 = \frac{1}{N^2} ||\boldsymbol{Z}(\omega_k, s)||^2$$

and

$$\psi_x(k, s) = -\left\{\tan^{-1}\left[\frac{-Z_2(\omega_k, s)}{Z_1(\omega_k, s)}\right] + s\omega_k \Delta t\right\} \qquad (3.7\text{-}17)$$

$$k = 0, 1, \ldots, N/2; \qquad s = 0, 1, \ldots, N-1$$

where $\omega_k = 2\pi k/(N\Delta t)$.

In conclusion, it is remarked that the time-varying spectra developed above can be used to display the manner in which the spectra of a discrete signal vary as the corresponding data sequence $X(m)$, $m = 0, 1, \ldots, N-1$ is being processed. This is best illustrated by means of a numerical example which follows.

Example 3.7-1

Consider the data sequence

$$x(m) = 3.97\, e^{-(m+1)/6} \sin\left(\frac{(2\pi)(4)(m+1)}{26}\right), \qquad m = 0, 1, \ldots, 25$$

which is obtained by sampling a 4 Hz damped sinusoid at the rate of 26 samples per second – that is, $\Delta t = 1/26$ sec. The time-varying power spectrum $|F_{x*}(f_k, s)|^2$, $s = 0, 1, \ldots, 25$ is desired at a set of 20 frequencies which are *equally spaced* on a *logarithmic scale*, and given by

$$\{f_k\} = \{1.38,\ 1.55,\ 1.74,\ 1.96,\ 2.20,\ 2.48,\ 2.79,\ 3.14,\ 3.53,\ 3.97,\ 4.47,\ 5.03,$$
$$5.65,\ 6.36,\ 7.15,\ 8.05,\ 9.06,\ 10.19,\ 11.46,\ 12.89\}$$

Computation of $|F_{x*}(f_k, s)|^2$, $s = 0, 1, \ldots, 25$ results in the "time-frequency-amplitude" plot shown in Fig. 3.4. To obtain this plot, the values of $|F_{x*}(f_k, s)|^2$ have been scaled by a convenient scale factor and subsequently converted to decibels. The dB value so obtained is denoted by $d(f_k, s)$. The manner in which the plot can be interpreted is best illustrated by the following examples.

(i) $d(3.97, 15) = 15$ and $d(1.96, 2) = 9$ implies that $|F(3.97, 15)|^2_{\text{dB}}$ is 6 dB greater than $|F(1.96, 2)|^2_{dB}$.

(ii) $d(2.48, 6) = 11$ and $d(10.19, 6) = -2$ implies that $|F(2.48, 6)|^2_{\text{dB}}$ is 13 dB greater than $|F(10.19, 6)|^2_{\text{dB}}$.

Frequency →

Time ($s\Delta t$) ↓ s	1.38	1.55	1.74	1.96	2.20	2.48	2.79	3.14	3.53	3.97	4.47	5.03	5.65	6.36	7.15	8.05	9.06	10.19	11.46	12.89
0	3	3	3	3	3	3	3	3	3	3	3	3	3	3	3	3	3	3	3	3
1	8	8	8	8	8	8	8	8	8	7	7	7	6	6	5	3	2	0	-5	-9
2	9	9	9	9	9	9	8	8	8	8	7	6	6	5	3	2	0	-2	-6	-9
3	8	8	8	8	8	9	9	9	9	9	9	9	8	7	5	2	-3	-9	-2	0
4	7	8	9	9	10	10	11	11	11	11	11	10	9	6	2	-6	-4	0	-1	-9
5	8	9	9	10	11	11	12	12	12	12	11	10	8	5	0	-3	-2	-3	-5	-5
6	7	8	9	9	10	11	11	12	12	12	12	11	9	5	-1	-3	0	-2	-9	-1
7	5	6	7	8	9	10	11	12	13	13	13	11	8	2	-2	1	-1	-8	-1	-9
8	4	5	6	7	9	10	12	13	13	13	13	11	7	2	0	0	-3	-3	-5	-4
9	4	5	6	7	9	10	12	13	13	13	13	11	7	1	0	0	-5	-2	-6	-3
10	6	6	6	7	8	9	11	12	13	14	13	11	6	1	2	-1	-2	-3	-3	-8
11	6	6	6	6	7	9	11	13	14	14	13	10	6	3	1	-2	-1	-5	-4	-4
12	6	6	6	6	7	9	11	13	14	14	13	10	6	3	1	-2	-1	-5	-4	-4
13	6	6	7	7	8	9	11	12	14	14	13	10	5	4	1	0	-3	-2	-6	-7
14	6	7	7	7	8	8	10	12	14	15	13	10	5	4	0	0	-3	-4	-4	-4
15	6	7	7	7	8	8	10	12	14	15	13	10	5	4	0	0	-3	-4	-4	-5
16	6	7	7	7	8	9	10	12	14	15	14	10	6	4	0	0	-2	-4	-6	-6
17	6	6	7	8	8	9	10	12	14	15	13	9	6	3	1	-1	-2	-3	-4	-4
18	6	6	7	8	8	9	10	12	14	15	13	9	6	3	1	-1	-2	-4	-5	-5
19	6	6	7	8	8	9	10	12	14	15	13	9	6	3	1	-1	-3	-4	-5	-5
20	6	6	7	7	8	9	10	12	14	15	13	9	6	3	1	0	-2	-4	-5	-5
21	6	6	7	7	8	9	10	12	14	15	13	9	6	3	1	0	-2	-4	-5	-5
22	6	6	7	7	8	9	10	12	14	15	13	9	6	3	0	0	-2	-3	-4	-5
23	6	6	7	7	8	9	10	12	14	15	13	9	6	3	0	-1	-2	-4	-5	-5
24	6	6	7	7	8	9	10	12	14	15	13	9	6	3	0	-1	-2	-4	-5	-5
25	6	6	7	7	8	9	10	12	14	15	13	9	6	3	0	-1	-2	-4	-5	-5

Fig. 3.4 Time-varying power spectrum of a 4 Hz damped sinusoid

For convenience, the minimum value of $d(f, s)$ has been constrained to -9. We observe that the power at the 3.97 Hz and adjacent frequencies is consistently higher than that at the remaining frequencies. Clearly, this is what one would expect since the discrete signal represents a 4 Hz damped sinusoid.

3.8 Summary

This chapter introduced the notion of the Fourier representation of data sequences $\{X(m)\}$ by means of the DFT. Some of the more important properties of the DFT were developed. The discrete analogs of convolution and correlation and their corresponding theorems were developed. It was shown that the DFT is directly proportional to the Fourier transform/series representation of $\{X(m)\}$.

Again, the definition of the DFT was extended to two dimensions. In conclusion, the notion of time-varying power and phase spectra was developed. It was demonstrated that such spectra could be used to display the manner in which the power and phase of a discrete/digital signal vary, as it is being processed.

Appendix 3.1

From Eq. (3.7-3) we have

$$F_{\hat{x}}(\omega_k) = \Delta t \sum_{m=0}^{N-1} X(N-1-m)\, e^{-i\omega_k m\,\Delta t}, \qquad k = 1, 2, \ldots, M$$

Introduction of the change of variable

$$\eta = N - 1 - m$$

leads to

$$F_{\hat{x}}(\omega_k) = \Delta t \left[\sum_{\eta=N-1}^{0} X(\eta)\, e^{i\eta\omega_k \Delta t}\, e^{-i(N-1)\omega_k \Delta t} \right] \tag{A3.1-1}$$

It is clear that

$$\sum_{\eta=N-1}^{0} X(\eta)\, e^{i\eta\omega_k \Delta t} = X(N-1)\, e^{i(N-1)\omega_k \Delta t} + \cdots + X(1)\, e^{i\omega_k \Delta t} + X(0)$$

$$= \sum_{m=0}^{N-1} X(m)\, e^{im\omega_k \Delta t} \tag{A3.1-2}$$

Substitution of Eq. (A3.1-2) in Eq. (A3.1-1) yields

$$F_{\hat{x}^*}(\omega_k) = \Delta t\, e^{-i(N-1)\omega_k \Delta t} \left[\sum_{m=0}^{N-1} X(m)\, e^{im\omega_k \Delta t} \right]$$

which implies that

$$F_{\hat{x}^*}(\omega_k) = e^{-i(N-1)\omega_k \Delta t}\, \bar{F}_{x^*}(\omega_k) \tag{A3.1-3}$$

where $\bar{F}_{x^*}(\omega_k)$ is the complex conjugate of $F_{x^*}(\omega_k)$. Again, since

$$F_{x^*}(\omega_k) = |F_{x^*}(\omega_k)|\, e^{i\psi_{x^*}(\omega_k)}$$

it follows that

$$\bar{F}_{x^*}(\omega_k) = |F_{x^*}(\omega_k)|\, e^{-i\psi_{x^*}(\omega_k)} \tag{A3.1-4}$$

Substitution of Eq. (A3.1-4) in Eq. (A3.1-3) leads to

$$F_{\hat{x}^*}(\omega_k) = |F_{x^*}(\omega_k)|\, e^{i[-\psi_{x^*}(\omega_k) - (N-1)\omega_k \Delta t]}$$

which implies that

$$|F_{\hat{x}*}(\omega_k)| = |F_{x*}(\omega_k)|$$

and

$$\psi_{\hat{x}*}(\omega_k) = -[\psi_{x*}(\omega_k) + (N-1)\,\omega_k \Delta t]$$

That is

$$|F_{x*}(\omega_k)| = |F_{\hat{x}*}(\omega_k)|$$

and

$$\psi_{x*}(\omega_k) = -[\psi_{\hat{x}*}(\omega_k) + (N-1)\,\omega_k \Delta t], \qquad k = 1, 2, \ldots, M$$

which are the desired results in Eq. (3.7-4).

References

1. Cochran, W. T. et al.: What is the Fast Fourier Transform? *Proc. IEEE* 55 (1967) 1664–1674.
2. Cooley, J. W., et al.: The Finite Fourier Transform. *IEEE Trans. Audio and Electroacoustics* AU-17 (1969) 77–85.
3. Gold, B., and Rader, C. M.: *Digital Processing of Signals.* New York, N.Y. McGraw-Hill, 1969.
4. Cooley, J. W., Lewis, P. A. W., and Welch, P. D.: The Fast Fourier Transform and Its Applications. IBM Res. Paper, RC-1743, 1967, IBM Watson Research Center, Yorktown Heights, New York.
5. Ahmed, N., and Rao, K. R.: Discrete Fourier and Hadamard Transforms. *Electronics Letters* 6 (1970) 221–224.
6. Jagadeesan, N.: n-dimensional Fast Fourier Transform. *Proc. of the 13th Midwest Symposium on Circuit Theory.* University of Minnesota, Minneapolis, Minn., May 7–8, 1970, pp. III 2.1–III 2.8.
7. Ahmed, N., Rao, K. R., and Tjoe, S. J.: Time-varying Fourier Transform. *Electronics Letters* 7 (1971) 535–536.
8. Arnold, C. R.: Spectral Estimation for Transient Waveforms. *IEEE Trans. Audio and Electroacoustics* AU-18 (1970) 248–257.
9. Ahmed, N., Natarajan, T., and Rao, K. R.: An Algorithm for the On-line Computation of Fourier Spectra. *International Journal of Computer Mathematics* 3 (1973) 361–370.
10. Andrews, H. C., and Pratt, W. K.: Digital Image Transform Processing. Proc. *Symposium on Applications of Walsh Functions,* 1970, pp. 183–194. This may be obtained from National Technical Information Service, Springfield, Va. 22151, order No. AD-707431.
11. Claire, E. J., Farber, S. M., and Green, R. R.: Practical Techniques for Transform Data Compression/Image Coding. Ibid. pp. 2–6.
12. Special Issue on Digital Picture Processing, *Proc. IEEE* 60, July 1972.
13. Huang, T. S., et al.: Image Processing, *Proc. IEEE* 59 (1971) 1586–1609.
14. Kinariwala, B. K., Kuo, F. F., and Tsao, Nai-Kuan: Linear Circuits and Computation. New York, N. Y.: John Wiley, 1973.

Problems

3-1 If $X(m) \leftrightarrow C_x(k)$, show that

(a)
$$X(-m) \leftrightarrow C_x(-k)$$

(b)
$$C_x(0) = \frac{1}{N} \sum_{m=0}^{N-1} X(m); \qquad X(0) = \sum_{k=0}^{N-1} C_x(k)$$

(c)
$$C_x(\pm k) = C_x(sN \pm k), \qquad s = 0, \pm 1, \pm 2, \ldots$$

3-2 If $\{A(m)\} = \{AA \cdots A\}$ is an N-periodic sequence with A being a constant, show that $\{C_a(k)\} = \{A\ 0\ 0 \cdots 0\}$.

3-3 The DFT coefficients of a real 8-periodic sequence $\{X(m)\}$ are as follows:

$$C_x(0) = 5, \quad C_x(1) = i, \quad C_x(2) = 1 + i, \quad C_x(3) = 2 + 3i, \quad C_x(4) = 2.$$

What are the values of $C_x(k)$, $k = 5, 6, 7$?

3-4 Consider a 4-periodic sequence $\{X(m)\} = \{X(0)\ X(1)\ X(2)\ X(3)\}$. Its DFT and IDFT are respectively defined as

$$C_x(k) = \frac{1}{4} \sum_{m=0}^{3} X(m)\, W^{km}, \qquad k = 0, 1, 2, 3 \tag{P3-4-1}$$

and

$$X(m) = \sum_{k=0}^{3} C_x(k)\, W^{-km}, \qquad m = 0, 1, 2, 3 \tag{P3-4-2}$$

(a) Write Eq. (P3-4-1) in matrix form as follows:

$$\begin{bmatrix} C_x(0) \\ C_x(1) \\ C_x(2) \\ C_x(3) \end{bmatrix} = \frac{1}{4} \mathbf{\Lambda} \begin{bmatrix} X(0) \\ X(1) \\ X(2) \\ X(3) \end{bmatrix}$$

where $\mathbf{\Lambda}$ is a (4×4) matrix.

(b) Verify that $\mathbf{\Lambda}$ is a *symmetric* matrix; that is

$$\mathbf{\Lambda}' = \mathbf{\Lambda}, \text{ prime denoting transpose.} \tag{P3-4-3}$$

(c) If $\mathbf{\Lambda}^*$ denotes the matrix obtained by replacing each element of $\mathbf{\Lambda}$ by its complex conjugate, verify that

$$\mathbf{\Lambda}^{*\prime}\mathbf{\Lambda} = 4\mathbf{I} \tag{P3-4-4}$$

where $\mathbf{I}$ is the (4×4) identity matrix.

Remark: The properties of $\mathbf{\Lambda}$ expressed by Eq. (P3-4-3) and Eq. (P3-4-4) can be shown to be true in general.

(d) Use the results of parts (c) to find $\mathbf{\Lambda}^{-1}$, the inverse of $\mathbf{\Lambda}$.

3-5 Given: $\{X(m)\} = \{X(0)\ X(1)\ X(2)\ X(3)\}$, and $\{C_x(k)\} = \{C_x(0)\ C_x(1)\ C_x(2)\ C_x(3)\}$. If $\{R(m)\} = \{X(1)\ X(0)\ X(3)\ X(2)\}$, find the DFT coefficients $C_r(k)$ in terms of $C_x(k)$, $k = 0, 1, 2, 3$.

Answer: $C_r(0) = C_x(0)$, $C_x(1) = -iC_x(3)$, $C_r(2) = -C_x(2)$, $C_r(3) = iC_x(1)$.

Hint: Show that

$$\begin{bmatrix} C_r(0) \\ C_r(1) \\ C_r(2) \\ C_r(3) \end{bmatrix} = \mathbf{\Lambda M \Lambda}^{-1} \begin{bmatrix} C_x(0) \\ C_x(1) \\ C_x(2) \\ C_x(3) \end{bmatrix} \tag{P3-5-1}$$

where $\mathbf{\Lambda}$ is defined in Prob. 3-4, and

$$\mathbf{M} = \begin{bmatrix} 0 & 1 & 0 & 0 \\ 1 & 0 & 0 & 0 \\ 0 & 0 & 0 & 1 \\ 0 & 0 & 1 & 0 \end{bmatrix}$$

The matrix product $\mathbf{\Lambda M \Lambda}^{-1}$ is called a *similarity transform* corresponding to $\mathbf{\Lambda}$.

3-6 Show that if the input $\{X(m)\}$ is real, then the DFT phase spectrum defined in Eq. (3.5-5) is an odd function about the point $k = N/2$, that is

$$\psi_x\left(\frac{N}{2} + l\right) = -\psi_x\left(\frac{N}{2} - l\right), \qquad l = 1, 2, \ldots, \frac{N}{2} - 1.$$

3-7 A "stretch function" is defined as follows [2]:

$$\text{Stretch}_s\,\{m : X\} = \begin{cases} X(m/s) & \text{for } m = ps, \quad p = 0, 1, \ldots, N-1 \\ 0, & \text{otherwise} \end{cases}$$

If $X(m) \leftrightarrow C_x(k)$; $m, k = 0, 1, \ldots, N-1$, show that

(a) $$\text{Stretch}_s\,\{m : X\} \leftrightarrow \frac{C_x(k)}{s}, \qquad k = 0, 1, \ldots, Ns - 1$$

(b) $$X(m) \leftrightarrow \text{Stretch}_s\,\{m : C_x(k)\}, \qquad m = 0, 1, \ldots, Ns - 1$$

3-8 Suppose a sequence $X(m)$, $m = 0, 1, \ldots, N-1$ is sampled at the points $0, s, 2s, \ldots, N/s - 1$, where N/s is an integer. Then a "sample function" is defined as follows [2]:

$\text{Sample}_s\,\{m : X\} = X(ms)$, $m = 0, 1, \ldots, N/s - 1$ and has a period of N/s. Now, if $X(m) \leftrightarrow C_x(k)$; $m, k = 0, 1, \ldots, N-1$, show that

$$\text{Sample}_s\,\{m : X\} \leftrightarrow \sum_{r=0}^{s-1} C_x\left(k + \frac{rN}{s}\right)$$

3-9 Show that

(a) $$\boldsymbol{L}(\omega) = \begin{bmatrix} \cos(\omega \Delta t) & -\sin(\omega \Delta t) \\ \sin(\omega \Delta t) & \cos(\omega \Delta t) \end{bmatrix}$$

is an orthonormal matrix, and

(b) $$\boldsymbol{L}(\omega)^m = \begin{bmatrix} \cos(m\omega \Delta t) & -\sin(m\omega \Delta t) \\ \sin(m\omega \Delta t) & \cos(m\omega \Delta t) \end{bmatrix}$$

3-10 Given the data sequence $\{X(m)\} = \{1 \;\; 2 \;\; -1 \;\; 3\}$. Show that its time-varying DFT power spectrum is as follows:

s	$\lvert C_x(0, s)\rvert^2$	$\lvert C_x(1, s)\rvert^2$	$\lvert C_x(2, s)\rvert^2$	$\lvert C_x(3, s)\rvert^2$
0	1/16	1/16	1/16	1/16
1	9/16	5/16	1/16	5/16
2	9/16	1/2	1/4	1/2
3	25/16	5/16	25/16	5/16

Chapter Four

Fast Fourier Transform

The main objective of this chapter is to develop a fast algorithm for efficient computation of the DFT. This algorithm, called the fast Fourier transform (FFT), significantly reduces the number of arithmetic operations and memory required to compute the DFT (or its inverse). Consequently, it has accelerated the application of Fourier techniques in digital signal processing in a number of diverse areas. A detailed development of the FFT is followed by some numerical examples which illustrate its applications.

4.1 Statement of the Problem

Let $\{X(m)\}$ denote the data sequence $X(m)$, $m = 0, 1, \ldots, N-1$ which is obtained by sampling a band-limited signal $x(t)$. We wish to obtain an *algorithm* to compute

$$C_x(k) = \frac{1}{N} \sum_{m=0}^{N-1} X(m) W^{km}, \qquad k = 0, 1, \ldots, N-1 \qquad (4.1\text{-}1)$$

where $W = e^{-i 2\pi/N}$, and $i = \sqrt{-1}$. It is recalled that Eq. (1) is the DFT of $\{X(m)\}$. The algorithm sought is called the *fast Fourier transform* (FFT) [1]. Since this algorithm was developed by Cooley and Tukey [2], it is also referred to as *Cooley-Tukey algorithm*. In what follows, we shall assume that $N = 2^n$, $n = 1, 2, \ldots, n_{max}$. Clearly, there is no loss of generality in doing so, as long as N is large enough to satisfy the sampling theorem – that is, $N \geq 2BL$ where B is the bandwidth of $x(t)$ in Hz, and L is its duration. The case when N is not necessarily of the form $N = 2^n$ is discussed elsewhere [1, 3, 4, 5].

4.2 Motivation to Search for an Algorithm

We consider the case when $\{X(m)\}$ is real-valued and $N = 8$. From the complex conjugate property we have

$$C_x(4 + l) = \bar{C}_x(4 - l), \qquad l = 1, 2, 3$$

Thus with respect to Eq. (4.1-1) it suffices to compute

$$C_x(k) = \frac{1}{8} \sum_{m=0}^{7} X(m) W^{km}, \qquad k = 0, 1, \ldots, 4 \tag{4.2-1}$$

where $W = e^{-i\pi/4}$.

From Eq. (1) it follows that

$$8\,C_x(k) = \sum_{m=0}^{7} X(m) \cos\left(\frac{mk\pi}{4}\right) - i \sum_{m=0}^{7} X(m) \sin\left(\frac{mk\pi}{4}\right) = A(k) - iB(k)$$

where

$$A(k) = \sum_{m=0}^{7} X(m) \cos\left(\frac{mk\pi}{4}\right),$$

and

$$B(k) = \sum_{m=0}^{7} X(m) \sin\left(\frac{mk\pi}{4}\right), \qquad k = 0, 1, \ldots, 4 \tag{4.2-2}$$

Using matrix notation, Eq. (2) becomes

$$\boldsymbol{A} = \mathbf{C}\boldsymbol{X}$$
$$\boldsymbol{B} = \mathbf{S}\boldsymbol{X}$$

where, $\boldsymbol{A}$ and $\boldsymbol{B}$ are 5-vectors, $\boldsymbol{X}$ is a 8-vector and $\mathbf{C}$ and $\mathbf{S}$ are (5×8) matrices as follows:

$$\mathbf{C} = \begin{bmatrix} 1 & 1 & 1 & 1 & 1 & 1 & 1 & 1 \\ 1 & 1/\sqrt{2} & 0 & -1/\sqrt{2} & -1 & -1/\sqrt{2} & 0 & 1/\sqrt{2} \\ 1 & 0 & -1 & 0 & 1 & 0 & -1 & 0 \\ 1 & -1/\sqrt{2} & 0 & 1/\sqrt{2} & -1 & 1/\sqrt{2} & 0 & -1/\sqrt{2} \\ 1 & -1 & 1 & -1 & 1 & -1 & 1 & -1 \end{bmatrix}$$

and

$$\mathbf{S} = \begin{bmatrix} 0 & 0 & 0 & 0 & 0 & 0 & 0 & 0 \\ 0 & 1/\sqrt{2} & 1 & 1/\sqrt{2} & 0 & -1/\sqrt{2} & -1 & -1/\sqrt{2} \\ 0 & 1 & 0 & -1 & 0 & 1 & 0 & -1 \\ 0 & 1/\sqrt{2} & -1 & 1/\sqrt{2} & 0 & -1/\sqrt{2} & 1 & -1/\sqrt{2} \\ 0 & 0 & 0 & 0 & 0 & 0 & 0 & 0 \end{bmatrix} \tag{4.2-3}$$

Inspection of Eq. (3) shows that there is a considerable repetition with respect to the elements of the matrices $\mathbf{C}$ and $\mathbf{S}$. There are two reasons for this.

1. The sinusoids form a family with strong characteristics which can be exploited only if sample points are properly chosen.

2. If N is the number of uniformly spaced sampled values of $x(t)$, then the repetition in the matrices corresponding to **C** and **S** increases with the number of factors in N. Thus the form $N = 2^n = 2 \cdot 2 \cdots 2$ is said to be "highly composite" [1, 6]. We will show later that this highly composite form of N yields substantial savings in computational time and memory.

4.3 Key to Developing the Algorithm

The key to developing the desired algorithm is to relate the repetition in the general forms of the above matrices **S** and **C** to powers of W, the principal N-th root of unity, and employ a notation which permits a generalization.

Notation. Each decimal value of m, $0 \leq m \leq N-1$, is expressed in binary form to obtain

$$m = m_{n-1}\, 2^{n-1} + m_{n-2}\, 2^{n-2} + \cdots + m_1 2^1 + m_0 2^0,$$

where

$$m_v = 0 \text{ or } 1, \qquad v = 0, 1, \ldots, n-1, \qquad n = \log_2 N$$

Similarly each decimal value of k, $0 \leq k \leq N/2$ is expressed as

$$k = k_{n-1}\, 2^{n-1} + k_{n-2}\, 2^{n-2} + \cdots + k_1 2^1 + k_0 2^0,$$

where

$$k_v = 0 \text{ or } 1, \qquad v = 0, 1, \ldots, n-1$$

Denoting the binary representation of $X(m)$ by $\tilde{X}(m)$, we obtain

$$\begin{aligned} X(m) &= \tilde{X}(m_{n-1}\, 2^{n-1} + m_{n-2}\, 2^{n-2} + \cdots + m_1 2^1 + m_0 2^0) \\ &= \tilde{X}(m_{n-1}, m_{n-2}, \ldots, m_1, m_0) \end{aligned} \tag{4.3-1}$$

Equation (1) leads to the useful relation

$$\begin{aligned} &\sum_{m=0}^{N-1} X(m) W^{km} \\ &= \sum_{m_0} \sum_{m_1} \cdots \sum_{m_{n-1}} \tilde{X}(m_{n-1}, m_{n-2}, \ldots, m_0) W^{k[m_{n-1} 2^{n-1} + m_{n-2} 2^{n-2} + \cdots + m_0 2^0]} \end{aligned} \tag{4.3-2}$$

For the purposes of illustration, we verify Eq. (2) for the case $N = 4$. For this case, the right hand side of Eq. (2) yields

$$\begin{aligned} \sum_{m_0} \sum_{m_1} \tilde{X}(m_1, m_0) W^{k[2m_1 + m_0]} &= \sum_{m_0} \{\tilde{X}(0, m_0) W^{km_0} + \tilde{X}(1, m_0) W^{k[2+m_0]}\} \\ &= \sum_{m_0} \tilde{X}(0, m_0) W^{km_0} + \sum_{m_0} \tilde{X}(1, m_0) W^{k[2+m_0]} \end{aligned}$$

That is

$$\sum_{m_0} \sum_{m_1} \tilde{X}(m_1, m_0) W^{k[2m_1+m_0]} = \tilde{X}(0,0) + \tilde{X}(0,1) W^k + \tilde{X}(1,0) W^{2k} + \tilde{X}(1,1) W^{3k} \tag{4.3-3}$$

We observe that the arguments of $X(m)$ in Eq. (3) correspond to the binary representation of the decimal values of m, taken in the *natural* order as shown in Table 4.3-1.

Table 4.3-1

m	m_1	m_0
0	0	0
1	0	1
2	1	0
3	1	1

Thus Eq. (3) yields

$$\sum_{m_0}\sum_{m_1} \tilde{X}(m_1, m_0) W^{k[2m_1+m_0]} = \sum_{m=0}^{3} X(m) W^{km}$$

which is Eq. (2) for $N = 4$.

4.4 Development of the Algorithm

The development of the algorithm [7, 8] is best explained for the case $N = 8$, and hence we have

$$C_x(k) = \frac{1}{8} \sum_{m=0}^{7} X(m) W^{km}, \qquad k = 0, 1, \ldots, 7 \tag{4.4-1}$$

where $W = e^{-i\pi/4}$. We express m in binary form to obtain

$$m = m_2 2^2 + m_1 2^1 + m_0 2^0 \tag{4.4-2}$$

From Eqs. (4.3-2), (1), and (2) it follows that

$$\begin{aligned} 8C_x(k) &= \sum_{m_0}\sum_{m_1}\sum_{m_2} \tilde{X}(m_2, m_1, m_0) W^{k[4m_2+2m_1+m_0]} \\ &= \sum_{m_0}\sum_{m_1}\sum_{m_2} \tilde{X}(m_2, m_1, m_0) W^{4km_2} W^{2km_1} W^{km_0} \end{aligned} \tag{4.4-3}$$

In Eq. (3) the innermost summation over m_2 is denoted by M_2 to obtain

$$M_2 = \sum_{m_2} \tilde{X}(m_2, m_1, m_0) W^{4km_2}$$

Substituting the binary form of k and noting that $W^4 = -1$, we obtain

$$M_2 = \sum_{m_2} \tilde{X}(m_2, m_1, m_0)(-1)^{m_2[4k_2+2k_1+k_0]}$$

Further, since $(-1)^{m_2[4k_2+2k_1]} = 1$, M_2 can be expressed as

$$M_2 = \sum_{m_2} \tilde{X}(m_2, m_1, m_0)(-1)^{k_0 m_2} \tag{4.4-4}$$

In Eq. (4), summation over m_2 results in a quantity which is a function of k_0, m_1, and m_0. Hence we introduce the notation

$$M_2 = \sum_{m_2} \tilde{X}(m_2, m_1, m_0)(-1)^{k_0 m_2} = \tilde{X}_1(k_0, m_1, m_0) \tag{4.4-5}$$

Substitution of Eq. (5) in Eq. (3) results in

$$8C_x(k) = \sum_{m_0} \sum_{m_1} \tilde{X}_1(k_0, m_1, m_0) W^{2km_1} W^{km_0} \tag{4.4-6}$$

Again, we consider the innermost summation in Eq. (6) and denote it by M_1. That is

$$\begin{aligned} M_1 &= \sum_{m_1} \tilde{X}_1(k_0, m_1, m_0) W^{2km_1} \\ &= \sum_{m_1} \tilde{X}_1(k_0, m_1, m_0)(-i)^{(4k_2+2k_1+k_0)m_1} \end{aligned} \tag{4.4-7}$$

In Eq. (7) the quantity $(-i)^{4k_2 m_1} = 1$, and hence it simplifies to yield

$$M_1 = \sum_{m_1} \tilde{X}_1(k_0, m_1, m_0)(-i)^{(2k_1+k_0)m_1} \tag{4.4-8}$$

The summation over m_1 in Eq. (8) results in a function of k_0, k_1, and m_0. Thus we have

$$M_1 = \tilde{X}_2(k_0, k_1, m_0) \tag{4.4-9}$$

Substituting Eq. (9) in Eq. (6) we obtain

$$8C_x(k) = \sum_{m_0} \tilde{X}_2(k_0, k_1, m_0) W^{m_0 k}$$

Corresponding to M_2 and M_1 we denote the summation over m_0 by M_0 to obtain

$$M_0 = \sum_{m_0} \tilde{X}_2(k_0, k_1, m_0) W^{m_0 k},$$

which yields

$$M_0 = \sum_{m_0} \tilde{X}_2(k_0, k_1, m_0) \left(\frac{1-i}{\sqrt{2}}\right)^{(4k_2+2k_1+k_0)m_0} \tag{4.4-10}$$

Inspection of Eq. (10) shows that no further simplification is possible. Again, the summation over m_0 yields a function of k_0, k_1, and k_2, which we denote by $\tilde{X}_3(k_0, k_1, k_2)$; that is

$$M_0 = \tilde{X}_3(k_0, k_1, k_2) \tag{4.4-11}$$

This is as far as we can go since

$$8C_x(k) = 8\tilde{C}_x(k_2, k_1, k_0) = \tilde{X}_3(k_0, k_1, k_2) \tag{4.4-12}$$

implying that Eqs. (5), (9), and (11) have all the information required to obtain the DFT coefficients $C_x(k)$. However, the following question remains:

what do these equations really do? As a first step towards answering this question, two observations are in order:

1. $N = 8$ results in $\log_2 N = 3$ such equations.

2. The coefficients (-1), $(-i)$, and $(1 - i)/\sqrt{2}$ in these equations are directly related to the roots of unity, $e^{-i2\pi}$. Clearly, (-1), $(-i)$, and $(1 - i)/\sqrt{2}$ are respectively the 2nd, 4th, and 8th primitive roots of unity. Thus, we introduce the notation

$$A_{2^r} = e^{-i2\pi/2^r}, \qquad r = 1, 2, \ldots, \log_2 N \tag{4.4-13}$$

where A_{2^r} is the 2^r-th primitive root of unity.

It is straightforward to show that A_{2^r} has the following properties;

(i) $$A_{2^r} = W^{N/2^r}, \qquad W = e^{-i2\pi/N} \tag{4.4-14a}$$

(ii) $$(A_{2^r})^{\lambda+l} = -(A_{2^r})^l; \qquad r = 1, 2, \ldots, \log_2 N,\ l = 0, 1, \ldots, 2^{r-1} \tag{4.4-14b}$$

where $\lambda = 2^{r-1}$

(iii) $$(A_N)^{N/2} = -1 \tag{4.4-14c}$$

Now, Eq. (5) is written in terms of A_{2^r} to obtain

$$\tilde{X}_1(k_0, m_1, m_0) = \sum_{m_2} \tilde{X}(m_2, m_1, m_0) A_2^{k_0 m_2}$$

That is

$$\tilde{X}_1(k_0, m_1, m_0) = \tilde{X}(0, m_1, m_0) + \tilde{X}(1, m_1, m_0) A_2^{k_0} \tag{4.4-15}$$

In Eq. (15) k_0 is either 0 or 1. Corresponding to each value of k_0 we generate 4 equations by varying m_1 and m_0 to obtain

Case 1: $k_0 = 0$

$$\left.\begin{aligned}
\tilde{X}_1(0, 0, 0) &= \tilde{X}(0, 0, 0) + \tilde{X}(1, 0, 0) \Rightarrow X_1(0) = X(0) + X(4)\\
\tilde{X}_1(0, 0, 1) &= \tilde{X}(0, 0, 1) + \tilde{X}(1, 0, 1) \Rightarrow X_1(1) = X(1) + X(5)\\
\tilde{X}_1(0, 1, 0) &= \tilde{X}(0, 1, 0) + \tilde{X}(1, 1, 0) \Rightarrow X_1(2) = X(2) + X(6)\\
\tilde{X}_1(0, 1, 1) &= \tilde{X}(0, 1, 1) + \tilde{X}(1, 1, 1) \Rightarrow X_1(3) = X(3) + X(7)
\end{aligned}\right\} \tag{4.4-16a}$$

Case 2: $k_0 = 1$

$$\left.\begin{aligned}
\tilde{X}_1(1, 0, 0) &= \tilde{X}(0, 0, 0) + A_2 \tilde{X}(1, 0, 0) \Rightarrow X_1(4) = X(0) - X(4)\\
\tilde{X}_1(1, 0, 1) &= \tilde{X}(0, 0, 1) + A_2 \tilde{X}(1, 0, 1) \Rightarrow X_1(5) = X(1) - X(5)\\
\tilde{X}_1(1, 1, 0) &= \tilde{X}(0, 1, 0) + A_2 \tilde{X}(1, 1, 0) \Rightarrow X_1(6) = X(2) - X(6)\\
\tilde{X}_1(1, 1, 1) &= \tilde{X}(0, 1, 1) + A_2 \tilde{X}(1, 1, 1) \Rightarrow X_1(7) = X(3) - X(7)
\end{aligned}\right\} \tag{4.4-16b}$$

This sequence of additions and subtractions are shown in Fig. 4.1, and designated by iteration #1 which corresponds to $r = 1$ in Eq. (13).

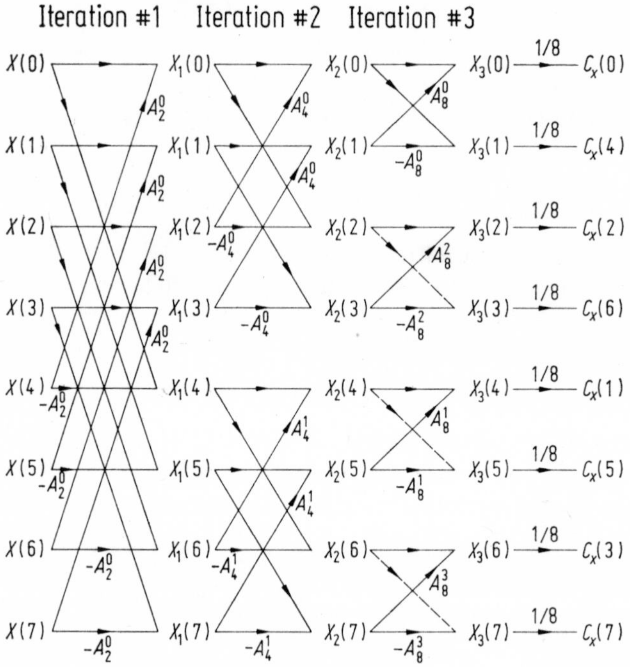

Fig. 4.1 FFT signal flow graph, $N = 8$

Again Eq. (9) in terms of A_4 yields

$$\widetilde{X}_2(k_0, k_1, m_0) = \widetilde{X}_1(k_0, 0, m_0) + \widetilde{X}_1(k_0, 1, m_0) A_4^{(2k_1+k_0)}$$

from which two equations are obtained for each set of values taken by (k_1, k_0). Since (k_1, k_0) take four sets of values, we consider four cases as follows:

Case 1: $(k_1, k_0) = (0, 0)$

$$\begin{aligned}\widetilde{X}_2(0, 0, 0) &= \widetilde{X}_1(0, 0, 0) + \widetilde{X}_1(0, 1, 0) \Rightarrow X_2(0) = X_1(0) + X_1(2)\\ \widetilde{X}_2(0, 0, 1) &= \widetilde{X}_1(0, 0, 1) + \widetilde{X}_1(0, 1, 1) \Rightarrow X_2(1) = X_1(1) + X_1(3)\end{aligned} \tag{4.4-17a}$$

Case 2: $(k_1, k_0) = (0, 1)$

$$\begin{aligned}\widetilde{X}_2(1, 0, 0) &= \widetilde{X}_1(1, 0, 0) + A_4 \widetilde{X}_1(1, 1, 0) \Rightarrow X_2(4) = X_1(4) + A_4 X_1(6)\\ \widetilde{X}_2(1, 0, 1) &= \widetilde{X}_1(1, 0, 1) + A_4 \widetilde{X}_1(1, 1, 1) \Rightarrow X_2(5) = X_1(5) + A_4 X_1(7)\end{aligned} \tag{4.4-17b}$$

Case 3: $(k_1, k_0) = (1, 0)$

$$\tilde{X}_2(0, 1, 0) = \tilde{X}_1(0, 0, 0) + A_4^2 \tilde{X}_1(0, 1, 0) \Rightarrow X_2(2) = X_1(0) - X_1(2)$$
$$\tilde{X}_2(0, 1, 1) = \tilde{X}_1(0, 0, 1) + A_4^2 \tilde{X}_1(0, 1, 1) \Rightarrow X_2(3) = X_1(1) - X_1(3) \tag{4.4-17c}$$

Case 4: $(k_1, k_0) = (1, 1)$

$$\tilde{X}_2(1, 1, 0) = \tilde{X}_1(1, 0, 0) + A_4^3 \tilde{X}_1(1, 1, 0) \Rightarrow X_2(6) = X_1(4) - A_4 X_1(6)$$
$$\tilde{X}_2(1, 1, 1) = \tilde{X}_1(1, 0, 1) + A_4^3 \tilde{X}_1(1, 1, 1) \Rightarrow X_2(7) = X_1(5) - A_4 X_1(7) \tag{4.4-17d}$$

The sequence of arithmetic operations in Eqs. (17a) through (17d) is represented by iteration #2 in Fig. 4.1, where iteration #2 corresponds to $r = 2$ in Eq. (13).

Finally Eq. (11) is written in terms of A_8 to obtain

$$\tilde{X}_3(k_0, k_1, k_2) = \tilde{X}_2(k_0, k_1, 0) + \tilde{X}_2(k_0, k_1, 1) A_8^{(4k_2+2k_1+k_0)} \tag{4.4-18}$$

which yields the following cases.

Case 1: $(k_2, k_1, k_0) = (0, 0, 0)$

$$\tilde{X}_3(0, 0, 0) = \tilde{X}_2(0, 0, 0) + \tilde{X}_2(0, 0, 1) \Rightarrow X_3(0) = X_2(0) + X_2(1) \tag{4.4-19a}$$

Case 2: $(k_2, k_1, k_0) = (0, 0, 1)$

$$\tilde{X}_3(1, 0, 0) = \tilde{X}_2(1, 0, 0) + A_8 \tilde{X}_2(1, 0, 1) \Rightarrow X_3(4) = X_2(4) + A_8 X_2(5) \tag{4.4-19b}$$

Case 3: $(k_2, k_1, k_0) = (0, 1, 0)$

$$\tilde{X}_3(0, 1, 0) = \tilde{X}_2(0, 1, 0) + A_8^2 \tilde{X}_2(0, 1, 1) \Rightarrow X_3(2) = X_2(2) + A_8^2 X_2(3) \tag{4.4-19c}$$

Case 4: $(k_2, k_1, k_0) = (0, 1, 1)$

$$\tilde{X}_3(1, 1, 0) = \tilde{X}_2(1, 1, 0) + A_8^3 \tilde{X}_2(1, 1, 1) \Rightarrow X_3(6) = X_2(6) + A_8^3 X_2(7) \tag{4.4-19d}$$

Case 5: $(k_2, k_1, k_0) = (1, 0, 0)$

$$\tilde{X}_3(0, 0, 1) = \tilde{X}_2(0, 0, 0) + A_8^4 \tilde{X}_2(0, 0, 1) \Rightarrow X_3(1) = X_2(0) - X_2(1) \tag{4.4-19e}$$

The sequence of arithmetic operations in Eqs. (19a) through (19e) is shown in Fig. 4.1 and designated by iteration #3 which corresponds to $r = 3$ in Eq. (13).

To obtain the desired $C_x(k)$ we recall that

$$8\tilde{C}_x(k_2, k_1, k_0) = \tilde{X}_3(k_0, k_1, k_2) \tag{4.4-20}$$

Note the *reversed* order in which the binary coefficients k_0, k_1, and k_2 appear in $\tilde{X}_3(\,)$ relative to the *natural* order in $\tilde{C}_x(\,)$. This is the so-called bit-reversal phenomenon. Thus the $X_3(\,)$ are related to the $C_x(\,)$ as follows:

$$\begin{aligned} X_3(0) &= 8C_x(0) \\ X_3(4) &= 8C_x(1) \\ X_3(2) &= 8C_x(2) \\ X_3(6) &= 8C_x(3) \\ X_3(1) &= 8C_x(4) \end{aligned} \tag{4.4-21}$$

Again[1], the remaining coefficients $C_x(k)$, $k = 5, 6, 7$ are obtained as

$$C_x(4 + l) = \bar{C}_x(4 - l), \qquad l = 1, 2, 3$$

That is, $C_x(5) = \bar{C}_x(3)$, $C_x(6) = \bar{C}_x(2)$, and $C_x(7) = \bar{C}_x(1)$. Alternately[2] $C_x(5)$, $C_x(6)$, and $C_x(7)$ can be computed by using Eqs. (12) and (18). This approach yields

$$\begin{aligned} 8C_x(5) &= X_3(5) = X_2(4) - A_8X_2(5) \\ 8C_x(6) &= X_3(3) = X_2(2) - A_8^2X_2(3) \\ 8C_x(7) &= X_3(7) = X_2(6) - A_8^3X_2(7) \end{aligned} \tag{4.4-22}$$

The arithmetic operations in Eq. (22) are shown under iteration #3 in Fig. 4.1 by means of dotted lines. Fig. 4.1 represents the FFT signal flow graph for $N = 8$.

Observations. With respect to the above FFT signal flow graph, we summarize the following observations.

1. The maximum value of the iteration index r is given by $n = \log_2 N = 3$, with $N = 8$.

2. In the r-th iteration, $r = 1, 2, \ldots, \log_2 N$, the multipliers are

$$A_{2^r}^s, \qquad s = 0, 1, \ldots, 2^{r-1} - 1; \tag{4.4-23}$$

$r = 1$ implies that A_2^0 is needed in iteration #1. Since $A_2^0 = 1$, only additions and subtractions are required in this iteration.
$r = 2$ implies that A_4^0 and A_4^1 are needed in iteration #2.
$r = 3$ implies that A_8^0, A_8^1, A_8^2, and A_8^3 are needed in iteration #3.

3. The r-th iteration consists of 2^{r-1} groups with $N/2^{r-1}$ members in each group. Each group takes *one* multiplier of the form $A_{2^r}^s$. Half the members of each group are associated with $+A_{2^r}^s$ while the remaining half is associated with $-A_{2^r}^s$.

[1] This can be done only if $\{X(m)\}$ is real.
[2] This approach is valid for both real and complex $\{X(m)\}$.

4. The first member $X_r(\ )$ of each group to which a multiplier A_{2r}^{s} is assigned in r-th iteration can be obtained as shown in Table 4.4-1 which is self-explanatory.

Table 4.4-1

s	$s = \tilde{s}(k_2, k_1, k_0)$ k_2	k_1	k_0	k_0	k_1	k_2	First member of group which takes A_{2r}^{s} $X_r(k) = \tilde{X}_r(k_0, k_1, k_2)$
0	0	0	0	0	0	0	$X_r(0)$
1	0	0	1	1	0	0	$X_r(4)$
2	0	1	0	0	1	0	$X_r(2)$
3	0	1	1	1	1	0	$X_r(6)$

5. The $C_x(k)$ corresponding to each $X_3(l)$, $l = 0, 1, 2, \ldots, 7$ are obtained as follows:

(a) Express the sequence $l = 0, 1, 2, \ldots, 7$ in binary form to obtain 000, 001, 010, 011, 100, 101, 110, 111.

(b) Reverse each of the 3-bit strings in (a) to obtain the sequence 000, 100, 010, 110, 001, 101, 011, 111.

(c) Write the binary sequence in (b) in decimal form to obtain 0, 4, 2, 6, 1, 5, 3, 7.

Then there is a one-to-one correspondence between the sequence $k = 0$, 4, 2, 6, 1, 5, 3, 7 and the sequence $l = 0, 1, 2, 3, 4, 5, 6, 7$.

6. All operations leading to $C_x(0)$ and $C_x(4)$ are real. Thus $C_x(0)$ and $C_x(4)$ are real numbers, which is in agreement with the DFT complex conjugate property.

7. Counting the real and imaginary parts separately, there are exactly 8 independent values involved in $C_x(k)$. This of course corresponds to $N = 8$, the number of data points $X(m)$ in the example.

8. The output of the r-th iteration can be stored in the same locations as those of its input. To see this, let us consider the output points $X_1(0) = X(0) + X(4)$ and $X_1(4) = X(0) - X(4)$. Clearly, $X_1(0)$ and $X_1(4)$ can be stored in two *temporary* locations $T1$ and $T2$ by computing $T1 = X(0) + X(4)$ and $T2 = X(0) - X(4)$. Now, since $X(0)$ and $X(4)$ are no longer needed for any further computations of the input to the first iteration, the contents of $T1$ and $T2$ can be stored back into the locations of $X(0)$ and $X(4)$. That is, $X_1(0)$ and $X_1(4)$ are stored in the locations of $X(0)$ and $X(4)$ respectively. Similarly the $X_1(m)$ are stored in the locations of $X(m)$, $m = 1$, 2, 3, 5, 6, 7 respectively. The same procedure can then be applied to store $X_2(m)$ in the locations of $X_1(m)$ and so on. This property of the flow graph in Fig. 4.1 is known as the "in-place" property [1]. Hence, essentially no additional storage locations are required as the data is being processed. The total number of storage locations required is $2N$, assuming complex data sequences as input. In addition, $2N$ storage locations are required to enable the bit-reversal process.

9. The total number of arithmetic operations (i.e., multiplication followed by an addition or subtraction) required to generate all the $C_x(k)$, $k = 0, 1, \ldots, N-1$, is approximately to $N \log_2 N$.

10. Clearly, the algorithm is valid regardless of whether the data $X(m)$, $m = 0, 1, \ldots, N-1$ is real or complex. Thus it can also be used to compute the IDFT with the following minor modifications which follow from Eqs. (3.1-4) and (4.1-1)

(i) Replace W by $\overline{W}$.
(ii) Omit the multiper $1/N$ following the last iteration.

We shall refer to this version of the algorithm as the *inverse fast Fourier transform* (IFFT).

11. The multipliers used in the signal flow graph can be expressed as powers of W rather than A_{2^r} using the relation $A_{2^r} = W^{N/2^r}$. We shall see shortly [Example 4.5-4] that the related signal flow graph can be constructed directly in terms of powers of W by means of a simple procedure.

4.5 Illustrative Examples

Example 4.5-1

Given the data sequence,

$$X(0) = 1, \quad X(1) = 2, \quad X(2) = 1, \quad X(3) = 1, \quad X(4) = 3, \quad X(5) = 2,$$
$$X(6) = 1, \quad \text{and } X(7) = 2,$$

use the FFT to compute the DFT coefficients $C_x(k)$, $k = 0, 1, \ldots, 7$.

Solution: Since $N = 8$, we use the signal flow graph in Fig. 4.1, where

$$A_{2^r} = e^{-i2\pi/2^r}; \qquad r = 1, 2, 3.$$

That is, $A_2 = -1$, $A_4 = -i$, $A_8 = (1-i)/\sqrt{2}$, $A_8^2 = -i$, and $A_8^3 = -(1+i)/\sqrt{2}$

Following the sequence of computations indicated in the flow graph, we obtain:

$X(0) = 1$	$X_1(0) = 4$	$X_2(0) = 6$	$X_3(0) = 13$	$\xrightarrow{1/8}$	$C_x(0)$
$X(1) = 2$	$X_1(1) = 4$	$X_2(1) = 7$	$X_3(1) = -1$	$\xrightarrow{1/8}$	$C_x(4)$
$X(2) = 1$	$X_1(2) = 2$	$X_2(2) = 2$	$X_3(2) = 2 - i$	$\xrightarrow{1/8}$	$C_x(2)$
$X(3) = 1$	$X_1(3) = 3$	$X_2(3) = 1$	$X_3(3) = 2 + i$	$\xrightarrow{1/8}$	$C_x(6) = \overline{C}_x(2)$
$X(4) = 3$	$X_1(4) = -2$	$X_2(4) = -2$	$X_3(4) = -1.293 + i0.707$	$\xrightarrow{1/8}$	$C_x(1)$
$X(5) = 2$	$X_1(5) = 0$	$X_2(5) = i$	$X_3(5) = -2.707 - i0.707$	$\xrightarrow{1/8}$	$C_x(5) = \overline{C}_x(3)$
$X(6) = 1$	$X_1(6) = 0$	$X_2(6) = -2$	$X_3(6) = -2.707 + i0.707$	$\xrightarrow{1/8}$	$C_x(3)$
$X(7) = 2$	$X_1(7) = -1$	$X_2(7) = -i$	$X_3(7) = -1.293 - i0.707$	$\xrightarrow{1/8}$	$C_x(7) = \overline{C}_x(1)$

Example 4.5-2

Given the data sequence,

$$X(0) = 1, \quad X(1) = 2, \quad X(2) = -1, \quad \text{and} \quad X(3) = 3.$$

(a) Use the FFT algorithm to compute the DFT coefficients $C_x(k)$, $k = 0, 1, 2, 3$.

(b) Verify that the algorithm can be used to recover the sequence $X(m)$, $m = 0, 1, 2, 3$ with the $C_x(k)$ obtained in (a) as the input.

Solution: (a) Here the number of iterations is 2, since $N = 4$. We compute

$$A_{2^r} = e^{-i2\pi/2^r}, \qquad r = 1, 2$$

to obtain

$$A_2 = -1, \quad \text{and} \quad A_4 = -i.$$

When $r = 1$, the multiplier is A_2^0 whereas A_4^0 and A_4^1 are the multipliers when $r = 2$.

Corresponding to Table 4.4-1 we obtain:

s	$s = \tilde{s}(k_1, k_0)$				First member of group which takes $A_{2^r}^s$
	k_1	k_0	k_0	k_1	$X_r(k) = \tilde{X}_r(k_0\, k_1)$
0	0	0	0	0	$X_r(0)$
1	0	1	1	0	$X_r(2)$

The above information is sufficient to construct the signal flow graph, $N = 4$ shown in Fig. 4.2.

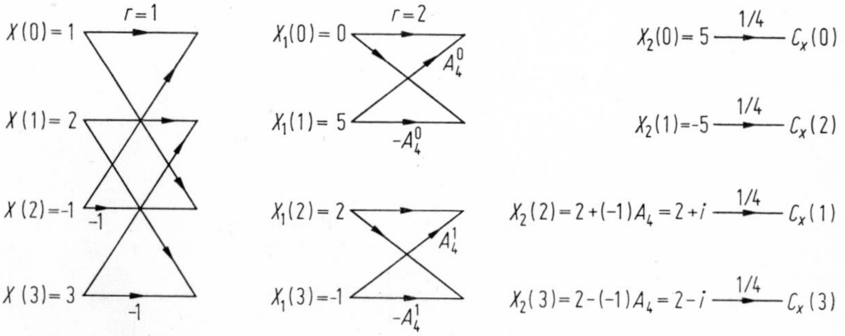

Fig. 4.2 Signal flow graph for Example 4.5-2

(b) Recall that the IDFT is given by

$$X(m) = \sum_{k=0}^{N-1} C_x(k) W^{-km}, \qquad m = 0, 1, \ldots, N-1$$

Figure 4.3 shows the corresponding signal flow graph where $\bar{A}_4 = i$ since $A_4 = -i$ in (a). The subscript x of $C_x(k)$ is dropped for convenience in Fig. 4.3.

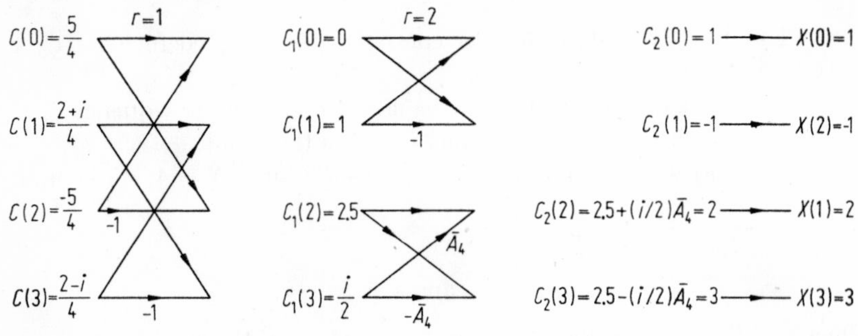

Fig. 4.3 Signal flow graph for Example 4.5-2

From Fig. 4.3 we have $X(0) = 1$, $X(1) = 2$, $X(2) = -1$, and $X(3) = 3$, which agrees with the sequence $X(m)$, $m = 0, 1, 2, 3$ in (a).

Example 4.5-3

For a sequence $X(m)$, $m = 0, 1, \ldots, 15$, develop the signal flow graph, $N = 16$.

Solution: $N = 16$ yields $n = 4$, which implies that the iteration index r takes the values 1, 2, 3, 4. Thus we compute the multipliers

$$A_{2^r}^s = [e^{-i2\pi/2^r}]^s, \qquad r = 1, 2, 3, 4; \qquad s = 0, 1, \ldots, 2^{r-1} - 1$$

to obtain Table 4.5-1.

Table 4.5-1

Iteration Number	Multipliers needed
1	A_2^0
2	A_4^0, A_4^1
3	$A_8^0, A_8^1, A_8^2, A_8^3$
4	$A_{16}^0, A_{16}^1, \ldots, A_{16}^7$

Table 4.5-2 shown below is constructed using the procedure used to construct Table 4.5-1.

Table 4.5-2

s	$s = \tilde{s}(k_3, k_2, k_1, k_0)$	k_0, k_1, k_2, k_3	First member of group which takes $A^s_{2^r}$ $X_r(k) = \tilde{X}_r(k_0, k_1, k_2, k_3)$
0	0 0 0 0	0 0 0 0	$X_r(0)$
1	0 0 0 1	1 0 0 0	$X_r(8)$
2	0 0 1 0	0 1 0 0	$X_r(4)$
3	0 0 1 1	1 1 0 0	$X_r(12)$
4	0 1 0 0	0 0 1 0	$X_r(2)$
5	0 1 0 1	1 0 1 0	$X_r(10)$
6	0 1 1 0	0 1 1 0	$X_r(6)$
7	0 1 1 1	1 1 1 0	$X_r(14)$

The resulting signal flow graph is shown in Fig. 4.4.

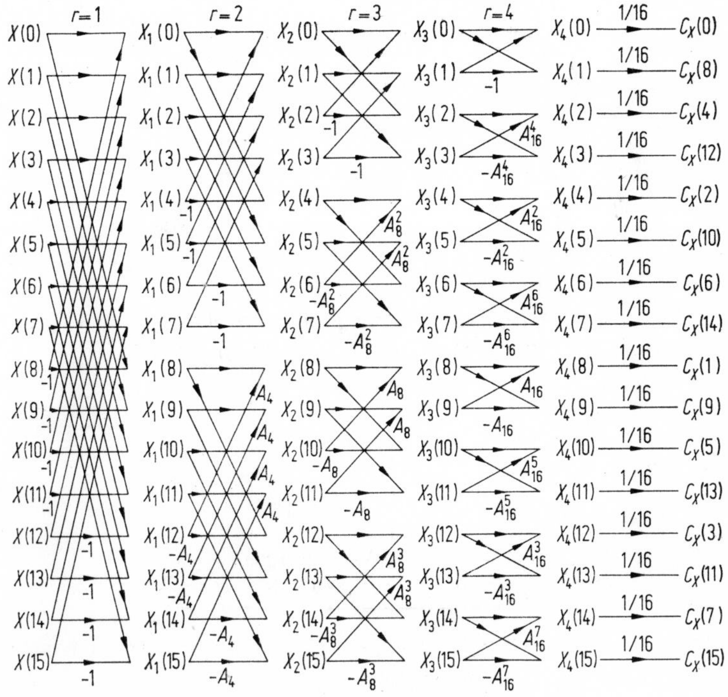

Fig. 4.4 FFT signal flow graph for Example 4.5-3

Example 4.5-4

(a) Apply Eq. (4.4-14a) to the signal flow graph, $N = 16$ in Fig. 4.4, and obtain the corresponding flow graph in terms of powers of W.

(b) Use the results of (a) to develop a set of rules which yields the signal flow graph, $N = 16$ in terms of powers of W, without resorting to Eq. (4.4-14a).

Solution: (a) $N = 16$ yields $n = 4$ and hence Eq. (4.4-14a) yields

$$A_{2^r} = W^{(16/2^r)}, \qquad r = 1, 2, 3, 4$$

where $W = e^{-i2\pi/16}$. Replacing the $A_{2^r}^s$ in Fig. 4.4 by means of this relation, we obtain the desired signal flow graph shown in Fig. 4.5.

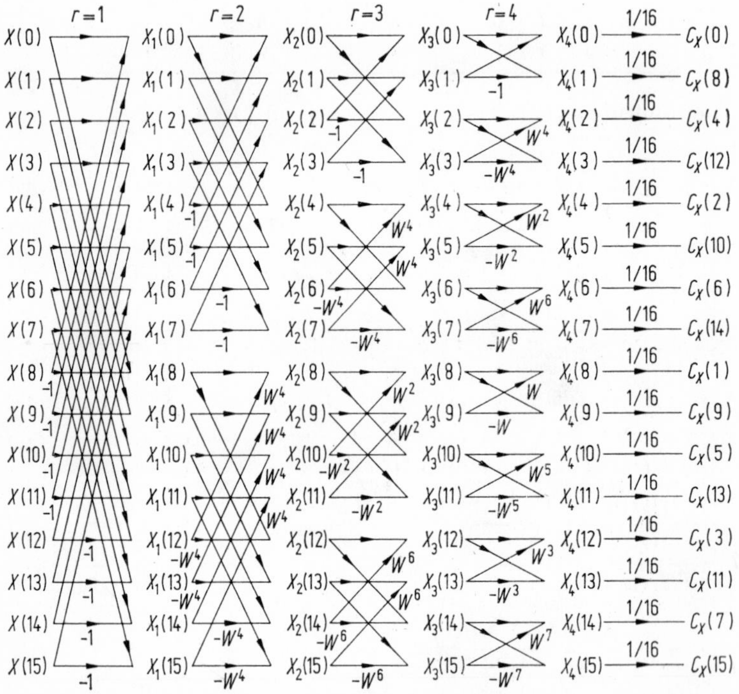

Fig. 4.5 FFT signal flow graph for Example 4.5-4

(b) Examination of Fig. 4.5 shows that the signal flow can be constructed by means of the following procedure:

Step 1: Express the sequence $l = 0, 1, \ldots, (N/2 - 1)$ in terms of $(n - 1)$-bit strings. This process yields the set S_1, where $S_1 = \{000, 001, 010, 011, 100, 101, 110, 111\}$

Step 2: Reverse each of the $(n - 1)$-bit strings in S_1 to obtain $S_2 = \{000, 100, 010, 110, 001, 101, 011, 111\}$

Step 3: Write the binary sequence in S_2 in decimal form to obtain
$S_3 = \{0, 4, 2, 6, 1, 5, 3, 7\}$

Step 4: Using S_3, form the set $S^* = \{W^0, W^4, W^2, W^6, W^1, W^5, W^3, W^7\}$

Step 5: Now, iteration r consists of 2^{r-1} groups, where $r = 1, 2, 3, 4$. The members of the set S^* obtained in Step 4 are assigned to these groups as follows:

Iteration Number	*Powers of W*
1	W^0
2	W^0, W^4
3	W^0, W^4, W^2, W^6
4	$W^0, W^4, W^2, W^6, W^1, W^5, W^3, W^7$

The above steps yield the signal flow graph, $N = 16$ in Fig. 4.5 without resorting to Eq. (4.4-14a). One can generalize this procedure for any $N = 2^n$; [see Prob. 4-4].

4.6 Shuffling

From the discussion in the previous sections, it is apparent that bit-reversal plays an important role in the FFT version we have considered. In general, bit-reversal is a time consuming operation. However, in cases where N is of the form $N = 2^n$, bit-reversal can be accomplished rapidly by means of a shuffling technique which involves decimal arithmetic only. The procedure to secure shuffling can be summarized as follows:

Step 1. Express N in terms of n factors

$$n_s = N/2^s, \qquad s = 1, 2, \ldots, n. \tag{4.6-1}$$

Step 2. Form the following table T_n

$$\begin{array}{lllll}
0 \\
n_1 \\
n_2 & (n_1 + n_2) \\
n_3 & (n_1 + n_3) & (n_2 + n_3) & (n_1 + n_2 + n_3) \\
n_4 & (n_1 + n_4) & (n_2 + n_4) & (n_1 + n_2 + n_4) \cdots \\
\vdots \\
n_n
\end{array}$$

That is, the k-th row, $k = 1, 2, \ldots, n + 1$ of T_n is obtained by adding n_{k-1} to each member of the preceding $(k - 1)$ rows. Then the desired bit-reversed sequence is given by L_n, where

$$L_n = \{0, n_1, n_2, (n_1 + n_2), n_3, (n_1 + n_3), \ldots, (n_1 + n_2 + \cdots + n_n)\} \quad (4.6\text{-}2)$$

As an illustrative example, we consider the case $N = 8$. Then Eq. (1) yields $n_1 = 4$, $n_2 = 2$, $n_3 = 1$. Forming T_3 we obtain

$$\begin{matrix} 0 \\ 4 \\ 2 & 6 \\ 1 & 5 & 3 & 7, \end{matrix}$$

which implies that $L_3 = \{0, 4, 2, 6, 1, 5, 3, 7\}$. This sequence can be looked upon as the "shuffled" version of the natural ordered sequence $\{0, 1, 2, 3, 4, 5, 6, 7\}$.

4.7 Operations Count and Storage Requirements

Operations count. From Eq. (4.1-1) it follows that the $C_x(k)$ can be computed by means of multiplying an $(N \times N)$ matrix, which consists of powers of W, by an $(N \times 1)$ data vector. The number of arithmetic operations (i.e., complex multiplication followed by an addition or subtraction) is approximately $2N^2$. We shall refer to this method of computing the DFT coefficients as

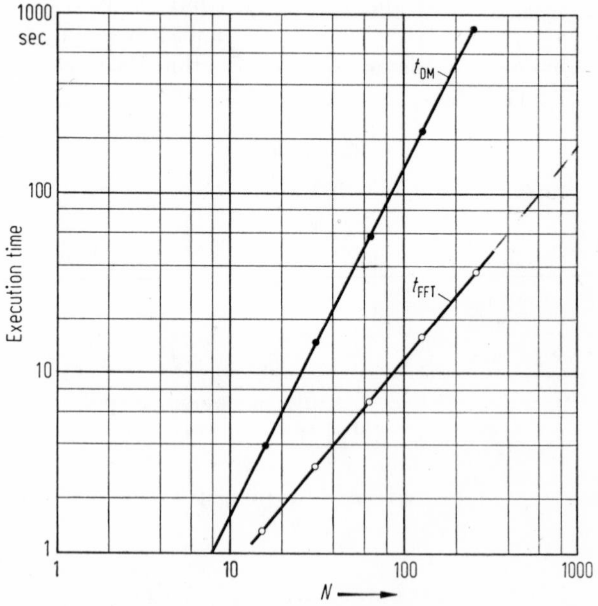

Fig. 4.6 Comparison of the direct method and FFT execution times

the "direct method". Use of the FFT results in $\log_2 N$ iterations, and the total number of arithmetic operations required is approximately $N \log_2 N$.

From the above discussion it is clear that the total number of arithmetic operations required by the direct method and the FFT are proportional to N^2 and N respectively. Hence as N increases, the FFT rapidly becomes more economical than the direct method. This is illustrated in Fig. 4.6, where t_{DM} and t_{FFT} denote the execution times associated with the direct method and the FFT respectively. From Fig. 4.6 we obtain

$$t_{DM} \simeq 0.0173\, N^{1.95}$$

and

$$t_{FFT} \simeq 5.528\, N^{1.167},$$

from which it follows that t_{DM} and t_{FFT} are approximately proportional to N^2 and N respectively. The above execution times were obtained using a 16-bit machine (NOVA 1200) and the programming language was BASIC. Again, it has been reported that using the IBM 7094 digital computer, the FFT required 5 seconds to compute all 8192 DFT coefficients, while the time taken by the direct method was of the order of half an hour [1].

Storage requirements. The FFT essentially requires $2N$ storage locations, as the data (assumed to be complex) is being processed. This is because the output of a particular iteration can be stored in the same locations as the input to that iteration. The shuffling process may require an additional $2N$ locations. Thus the FFT requires approximately $4N$ storage locations. In contrast, the direct method requires approximately $2N^2$ locations since it is necessary to store N^2 values of the powers of W. The FFT storage requirements are hence significantly less severe than those for the direct method.

Round-off error. The FFT significantly reduces the round-off error associated with the arithmetic operations. Compared to the direct method the FFT reduces the round-off error approximately by a factor of $(\log_2 N)/N$; [1, 9].

4.8 Some Applications

In this section we discuss some applications of the FFT and illustrate them by means of examples [1, 5, 10–12].

1. Computation of Fourier amplitude and phase spectra. Let $x(t)$ be an exponentially decaying signal, such that

$$x(t) = e^{-t}, \qquad 0 \leq t \leq 4 \text{ msec}$$

and suppose its DFT amplitude and phase spectra are desired.

To use the FFT, $x(t)$ has to be sampled to obtain N sampled values, such that $N = 2^n$. For example, we consider the case $N = 32$ which results in the exponentially decaying sequence

$$X(m) = x(mT), \qquad m = 0, 1, \ldots, 31 \tag{4.8-1}$$

where T is the sampling interval, and $X(0) = 1$. The amplitude and phase spectra are given by

$$p(k) = |C_x(k)| \tag{4.8-2}$$

and

$$\psi_x(k) = \tan^{-1}\left\{\frac{I_x(k)}{R_x(k)}\right\}, \qquad k = 0, 1, \ldots, 31 \tag{4.8-3}$$

where $R_x(k)$ and $I_x(k)$ are the real and imaginary parts of $C_x(k)$. Use of the FFT results in the spectra shown in Fig. 4.7 where the amplitude spectrum has been normalized. In Fig. 4.7 we observe that the amplitude spectrum has $N/2 + 1 = 17$ independent spectral points $C_x(0)$, $|C_x(1)|, \ldots,$ $|C_x(15)|$, $C_x(16)$, due to complex conjugate theorem [see Eq. (3.2-2)].

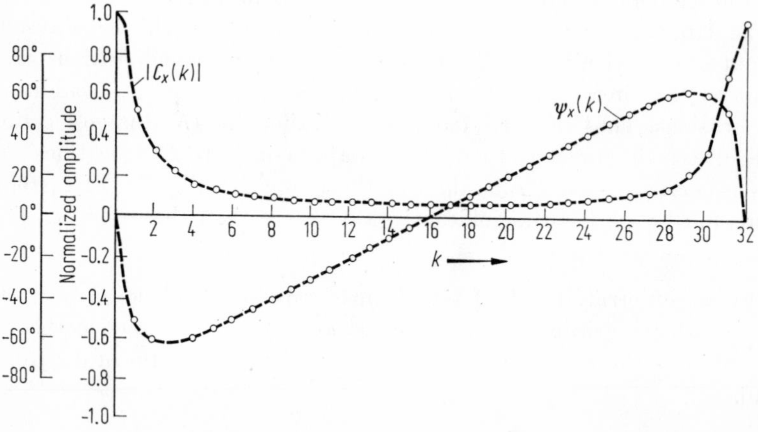

Fig. 4.7 DFT amplitude and phase spectra of an exponentially decaying sequence

Again, $\psi_x(k)$ is an odd function about $k = N/2$ with $N = 32$ [see Prob. 3-6]. The variable k is sometimes referred to as the "frequency number". Since the fundamental frequency f_0 is 250 Hz (= 1/4 msec), the bandwidth of $x(t)$ has been assumed to be

$$B = \frac{N}{2} f_0 = 4000 \text{ Hz}.$$

Using the information in Fig. 4.7, we obtain Fig. 4.8 that shows $|F_{x*}(k\omega_0)|$ and $\psi_{x*}(k\omega_0)$, $\omega_0 = 2\pi f_0$, which are the Fourier amplitude and phase spectra of the sequence $X(m)$, $m = 0, 1, \ldots, 31$.

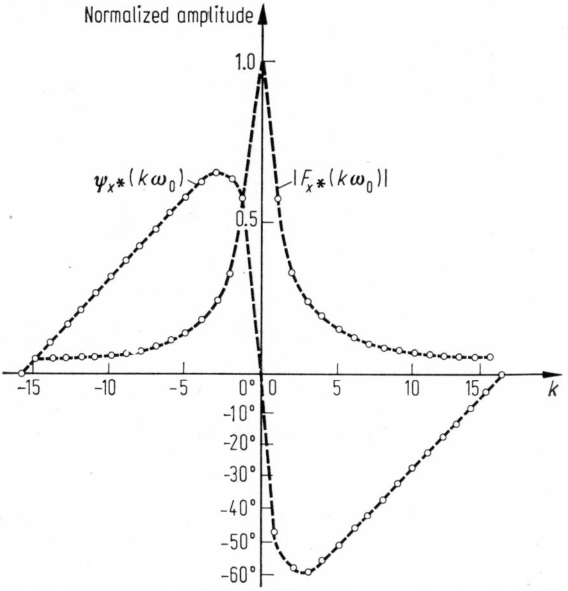

Fig. 4.8 Fourier amplitude and phase spectra of an exponentially decaying sequence

2. Computation of correlation sequences. If $\{X(m)\}$ and $\{Y(m)\}$ are two real-valued N-periodic sequences, then the sequence $\{\hat{Z}(m)\}$ obtained by their correlation is given by

$$\hat{Z}(m) = \frac{1}{N} \sum_{h=0}^{N-1} X(h) Y(m+h), \qquad m = 0, 1, \ldots, N-1 \qquad (4.8\text{-}4)$$

The correlation theorem [see Eq. (3.2-16)] yields

$$C_{\hat{z}}(k) = \bar{C}_x(k) C_y(k), \qquad k = 0, 1, \ldots, N-1 \qquad (4.8\text{-}5)$$

from which it follows that $\{\hat{Z}(m)\}$ can be computed using the FFT, as indicated in Fig. 4.9. As an illustrative example, the $\{\hat{Z}(m)\}$ sequence which

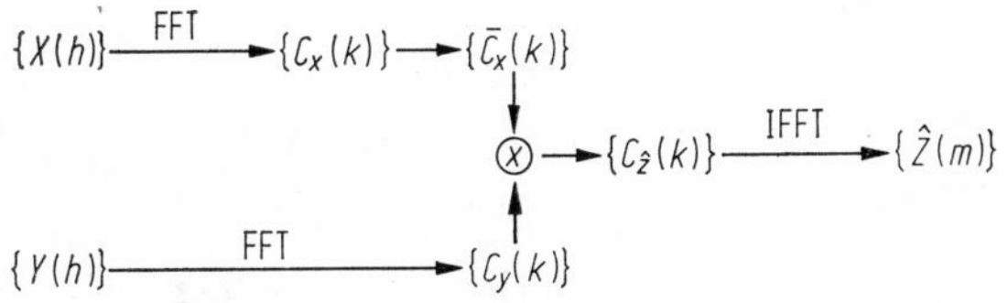

Fig. 4.9 Sequence of computations to obtain $\{\hat{Z}(m)\}$

results by correlating the exponentially decaying sequence [see Eq. (1)], with itself (i.e., autocorrelation) is shown in Fig. 4.10.

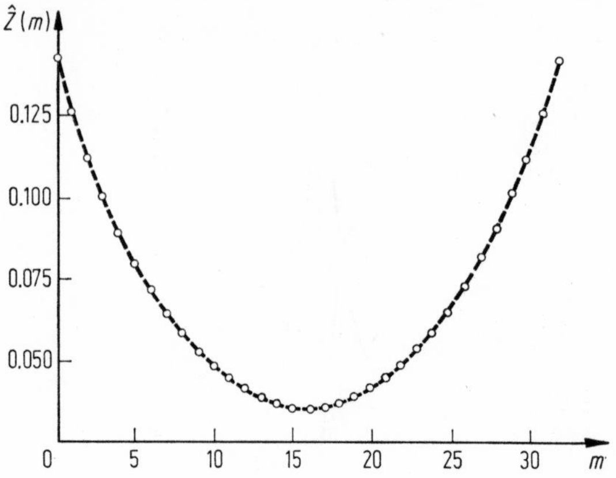

Fig. 4.10 Autocorrelation of exponentially decaying sequence

3. Computation of convolution sequences. The convolution sequence $\{Z(m)\}$ corresponding to two real-valued N-periodic sequences $\{X(m)\}$ and $\{Y(m)\}$ is given by

$$Z(m) = \sum_{h=0}^{N-1} X(h)Y(m-h), \qquad m = 0, 1, \ldots, N-1 \tag{4.8-6}$$

From the convolution theorem [see Eq. (3.2-11)], we have

$$C_z(k) = C_x(k)C_y(k), \qquad k = 0, 1, \ldots, N-1 \tag{4.8-7}$$

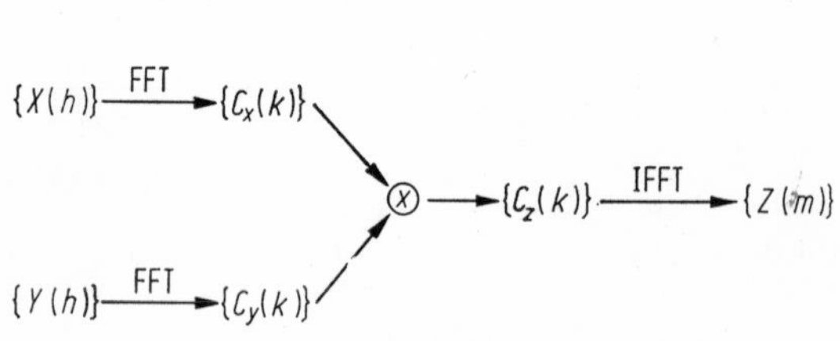

Fig. 4.11 Sequence of computations to obtain $\{Z(m)\}$

From Eq. (7) it follows that the FFT can be used to compute $\{Z(m)\}$ as indicated in Fig. 4.11. The convolution of the exponentially decaying sequence [see Eq. (1)] with itself is shown in Fig. 4.12.

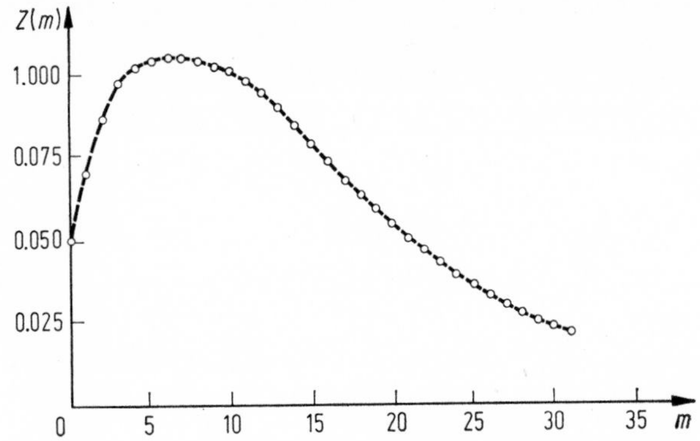

Fig. 4.12 Convolution of exponentially decaying sequence with itself

Particular case. We consider the case when the data sequences to be convolved are *aperiodic*[1], and are denoted by $\tilde{X}(m)$, $m = 0, 1, \ldots, M$ and $\tilde{Y}(m)$, $m = 0, 1, \ldots, P$. The convolution of these sequences is given by

$$\tilde{Z}(m) = \frac{1}{M+P+1} \sum_{h=0}^{M+P} \tilde{X}(h)\tilde{Y}(m-h), \qquad m = 0, 1, \ldots, M+P \tag{4.8-8}$$

where $\tilde{Y}(l) = 0$, $l < 0$. The $\tilde{Z}(m)$ can be computed using the FFT as follows [13].

(1) Let N be the smallest integer power of 2 greater than $(M + P)$.

(2) Form N-periodic sequences $\{X(m)\}$ and $\{Y(m)\}$ such that

$$X(m) = \begin{cases} \tilde{X}(m), & m = 0, 1, \ldots, M \\ 0, & m = M+1, M+2, \ldots, N-1 \end{cases} \tag{4.8-9}$$

and

$$Y(m) = \begin{cases} \tilde{Y}(m), & m = 0, 1, \ldots, P \\ 0, & m = P+1, P+2, \ldots, N-1 \end{cases} \tag{4.8-10}$$

(3) Use the FFT to compute the convolution sequence $\{Z(m)\}$ of the *periodic* sequences $\{X(m)\}$ and $\{Y(m)\}$ where

$$Z(m) = \frac{1}{N} \sum_{h=0}^{N-1} X(h)Y(m-h), \qquad m = 0, 1, \ldots, N-1 \tag{4.8-11}$$

(4) Obtain the desired $\tilde{Z}(m)$ using the relation

$$\tilde{Z}(m) = \frac{N}{M+P+1} Z(m), \qquad m = 0, 1, \ldots, M+P \tag{4.8-12}$$

[1] To distinguish between convolution/correlation associated between periodic and aperiodic sequences, the convolution/correlation associated with periodic sequences is sometimes referred to as cyclic or circular convolution/correlation.

For the purposes of illustration, we consider the case $M = 3$ and $P = 2$. Then it follows that $N = 8$ and Eqs. (9) and (10) yield

$$X(m) = \begin{cases} \tilde{X}(m), & m = 0, 1, 2, 3 \\ 0, & m = 4, 5, 6, 7 \end{cases} \tag{4.8-13}$$

and

$$Y(m) = \begin{cases} \tilde{Y}(m), & m = 0, 1, 2 \\ 0, & m = 3, \ldots, 7 \end{cases} \tag{4.8-14}$$

Substitution of Eqs. (13) and (14) in Eq. (6) leads to the matrix equation [see Eq. (3.3-8)]

$$\begin{bmatrix} Z(0) \\ Z(1) \\ Z(2) \\ Z(3) \\ Z(4) \\ Z(5) \\ Z(6) \\ Z(7) \end{bmatrix} = \frac{1}{8} \begin{bmatrix} \tilde{X}(0) & \tilde{X}(1) & \tilde{X}(2) & \tilde{X}(3) & 0 & 0 & 0 & 0 \\ \tilde{X}(1) & \tilde{X}(2) & \tilde{X}(3) & 0 & 0 & 0 & 0 & \tilde{X}(0) \\ \tilde{X}(2) & \tilde{X}(3) & 0 & 0 & 0 & 0 & \tilde{X}(0) & \tilde{X}(1) \\ \tilde{X}(3) & 0 & 0 & 0 & 0 & \tilde{X}(0) & \tilde{X}(1) & \tilde{X}(2) \\ 0 & 0 & 0 & 0 & \tilde{X}(0) & \tilde{X}(1) & \tilde{X}(2) & \tilde{X}(3) \\ 0 & 0 & 0 & \tilde{X}(0) & \tilde{X}(1) & \tilde{X}(2) & \tilde{X}(3) & 0 \\ 0 & 0 & \tilde{X}(0) & \tilde{X}(1) & \tilde{X}(2) & \tilde{X}(3) & 0 & 0 \\ 0 & \tilde{X}(0) & \tilde{X}(1) & \tilde{X}(2) & \tilde{X}(3) & 0 & 0 & 0 \end{bmatrix} \begin{bmatrix} \tilde{Y}(0) \\ 0 \\ 0 \\ 0 \\ 0 \\ 0 \\ \tilde{Y}(2) \\ \tilde{Y}(1) \end{bmatrix} \tag{4.8-15}$$

The sequence $Z(m)$, $m = 0, 1, \ldots, 7$ is computed using an 8-point FFT and subsequently the desired $\tilde{Z}(m)$, $m = 0, 1, \ldots, 5$ is obtained from Eq. (12). From the above discussion it is apparent that the FFT can also be used with a similar procedure to compute the correlation of two aperiodic sequences.

4. Waveform synthesis. The FFT can be used for the synthesis of the L-periodic extension of a digital signal $x^*(t)$ [see Fig. 4.13], whose Fourier amplitude and phase spectra have been specified. Suppose we are given

$$|F_{x^*}(k\omega_0)| \quad \text{and} \quad \psi_{x^*}(k\omega_0), \qquad k = 0, 1, \ldots, N/2$$

where N is an integer power of 2, and $\omega_0 = 2\pi/L$.

From the discussion in Sec. 3.4 it can be seen that $x^*(t)$ can be obtained as follows:

(1) $$F_{x^*}(k\omega_0) = |F_{x^*}(k\omega_0)|\, e^{i\psi_{x^*}(k\omega_0)}, \qquad k = 0, 1, \ldots, N/2 \tag{4.8-16}$$

(2) Define $\quad C_x(k) = (1/L)\, F_{x^*}(k\omega_0), \qquad k = 0, 1, \ldots, N-1$

where $\quad C_x(N/2 + l) = \bar{C}_x(N/2 - l), \qquad l = 1, 2, \ldots, N/2 - 1$

(3) Use the IFFT to compute

$$X(m) = \sum_{k=0}^{N-1} C_x(k) W^{-km}, \qquad m = 0, 1, \ldots, N-1$$

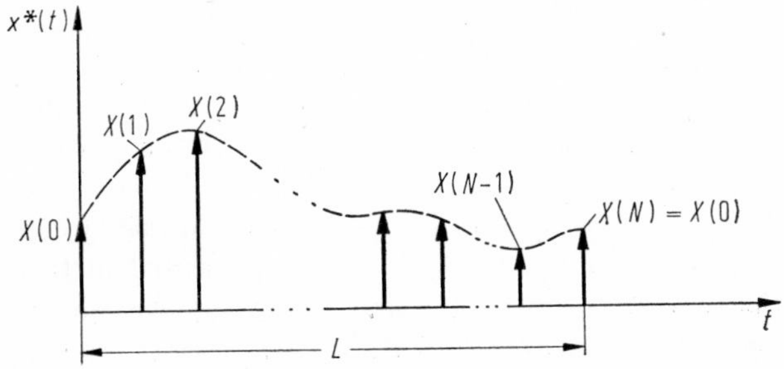

Fig. 4.13 L-periodic extension of waveform to be synthesized

For the purposes of illustration, an example is considered next. The values of $F_{x^*}(k\omega_0)$, $k = 1, 2, \ldots, 16$ are given in the table below.

k	$F_{x^*}(k\omega_0) \cdot 10^3$	k	$F_{x^*}(k\omega_0) \cdot 10^3$
0	16.0000	9	1.1312
1	5.3920	10	0
2	0	11	−0.9408
3	−3.3920	12	0
4	0	13	0.7840
5	2.0320	14	0
6	0	15	−0.6784
7	−1.4544	16	0
8	0		

With $N = 32$ and assuming that $L = 32$ msec, the steps listed above yield the $x^*(t)$ shown in Fig. 4.14.

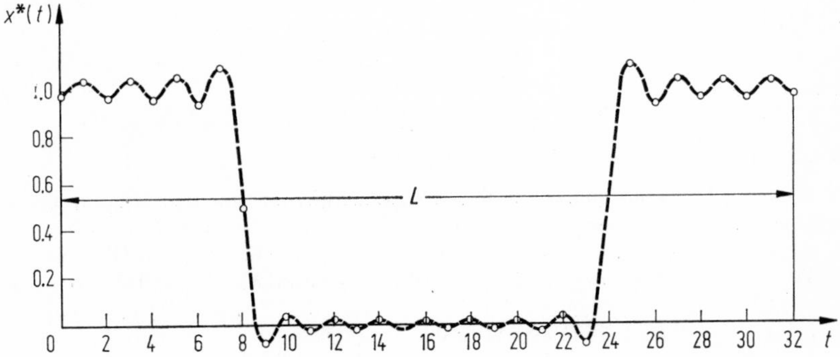

Fig. 4.14 Synthesized waveform

4.9 Summary

The most important aspect of this chapter is the development of the FFT. We restricted our attention to the 1-dimensional case where the data is in the form of a vector. From the discussion in Sec. 3.6 it is apparent that the 1-dimensional FFT can be used $N_1 N_2$ times to compute the 2-dimensional DFT, where the data is in the form of a $(N_1 \times N_2)$ matrix. Alternately the 2-dimensional FFT can be computed as a 1-dimensional FFT, where the data matrix is rearranged in the form of a $(N_1 N_2 \times 1)$ vector [14]. This approach can also be extended to the m-dimensional case, where $m \geq 2$.

The development of the FFT which was presented in this chapter is based on the derivation by Cooley and Tukey [2]. Alternately it can be derived by matrix partitioning [22] or matrix factorization [1, 23, 24] techniques. However, several other useful versions of the FFT are available [8]. These are attributed to Brenner [15], Fisher [16], Singleton [3], Sande [17], and Bluestein [18]. A quantitative study of the four versions of the FFT described in [2, 3, 15, 16] has been made on the basis of program execution time, storage, and accuracy [19]. The study was restricted to FFT sizes which are integer powers of 2, from 16 through 8192.

In conclusion, some applications of the FFT were considered and illustrated by means of numerical examples. A discussion of several other applications related to the area of digital signal processing is available in [1, 8, 11, 20, 21].

Appendix 4.1

This appendix provides a listing of a subroutine (p. 79) for computing the DFT. An interesting derivation of the FFT version used in this subroutine is available in [24].

References

1. Cochran, W. T., et al.: What is the Fast Fourier Transform? *Proc. IEEE* 55 (1967) 1664–74.
2. Cooley, J. W., and Tukey, J. W.: An Algorithm for Machine Computation of Complex Fourier Series. *Mathematics of Computation* 19 (1965) 297–301.
3. Singleton, R. C.: An Algorithm for Computing the Mixed Radix Fast Fourier Transform. *IEEE Trans. Audio and Electroacoustics* AU-17 (1969) 99–103.
4. Rader, C.: Discrete Fourier Transforms when the Number of Data Samples is Prime. *Proc. IEEE* 56 (1968) 1107–08.
5. Gold, B., and Rader, C. M.: *Digital Processing of Signals*. New York: McGraw-Hill, 1969.

```
      SUBROUTINE FFT(X,N,INV)
C*************************************************************************
C
C     THIS PROGRAM IMPLEMENTS THE FFT ALGORITHM TO COMPUTE THE DISCRETE
C     FOURIER COEFFICIENTS OF A DATA SEQUENCE OF N POINTS
C
C     CALLING SEQUENCE FROM THE MAIN PROGRAM:
C     CALL FFT(X,N,INV)
C          N: NUMBER OF DATA POINTS
C          X: COMPLEX ARRAY CONTAINING THE DATA SEQUENCE. IN THE END DFT
C             COEFFS. ARE RETURNED IN THE ARRAY. MAIN PROGRAM SHOULD
C             DECLARE II AS--    COMPLEX  X(512)
C        INV: FLAG FOR INVERSE
C             INV=0  FOR FORWARD TRANSFORM
C             INV=1  FOR INVERSE TRANSFORM
C
C*************************************************************************
      COMPLEX X(512),W,T,CMPLX
C
C     CALCULATE THE # OF ITERATIONS (LOG. N TO THE BASE 2)
C
      ITER=0
      IREM=N
   10 IREM=IREM/2
      IF (IREM.EQ.0) GO TO 20
      ITER=ITER+1
      GO TO 10
   20 CONTINUE
      SIGN=-1
      IF (INV.EQ.1) SIGN=1.
      NXP2=N
      DO 50 IT=1,ITER
C
C     COMPUTATION FOR EACH ITERATION
C     NXP: NUMBER OF POINTS IN A PARTITION
C     NXP2: NXP/2
C
      NXP=NXP2
      NXP2=NXP/2
      WPWR=3.141592/FLOAT(NXP2)
      DO 40 M=1,NXP2
C
C     CALCULATE THE MULTIPLIER
C
      ARG=FLOAT(M-1)*WPWR
      W=CMPLX(COS(ARG),SIGN*SIN(ARG))
      DO 40 MXP=NXP,N,NXP
C
C     COMPUTATION FOR EACH PARTITION
C
      J1=MXP-NXP+M
      J2=J1+NXP2
      T=X(J1)-X(J2)
      X(J1)=X(J1)+X(J2)
   40 X(J2)=T
   50 CONTINUE
C
C     UNSCRAMBLE THE BIT-REVERSED DFT COEFFS.
C
      N2=N/2
      N1=N-1
      J=1
      DO 65 I=1,N1
      IF(I.GE.J) GO TO 55
      T=X(J)
      X(J)=X(I)
      X(I)=T
   55 K=N2
   60 IF(K.GE.J) GO TO 65
      J=J-K
      K=K/2
      GO TO 60
   65 J=J+K
      IF (INV.EQ.1) GO TO 75
      DO 70 I=1,N
   70 X(I)=X(I)/FLOAT(N)
   75 CONTINUE
      RETURN
      END
```

6. Cooley, J. W. et al.: Historical Notes on the Fast Fourier Transform. *Proc. IEEE* 55 (1967) 1675–77.
7. Ohnsorg, F. R.: The Tukey-Cooley Algorithm. Honeywell Inter-office Correspondence, MR-9959, 1967, Systems and Research Center, St. Paul Minn., 55113, USA.
8. Bergland, G. D.: A Guided Tour of the Fast Fourier Transform. *IEEE Spectrum* 6, July 1969, 41–52.
9. Kaneko, T., and Liu, B.: Accumulation of Round-off Error in Fast Fourier Transforms. *Journal of ACM* 17 (1970) 637–654.
10. Cooley, J. W., Lewis, P. A. W., and Welch, P. D.: The Fast Fourier Transform and its Applications. *IBM Research Report* RC-1743, IBM Watson Research Center, Yorkstown, N. Y.
11. Brigham, E. O., and Morrow, R. E.: The Fast Fourier Transform. *IEEE Spectrum* 4, (1967) 63–70.
12. Campbell, M. D., Houts, R. D., and Reinhard, E. A.: A Computer Utility Incorporating the FFT Algorithm for a Signal and System Theory Course. *IEEE Trans. Education* E-16 (1973) 42–47.
13. Stockham, Jr., T. G.: High Speed Convolution and Correlation. *1966 Spring Joint Computer Conference*, AFIPS Proc., 1966, 229–233.
14. Jagadeesan, M.: n-Dimensional Fast Fourier Transform. *Proc. Thirteenth Midwest Symposium on Circuit Theory*, 1970, III.2.1–III.2.8.
15. Brenner, N. M.: Three Fortran Programs that Perform the Cooley-Tukey Fourier Transform. MIT Lincoln Lab. Publication AD 657019, 1967.
16. Fisher, J. R.: Fortran Program for Fast Fourier Transform. Naval Research Laboratory Report No. 7041, Washington, D.C., 1970.
17. Sande, G.: Arbitrary Radix One-dimensional Fast Fourier Transform Subroutines, University of Chicago, Illinois, 1968.
18. Bluestein, L. I.: Several Fourier Transform Algorithms. NEREM Record, 10, 1968, 218–219, published by the Boston section of the IEEE.
19. Ferrie, J. R., and Nuttall, A. H.: Comparison of Four Fast Fourier Transform Algorithms. Navy Underwater Systems Center, Newport, Rhode Island 02840, NUSC Report No. 4113, 1971.
20. Singleton, R. C.: A Short Bibliography on the Fast Fourier Transform. *IEEE Trans. Audio and Electroacoustics* AU-17 (1969) 166–169.
21. Special Issues on Fast Fourier Transform. *IEEE Trans. Audio and Electroacoustics* AU-15, June 1967, and AU-17, June 1969.
22. Ahmed, N., and Cheng, S. M.: On Matrix Partitioning and a Class of Algorithms. *IEEE Trans. Education* E-13 (1970) 103–105.
23. Theilheimer, F.: A Matrix Version of the Fast Fourier Transform. *IEEE Trans. Audio and Electroacoustics* AU-17 (1969) 158–161.
24. Kahaner, D. K.: Matrix Description of the Fast Fourier Transform. *IEEE Trans. Audio and Electroacoustics* AU-18 (1970) 442–452.
25. White Jr., Andrew, and Gray, Paul E.: Fast Fourier Transform: A Pragmatic Approach. Dept. of Electrical Engineering, North Carolina Agricultural and Technical State University, Greensboro, North Carolina, USA. Monograph EE-M-NO. 2, April 1973.
26. Brigham, E. O.: *The Fast Fourier Transform*. Englewood Cliffs, New Jersey: Prentice-Hall, 1974.

Problems

4-1 Consider two 4-periodic sequences

$$\{X(m)\} = \{1\ \ 2\ \ -1\ \ 3\} \quad \text{and} \quad \{Y(m)\} = \{Y(0)\,Y(1)\,Y(2)\,Y(3)\},$$

such that

$$\begin{bmatrix} -1 \\ 1 \\ 4 \\ 1 \end{bmatrix} = \frac{1}{4} \begin{bmatrix} 1 & 2 & -1 & 3 \\ 2 & -1 & 3 & 1 \\ -1 & 3 & 1 & 2 \\ 3 & 1 & 2 & -1 \end{bmatrix} \begin{bmatrix} Y(0) \\ Y(3) \\ Y(2) \\ Y(1) \end{bmatrix}$$

Find $\{Y(m)\}$ *without inverting* the (4×4) matrix in the above matrix equation.

4-2 Given:

$$\{X(m)\} = \{1\ \ 2\ \ -1\ \ 3\}, \quad \text{and} \quad \{C_x(k)\} = \left\{\frac{5}{4} \quad \frac{(2+i)}{4} \quad \frac{-5}{4} \quad \frac{(2-i)}{4}\right\}.$$

Consider the transformation

$$\begin{bmatrix} R(0) \\ R(1) \\ R(2) \\ R(3) \end{bmatrix} = \frac{1}{4} \begin{bmatrix} 1 & 2 & -1 & 3 \\ 3 & 1 & 2 & -1 \\ -1 & 3 & 1 & 2 \\ 2 & -1 & 3 & 1 \end{bmatrix} \begin{bmatrix} 1 \\ 2 \\ -1 \\ 3 \end{bmatrix}$$

Find $\{C_r(k)\}$ *without* evaluating the matrix product.

4-3 (a) Express the multipliers A^s_{2r} in Fig. 4.1 in terms of powers of W. Use the procedure outlined in Example 4.5-4.

(b) Use the signal flow graph obtained in (a) to verify that the DFT coefficient sequence $\{C_x(k)\}$ and the data sequence $\{X(m)\}$ are related by a matrix product as follows. (Here $W = e^{-i2\pi/8}$.)

$$\begin{bmatrix} C_x(0) \\ C_x(4) \\ C_x(2) \\ C_x(6) \\ C_x(1) \\ C_x(5) \\ C_x(3) \\ C_x(7) \end{bmatrix} = \frac{1}{8} \begin{bmatrix} 1 & 1 & & & & & & \mathbf{0} \\ 1 & -1 & & & & & & \\ & & 1 & W^2 & & & & \\ & & 1 & -W^2 & & & & \\ & & & & 1 & W & & \\ & & & & 1 & -W & & \\ & & & & & & 1 & W^3 \\ \mathbf{0} & & & & & & 1 & -W^3 \end{bmatrix}$$

$$\times \begin{bmatrix} 1 & 0 & 1 & 0 & & & & \\ 0 & 1 & 0 & 1 & & \mathbf{0}_4 & & \\ 1 & 0 & -1 & 0 & & & & \\ 0 & 1 & 0 & -1 & & & & \\ & & & & 1 & 0 & W^2 & 0 \\ & \mathbf{0}_4 & & & 0 & 1 & 0 & W^2 \\ & & & & 1 & 0 & -W^2 & 0 \\ & & & & 0 & 1 & 0 & -W^2 \end{bmatrix} \begin{bmatrix} 1 & 0 & 0 & 0 & 1 & 0 & 0 & 0 \\ 0 & 1 & 0 & 0 & 0 & 1 & 0 & 0 \\ 0 & 0 & 1 & 0 & 0 & 0 & 1 & 0 \\ 0 & 0 & 0 & 1 & 0 & 0 & 0 & 1 \\ 1 & 0 & 0 & 0 & -1 & 0 & 0 & 0 \\ 0 & 1 & 0 & 0 & 0 & -1 & 0 & 0 \\ 0 & 0 & 1 & 0 & 0 & 0 & -1 & 0 \\ 0 & 0 & 0 & 1 & 0 & 0 & 0 & -1 \end{bmatrix} \begin{bmatrix} X(0) \\ X(1) \\ X(2) \\ X(3) \\ X(4) \\ X(5) \\ X(6) \\ X(7) \end{bmatrix}$$

4-4 In Example 4.5-4, a set of rules was developed to obtain the FFT signal flow graph in terms of powers of W, for $N = 16$. Develop the corresponding set of rules for the general case when $N = 2^n$, $n = 1, 2, \ldots, n_{max}$.

4-5 Evaluate the 2-dimensional DFT of the data matrix

$$\mathbf{X} = \begin{bmatrix} 1 & 3 \\ 1 & 1 \\ 2 & 2 \\ 1 & 2 \end{bmatrix}$$

using the definition of the 2-dimensional DFT [see Sec. 3.6].
Answer:

$$\frac{1}{8}\begin{bmatrix} 13 & -3 \\ i & -2-i \\ 3 & -1 \\ -i & -2+i \end{bmatrix}$$

4-6 Extend the 1-dimensional FFT program given in Appendix 4-1 so that it is capable of computing the 2-dimensional DFT. Subsequently use it to compute the 2-dimensional DFT of the data matrix

$$\mathbf{X} = \begin{bmatrix} 1 & \varrho & \varrho^2 & \varrho^3 \\ \varrho & 1 & \varrho & \varrho^2 \\ \varrho^2 & \varrho & 1 & \varrho \\ \varrho^3 & \varrho^2 & \varrho & 1 \end{bmatrix}$$

where $\varrho = 0.9$.

4-7 The relationship between the matrix factors described in Prob. 4-3b and the multipliers in the signal flow graph obtained in Prob. 4-3a can be easily observed. From the FFT signal flow graph described in Fig. 4.5 for $N = 16$, *write down* the matrix factors of the transform matrix.

4-8 For $N = 32$, $W = e^{-i2\pi/32}$ the DFT transform matrix can be expressed as a product of matrix factors as follows.

$$\left(\operatorname{diag}\left[\begin{bmatrix} 1 & 1 \\ 1 & -1 \end{bmatrix}, \begin{bmatrix} 1 & W^8 \\ 1 & -W^8 \end{bmatrix}, \begin{bmatrix} 1 & W^4 \\ 1 & -W^4 \end{bmatrix}, \begin{bmatrix} 1 & W^{12} \\ 1 & -W^{12} \end{bmatrix}, \begin{bmatrix} 1 & W^2 \\ 1 & -W^2 \end{bmatrix}, \begin{bmatrix} 1 & W^{10} \\ 1 & -W^{10} \end{bmatrix},\right.\right.$$

$$\begin{bmatrix} 1 & W^6 \\ 1 & -W^6 \end{bmatrix}, \begin{bmatrix} 1 & W^{14} \\ 1 & -W^{14} \end{bmatrix}, \begin{bmatrix} 1 & W^1 \\ 1 & -W^1 \end{bmatrix}, \begin{bmatrix} 1 & W^9 \\ 1 & -W^9 \end{bmatrix}, \begin{bmatrix} 1 & W^5 \\ 1 & -W^5 \end{bmatrix}, \begin{bmatrix} 1 & W^{13} \\ 1 & -W^{13} \end{bmatrix},$$

$$\left.\left.\begin{bmatrix} 1 & W^3 \\ 1 & -W^3 \end{bmatrix}, \begin{bmatrix} 1 & W^{11} \\ 1 & -W^{11} \end{bmatrix}, \begin{bmatrix} 1 & W^7 \\ 1 & -W^7 \end{bmatrix}, \begin{bmatrix} 1 & W^{15} \\ 1 & -W^{15} \end{bmatrix}\right]\right) \times$$

$$\left(\operatorname{diag}\left[\begin{bmatrix} \mathbf{I}_2 & \mathbf{I}_2 \\ \mathbf{I}_2 & -\mathbf{I}_2 \end{bmatrix}, \begin{bmatrix} \mathbf{I}_2 & \mathbf{I}_2W^8 \\ \mathbf{I}_2 & -\mathbf{I}_2W^8 \end{bmatrix}, \begin{bmatrix} \mathbf{I}_2 & \mathbf{I}_2W^4 \\ \mathbf{I}_2 & -\mathbf{I}_2W^4 \end{bmatrix}, \begin{bmatrix} \mathbf{I}_2 & \mathbf{I}_2W^{12} \\ \mathbf{I}_2 & -\mathbf{I}_2W^{12} \end{bmatrix}, \begin{bmatrix} \mathbf{I}_2 & \mathbf{I}_2W^2 \\ \mathbf{I}_2 & -\mathbf{I}_2W^2 \end{bmatrix},\right.\right.$$

$$\left.\left.\begin{bmatrix} \mathbf{I}_2 & \mathbf{I}_2W^{10} \\ \mathbf{I}_2 & -\mathbf{I}_2W^{10} \end{bmatrix}, \begin{bmatrix} \mathbf{I}_2 & \mathbf{I}_2W^6 \\ \mathbf{I}_2 & -\mathbf{I}_2W^6 \end{bmatrix}, \begin{bmatrix} \mathbf{I}_2 & \mathbf{I}_2W^{14} \\ \mathbf{I}_2 & -\mathbf{I}_2W^{14} \end{bmatrix}\right]\right) \times$$

$$\left(\operatorname{diag}\left[\begin{bmatrix} \mathbf{I}_4 & \mathbf{I}_4 \\ \mathbf{I}_4 & -\mathbf{I}_4 \end{bmatrix}, \begin{bmatrix} \mathbf{I}_4 & \mathbf{I}_4W^8 \\ \mathbf{I}_4 & -\mathbf{I}_4W^8 \end{bmatrix}, \begin{bmatrix} \mathbf{I}_4 & \mathbf{I}_4W^4 \\ \mathbf{I}_4 & -\mathbf{I}_4W^4 \end{bmatrix}, \begin{bmatrix} \mathbf{I}_4 & \mathbf{I}_4W^{12} \\ \mathbf{I}_4 & -\mathbf{I}_4W^{12} \end{bmatrix}\right]\right) \times$$

$$\left(\operatorname{diag}\left[\begin{bmatrix} \mathbf{I}_8 & \mathbf{I}_8 \\ \mathbf{I}_8 & -\mathbf{I}_8 \end{bmatrix}, \begin{bmatrix} \mathbf{I}_8 & \mathbf{I}_8W^8 \\ \mathbf{I}_8 & -\mathbf{I}_8W^8 \end{bmatrix}\right]\right) \begin{bmatrix} \mathbf{I}_{16} & \mathbf{I}_{16} \\ \mathbf{I}_{16} & -\mathbf{I}_{16} \end{bmatrix}$$

where $\mathbf{I}_m$ denotes the $(m \times m)$ identity matrix.

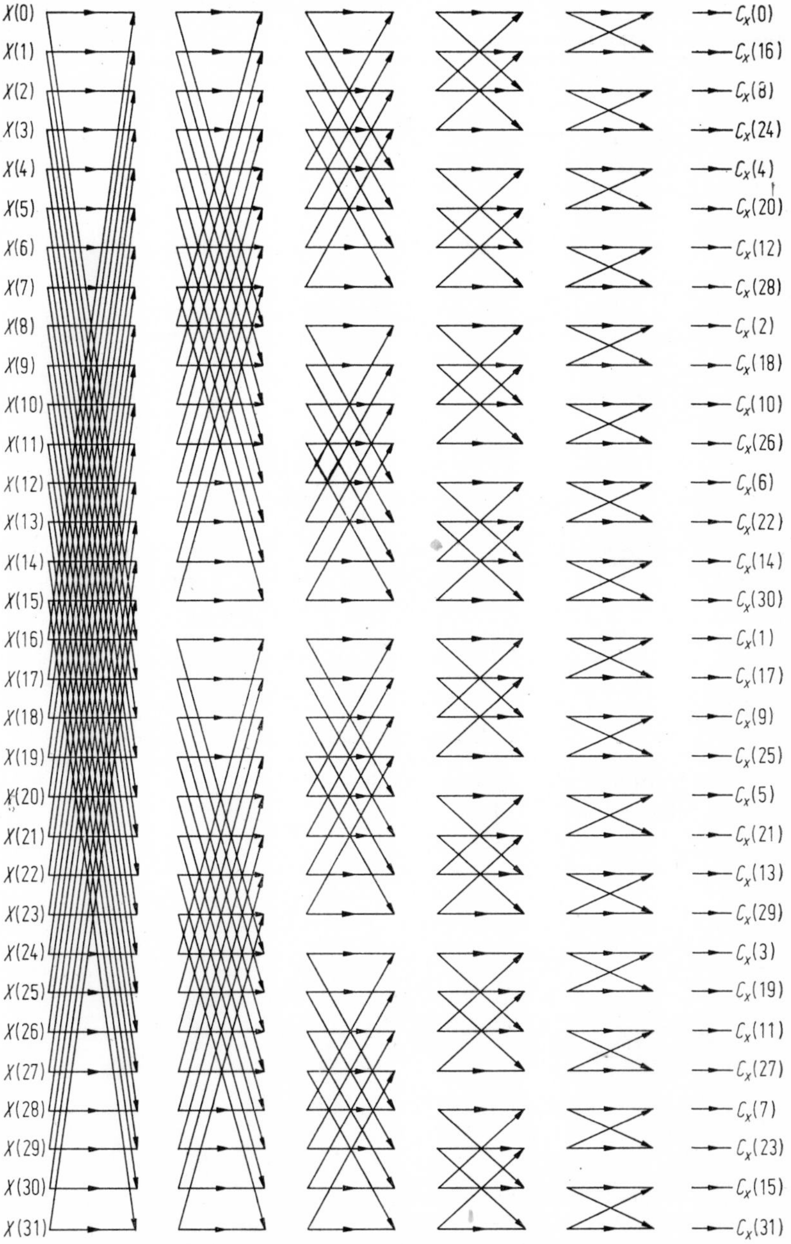

Fig. P4-8-1 FFT signal flow graph for $N = 32$

Using these matrix factors, fill in the multipliers in the FFT signal flow graph shown in Fig. P4-8-1 for $N = 32$.

4-9 Jagadeesan [14] has presented a technique, which uses the one dimensional transform for computing the n-dimensional FFT. Use his method to solve Prob. 4-5. Fill in the multipliers (powers of W) in Fig. P4-9-1.

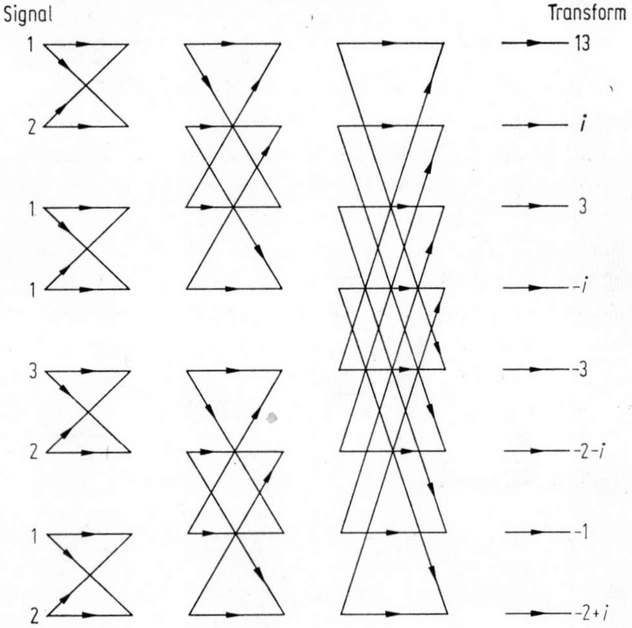

Fig. P4-9-1 FFT signal flow graph for evaluating the 2-dimensional DFT [14]

Chapter Five

A Class of Orthogonal Functions

The purpose of this chapter is to introduce a class of nonsinusoidal orthogonal functions which consists of: (1) Rademacher functions, (2) Haar functions, and (3) Walsh functions. These orthogonal functions consist of either square or rectangular waves. Individual functions belonging to such sets of functions are distinguished by means of a parameter called *sequency*. Some aspects pertaining to notation for representing nonsinusoidal orthogonal functions are also discussed.

The motivation for studying the above class of functions is that they will be used in connection with the development of the Haar, Walsh-Hadamard, and modified Walsh-Hadamard transforms. Such a development will be undertaken in the next two chapters.

5.1 Definition of Sequency

The term *frequency* is applied to a set of sinusoidal (periodic) functions whose zero-crossings are uniformly spaced over an interval. It is a parameter f that distinguishes the individual functions which belong to the sets $\{\cos 2\pi ft\}$ and $\{\sin 2\pi ft\}$, and is interpreted as the number of complete cycles (or one-half the number of zero-crossings) generated by a sinusoidal function per unit time.

The generalization of frequency is achieved by defining *generalized frequency* as one-half the average number of zero-crossings per unit time [1]. Harmuth [2] introduced the term *sequency* to describe generalized frequency, and applied it to distinguish functions whose zero-crossings are not uniformly spaced over an interval, and which are not neccessarily periodic. The definition of sequency coincides with that of frequency when applied to sinusoidal functions. Applying the above definition to periodic and aperiodic functions, we obtain:

(i) The sequency of a periodic function equals one-half the number of sign changes per period.
(ii) The sequency of an aperiodic function equals one-half the number of sign changes per unit time, if this limit exists.

For the purposes of illustration, consider the continuous functions $f_1(t)$ and $f_2(t)$ in Fig. 5.1 which are defined in the half-open interval $[-0.5, 0.5)$.

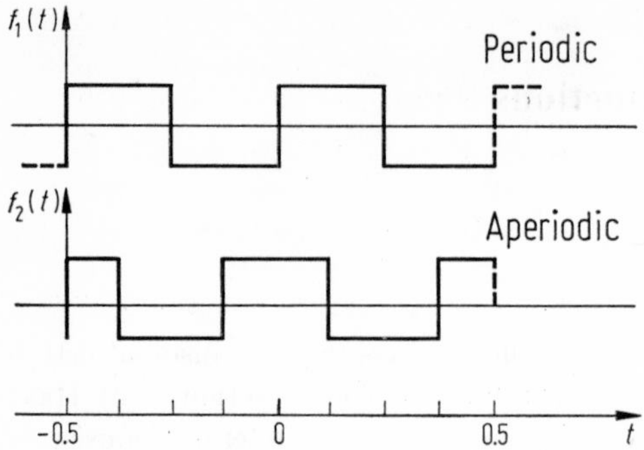

Fig. 5.1 Pertaining to definition of sequency of continuous functions

Since each function has 4 zero-crossings in the interval, the sequency of each equals 2. Analogous to frequency which is expressed in cycles per second or Hertz, sequency is expressed in terms of zero-crossings per second, for which one may use the abbreviation "zps".

The above definition of sequency can be applied with minor modification to the corresponding discrete function $f^*(t)$, which is obtained by sampling $f(t)$ at equidistant points. If the number of sign changes per unit time of $f^*(t)$ equals η, then the sequency of $f^*(t)$ is defined as $\eta/2$ or $(\eta + 1)/2$, when η is even or odd respectively. For example, consider $f_1^*(t)$ and $f_2^*(t)$ shown in Fig. 5.2. These discrete functions are obtained by sampling $f_1(t)$ and $f_2(t)$ in Fig. 5.1 at a set of 8 equidistant points. From Fig. 5.2 it follows that $\eta_1 = 3$ and $\eta_2 = 4$. Thus the sequency of each of the functions $f_1^*(t)$ and $f_2^*(t)$ equals 2, as was the case with $f_1(t)$ and $f_2(t)$.

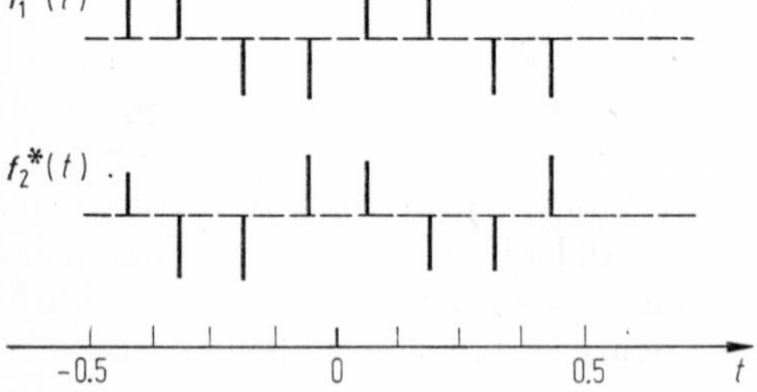

Fig. 5.2 Pertaining to definition of sequency of discrete functions

5.2 Notation

In the case of trigonometric, exponential and logarithmic functions, many of the notations accepted as "standard" consist of three letters. Examples of these include sin, cos, exp, erf and log. Thus it is reasonable to introduce a similar scheme to denote nonsinusoidal complete orthogonal functions [1].

In addition, it is necessary to distinguish continuous functions from the corresponding discrete functions. A possible notation is proposed in Table 5.2-1 as suggested by Harmuth [2]. The complete set of Walsh functions, defined on the unit interval [0, 1) can be divided into two groups of even and odd functions about the point $t = 0.5$. These even and odd functions are analogous to the sine and cosine functions respectively; hence they are denoted by sal (*s*ine W*al*sh) and cal (*c*osine W*al*sh) respectively. A rigorous mathematical development of the sal-cal system has been carried out by Pichler [3].

Table 5.2-1 Notation for continuous and discrete functions

Name of function	Notation	
	Continuous functions	Discrete functions
Rademacher	rad	Rad
Haar	har	Har
Walsh	wal	Wal
"cosine Walsh"	cal	Cal
"sine Walsh"	sal	Sal

5.3 Rademacher and Haar Functions

Rademacher functions are an incomplete set of orthonormal functions which were developed in 1922 [4]. The Rademacher function of index m, denoted by $\text{rad}(m, t)$, is a train of rectangular pulses with 2^{m-1} cycles in the half-open interval [0, 1), taking the values $+1$ or -1 (see Fig. 5.3). An exception is $\text{rad}(0, t)$, which is the unit pulse.

Rademacher functions are periodic with period 1, i.e.,

$$\text{rad}(m, t) = \text{rad}(m, t + 1)$$

They are also periodic over shorter intervals [5] such that

$$\text{rad}(m, t + n\,2^{1-m}) = \text{rad}(m, t), \qquad m = 1, 2, \ldots; \quad n = \pm 1, \pm 2, \ldots$$

Rademacher functions can be generated using the recurrence relation

$$\text{rad}(m, t) = \text{rad}(1, 2^{m-1}t)$$

with

$$\text{rad}(1, t) = \begin{cases} 1, & t \in [0, 1/2) \\ -1, & t \in [1/2, 1) \end{cases} \tag{5.3-1}$$

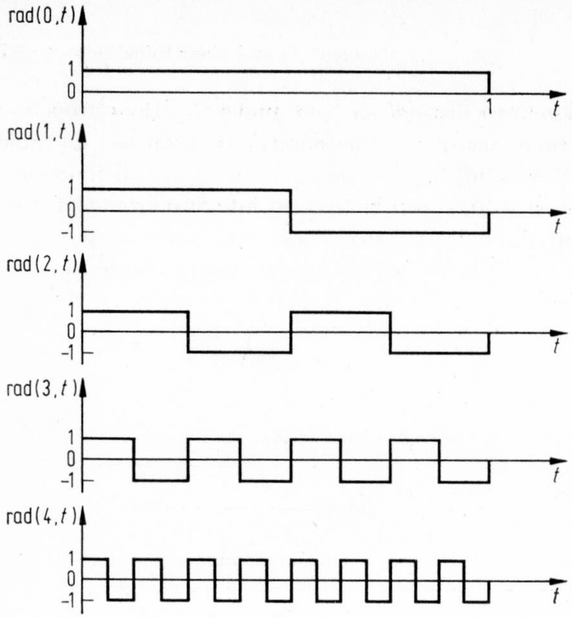

Fig. 5.3 Rademacher functions

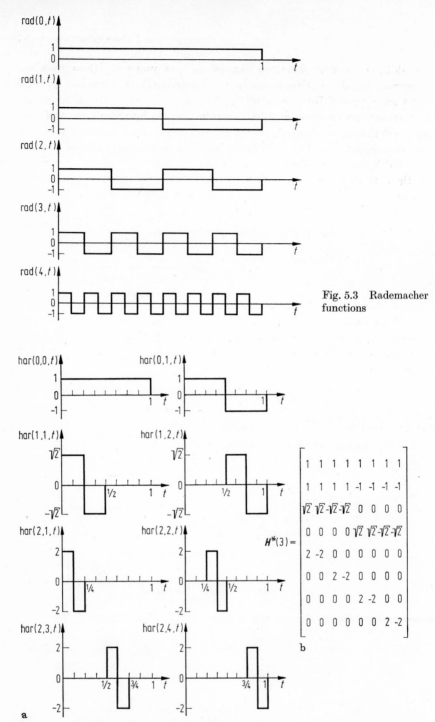

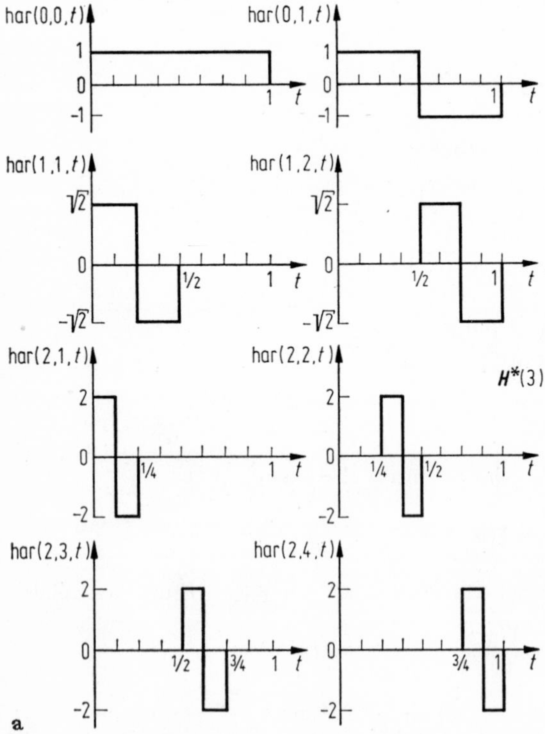

a

$$H^*(3) = \begin{bmatrix} 1 & 1 & 1 & 1 & 1 & 1 & 1 & 1 \\ 1 & 1 & 1 & 1 & -1 & -1 & -1 & -1 \\ \sqrt{2} & \sqrt{2} & -\sqrt{2} & -\sqrt{2} & 0 & 0 & 0 & 0 \\ 0 & 0 & 0 & 0 & \sqrt{2} & \sqrt{2} & -\sqrt{2} & -\sqrt{2} \\ 2 & -2 & 0 & 0 & 0 & 0 & 0 & 0 \\ 0 & 0 & 2 & -2 & 0 & 0 & 0 & 0 \\ 0 & 0 & 0 & 0 & 2 & -2 & 0 & 0 \\ 0 & 0 & 0 & 0 & 0 & 0 & 2 & -2 \end{bmatrix}$$

b

Fig. 5.4 a Continuous Haar functions, $N = 8$; b Discrete Haar functions, $N = 8$

The set of *Haar* functions $\{\text{har}(n, m, t)\}$ is periodic, orthonormal, and complete [5, 15–17], and was proposed in 1910 by Haar [6]. Fig. 5.4a shows the set of first 8 Haar functions.

A recurrence relation which enables one to generate $\{\text{har}(n, m, t)\}$ is given by [7]

$$\text{har}(0, 0, t) = 1, \qquad t \in [0, 1)$$

$$\text{har}(r, m, t) = \begin{cases} 2^{r/2}, & \dfrac{m-1}{2^r} \leq t < \dfrac{m-1/2}{2^r} \\ -2^{r/2}, & \dfrac{m-1/2}{2^r} \leq t < \dfrac{m}{2^r} \\ 0, & \text{elsewhere for } t \in [0, 1) \end{cases} \tag{5.3-2}$$

where $0 \leq r < \log_2 N$ and $1 \leq m \leq 2^r$.

Sampling of the set of Haar functions in Fig. 5.4a results in the array shown in Fig. 5.4b, each row of which is a discrete Haar function $\text{Har}(r, m, t)$. Arrays so obtained are used in connection with the Haar transform and may be denoted by $\mathbf{H}^*(n)$, where $n = \log_2 N$.

5.4 Walsh Functions

The incomplete set of Rademacher functions was completed by Walsh [8] in 1923, to form the complete orthonormal set of rectangular functions, now known as Walsh functions. The set of Walsh functions is generally classified into three groups. These groups differ from one another in that the *order* in which individual functions appear is different. The three types of orderings are: (1) sequency or Walsh ordering, (2) dyadic or Paley ordering, and (3) natural or Hadamard ordering. In what follows, we discuss some aspects pertaining to each of these orderings.

(1) Sequency or Walsh Ordering. This is the ordering which was originally employed by Walsh [8]. We denote the set of Walsh functions belonging to this set by

$$S_w = \{\text{wal}_w(i, t), \ i = 0, 1, \ldots, N-1\} \tag{5.4-1}$$

where

$$N = 2^n, \qquad n = 1, 2, 3, \ldots,$$

the subscript "w" denotes Walsh ordering, and i denotes the i-th member of S_w.

If s_i represents the sequency of $\mathrm{wal}_w(i, t)$, then s_i is given by

$$s_i = \begin{cases} 0, & i = 0 \\ i/2, & i \text{ even} \\ (i+1)/2, & i \text{ odd} \end{cases} \tag{5.4-2}$$

The cal and sal functions corresponding to $\mathrm{wal}_w(i, t)$ are denoted as

$$\begin{aligned} \mathrm{cal}\,(s_i, t) &= \mathrm{wal}_w\,(i, t), \quad i \text{ even} \\ \mathrm{sal}\,(s_i, t) &= \mathrm{wal}_w\,(i, t), \quad i \text{ odd} \end{aligned} \tag{5.4-3}$$

Using the above notation, the first 8 of the Walsh functions are shown in Fig. 5.5a. It is apparent from Fig. 5.5a that the sequency of a Walsh function is greater than or equal to that of the preceding Walsh function and has exactly one more zero-crossing in the open interval $t \in (0, 1)$. Hence the alternate name "sequency ordering". S_w can be generated using the set of Rademacher functions and the Gray code [9], (see Prob. 5-1).

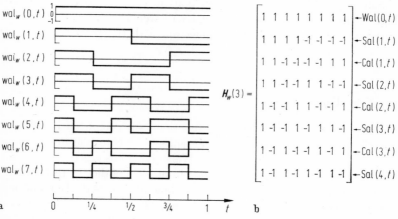

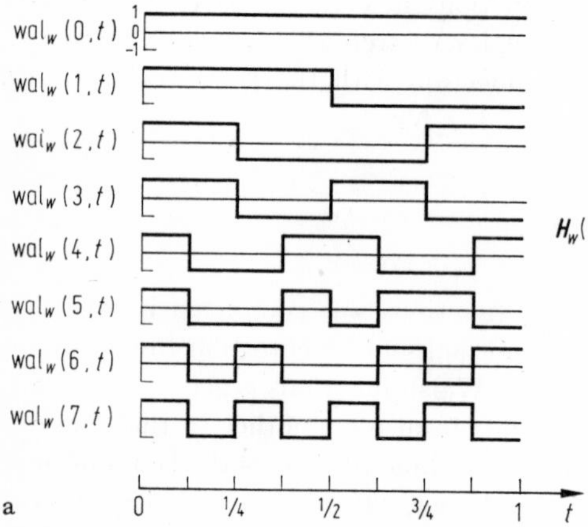

a

$$H_w(3) = \begin{bmatrix} 1 & 1 & 1 & 1 & 1 & 1 & 1 & 1 \\ 1 & 1 & 1 & 1 & -1 & -1 & -1 & -1 \\ 1 & 1 & -1 & -1 & -1 & -1 & 1 & 1 \\ 1 & 1 & -1 & -1 & 1 & 1 & -1 & -1 \\ 1 & -1 & -1 & 1 & 1 & -1 & -1 & 1 \\ 1 & -1 & -1 & 1 & -1 & 1 & 1 & -1 \\ 1 & -1 & 1 & -1 & -1 & 1 & -1 & 1 \\ 1 & -1 & 1 & -1 & 1 & -1 & 1 & -1 \end{bmatrix} \begin{matrix} \leftarrow \mathrm{Wal}(0,t) \\ \leftarrow \mathrm{Sal}(1,t) \\ \leftarrow \mathrm{Cal}(1,t) \\ \leftarrow \mathrm{Sal}(2,t) \\ \leftarrow \mathrm{Cal}(2,t) \\ \leftarrow \mathrm{Sal}(3,t) \\ \leftarrow \mathrm{Cal}(3,t) \\ \leftarrow \mathrm{Sal}(4,t) \end{matrix}$$

b

Fig. 5.5a Walsh-ordered continuous Walsh functions, $N = 8$
Fig. 5.5b Walsh-ordered discrete Walsh functions, $N = 8$

Discrete case. Sampling of the Walsh functions in Fig. 5.5a at 8 equidistant points results in the (8×8) matrix shown in Fig. 5.5b. In general an $(N \times N)$ matrix would be obtained. We denote such matrices by $\mathbf{H}_w(n)$, $n = \log_2 N$, since they can be obtained by reordering the rows of a class of matrices called Hadamard matrices.

Let u_i and v_i denote the i-th bit in the binary representations of the integers u and v respectively; that is

$$(u)_{\text{decimal}} = (u_{n-1}\, u_{n-2} \cdots u_1\, u_0)_{\text{binary}}$$

and

$$(v)_{\text{decimal}} = (v_{n-1}\, v_{n-2} \cdots v_1\, v_0)_{\text{binary}}$$

Then the elements $h_{uv}^{(w)}$ of $\mathbf{H}_w(n)$ can be generated as follows [10]:

$$h_{uv}^{(w)} = (-1)^{\sum_{i=0}^{n-1} r_i(u) v_i}, \qquad u, v = 0, 1, \ldots, N-1 \tag{5.4-4}$$

where

$$\begin{aligned} r_0(u) &= u_{n-1}, \\ r_1(u) &= u_{n-1} + u_{n-2}, \\ r_2(u) &= u_{n-2} + u_{n-3}, \\ &\vdots \\ r_{n-1}(u) &= u_1 + u_0 \end{aligned}$$

(2) Dyadic or Paley Ordering. The dyadic type of ordering was introduced by Paley [11]. Walsh functions are elements of the dyadic group and can be ordered using the Gray code [9, 12]. This set of Walsh functions is denoted as

$$S_p = \{\text{wal}_p(i, t),\ i = 0, 1, \ldots, N-1\} \tag{5.4-5}$$

where the subscript p denotes Paley ordering and i denotes the i-th member of S_p. The set S_p is related to the Walsh-ordered set S_w by the relation

$$\text{wal}_p(i, t) = \text{wal}_w[b(i), t] \tag{5.4-6}$$

where $b(i)$ represents the *Gray code-to-binary conversion*[1] of i. An algorithm to evaluate the conversion in Eq. (6) along with its implementation is considered in [12]. For the purposes of illustration, we evaluate Eq. (6) for the case $N = 8$. The corresponding results are summarized in Table 5.4-1.

Table 5.4-1 Relationship between the Walsh-ordered and Paley-ordered Walsh functions

i_{decimal}	i_{binary}	$b(i)_{\text{binary}}$	$b(i)_{\text{decimal}}$	Eq. (5.4-6)
0	000	000	0	$\text{wal}_p(0, t) = \text{wal}_w(0, t)$
1	001	001	1	$\text{wal}_p(1, t) = \text{wal}_w(1, t)$
2	010	011	3	$\text{wal}_p(2, t) = \text{wal}_w(3, t)$
3	011	010	2	$\text{wal}_p(3, t) = \text{wal}_w(2, t)$
4	100	111	7	$\text{wal}_p(4, t) = \text{wal}_w(7, t)$
5	101	110	6	$\text{wal}_p(5, t) = \text{wal}_w(6, t)$
6	110	100	4	$\text{wal}_p(6, t) = \text{wal}_w(4, t)$
7	111	101	5	$\text{wal}_p(7, t) = \text{wal}_w(5, t)$

Applying the information in Table 5.4-1 to the functions $\text{wal}_w(i, t)$, $i = 0, 1, \ldots, 7$ in Fig. 5.5a, we obtain the first 8 Walsh functions $\text{wal}_p(i, t)$ shown in Fig. 5.6a.

[1] For a discussion on the binary-to-Gray code conversion and Gray code-to-binary conversion, see Appendix 5.1.

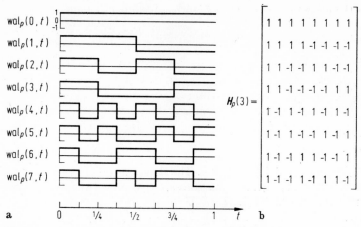

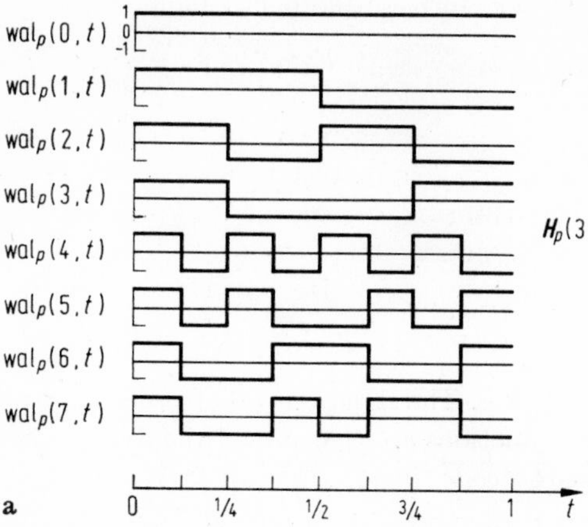

a

$$
H_p(3) = \begin{bmatrix}
1 & 1 & 1 & 1 & 1 & 1 & 1 & 1 \\
1 & 1 & 1 & 1 & -1 & -1 & -1 & -1 \\
1 & 1 & -1 & -1 & 1 & 1 & -1 & -1 \\
1 & 1 & -1 & -1 & -1 & -1 & 1 & 1 \\
1 & -1 & 1 & -1 & 1 & -1 & 1 & -1 \\
1 & -1 & 1 & -1 & -1 & 1 & -1 & 1 \\
1 & -1 & -1 & 1 & 1 & -1 & -1 & 1 \\
1 & -1 & -1 & 1 & -1 & 1 & 1 & -1
\end{bmatrix}
$$

b

Fig. 5.6a Paley-ordered continuous Walsh functions, $N = 8$
Fig. 5.6b Paley-ordered discrete Walsh functions, $N = 8$

Discrete case. By sampling the Walsh functions shown in Fig. 5.6a, we obtain the (8×8) matrix shown in Fig. 5.6b. This matrix can also be obtained by rearranging the rows of the (8×8) Hadamard matrix. The matrices associated with the Paley-ordered Walsh functions are denoted by $\mathbf{H}_p(n)$, $n = \log_2 N$.

The elements $h_{uv}^{(p)}$ of $\mathbf{H}_p(n)$ can be generated using the relation

$$h_{uv}^{(p)} = (-1)^{\sum_{i=0}^{n-1} u_{n-1-i} v_i}, \qquad u, v = 0, 1, \ldots, N-1 \tag{5.4-7}$$

(3) Natural or Hadamard Ordering. This set of Walsh functions is denoted by

$$S_h = \{\mathrm{wal}_h(i, t), \quad i = 0, 1, \ldots, N-1\} \tag{5.4-8}$$

where the subscribt h denotes Hadamard ordering and i denotes the i-th member of S_h. The functions belonging to S_h are related to the Walsh-

Table 5.4-2 Relationship between the Walsh-ordered and Hadamard-ordered Walsh functions

i	i_{binary}	$\langle i\rangle_{\text{binary}}$	$b(\langle i\rangle)_{\text{binary}}$	$b(\langle i\rangle)_{\text{decimal}}$	Eq. (5.4-9)
0	000	000	000	0	$\mathrm{wal}_h(0, t) = \mathrm{wal}_w(0, t)$
1	001	100	111	7	$\mathrm{wal}_h(1, t) = \mathrm{wal}_w(7, t)$
2	010	010	011	3	$\mathrm{wal}_h(2, t) = \mathrm{wal}_w(3, t)$
3	011	110	100	4	$\mathrm{wal}_h(3, t) = \mathrm{wal}_w(4, t)$
4	100	001	001	1	$\mathrm{wal}_h(4, t) = \mathrm{wal}_w(1, t)$
5	101	101	110	6	$\mathrm{wal}_h(5, t) = \mathrm{wal}_w(6, t)$
6	110	011	010	2	$\mathrm{wal}_h(6, t) = \mathrm{wal}_w(2, t)$
7	111	111	101	5	$\mathrm{wal}_h(7, t) = \mathrm{wal}_w(5, t)$

ordered functions by the relation

$$\mathrm{wal}_h(i, t) = \mathrm{wal}_w[b\,(\langle i \rangle), t] \tag{5.4-9}$$

where $\langle i \rangle$ is obtained by the *bit-reversal* of i and, $b(\langle i \rangle)$ is the *Gray code-to-binary conversion* of $\langle i \rangle$. For the purposes of illustration, Eq. (9) is evaluated for $N = 8$ and the results are summarized in Table 5.4-2, (p. 92).
The above table along with the functions $\mathrm{wal}_w(i, t)$, $i = 0, 1, \ldots, 7$ in Fig. 5 5a yields the first 8 Walsh functions $\mathrm{wal}_h(i, t)$ shown in Fig. 5.7a.

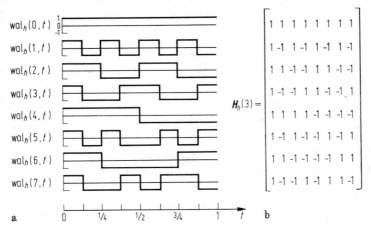

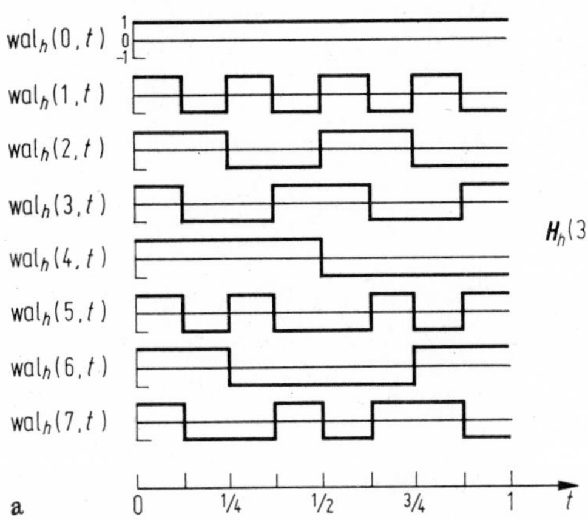

$$H_h(3) = \begin{bmatrix} 1 & 1 & 1 & 1 & 1 & 1 & 1 & 1 \\ 1 & -1 & 1 & -1 & 1 & -1 & 1 & -1 \\ 1 & 1 & -1 & -1 & 1 & 1 & -1 & -1 \\ 1 & -1 & -1 & 1 & 1 & -1 & -1 & 1 \\ 1 & 1 & 1 & 1 & -1 & -1 & -1 & -1 \\ 1 & -1 & 1 & -1 & -1 & 1 & -1 & 1 \\ 1 & 1 & -1 & -1 & -1 & -1 & 1 & 1 \\ 1 & -1 & -1 & 1 & -1 & 1 & 1 & -1 \end{bmatrix}$$

b

Fig. 5.7a Hadamard-ordered continuous Walsh functions, $N = 8$
Fig. 5.7b Hadamard-ordered discrete Walsh functions, $N = 8$

Discrete case. Sampling of the Walsh functions in Fig. 5.7a results in the (8×8) Hadamard matrix shown in Fig. 5.7b. In general an $(N \times N)$ matrix $\mathbf{H}_h(n)$ would be obtained, where $n = \log_2 N$. This class of Hadamard matrices can be partitioned in the form

$$\mathbf{H}_h(n) = \begin{bmatrix} \mathbf{H}_h(n-1) & \mathbf{H}_h(n-1) \\ \mathbf{H}_h(n-1) & -\mathbf{H}_h(n-1) \end{bmatrix} \tag{5.4-10}$$

and are considered to be in "natural" form [10]. Hence the name natural ordering.

The elements $h_{uv}^{(h)}$ of $\mathbf{H}_h(n)$ can be generated using the relation

$$h_{uv}^{(h)} = (-1)^{\sum_{i=0}^{n-1} u_i v_i}, \qquad u, v = 0, 1, \ldots, N-1 \tag{5.4-11}$$

5.5 Summary

Rademacher, Haar, and Walsh functions were considered in this chapter. The definition of sequency was introduced and was used as a parameter to distinguish individual functions which belong to sets of nonsinusoidal functions. It was shown that Walsh functions can be classified into three groups: (1) Walsh-ordered, (2) Paley-ordered and (3) Hadamard-ordered. These three types of orderings were related using the Gray code.

Sampling of finite sets of Haar and Walsh functions at a set of equidistant points resulted in Haar and Hadamard matrices respectively. The rows of the matrices so obtained will serve as basis vectors for defining the Haar and Walsh-Hadamard transforms respectively.

Appendix 5.1 Elements of the Gray Code

In certain practical applications, e.g., analog-to-digital conversion, it is desirable to use codes in which all successive code words differ in exactly one digit. Codes that possess such a property are called *cyclic codes*. An important cyclic code is the Gray code. A 4-bit Gray code is shown in Table A5.1-1. The property which makes this cyclic code useful is that a binary number can easily be converted into the Gray code using half adders.

Table A5.1-1 4-bit Gray code

Decimal number	Gray				Binary			
	g_3	g_2	g_1	g_0	b_3	b_2	b_1	b_0
0	0	0	0	0	0	0	0	0
1	0	0	0	1	0	0	0	1
2	0	0	1	1	0	0	1	0
3	0	0	1	0	0	0	1	1
4	0	1	1	0	0	1	0	0
5	0	1	1	1	0	1	0	1
6	0	1	0	1	0	1	1	0
7	0	1	0	0	0	1	1	1
8	1	1	0	0	1	0	0	0
9	1	1	0	1	1	0	0	1
10	1	1	1	1	1	0	1	0
11	1	1	1	0	1	0	1	1
12	1	0	1	0	1	1	0	0
13	1	0	1	1	1	1	0	1
14	1	0	0	1	1	1	1	0
15	1	0	0	0	1	1	1	1

Binary-to-Gray code conversion. Let $g_{n-1}\, g_{n-2} \cdots g_2\, g_1\, g_0$ denote a code word in the n-bit Gray code corresponding to the binary number

$b_{n-1}\, b_{n-2} \cdots b_2\, b_1\, b_0$. Then, the g_i are given by

$$g_i = b_i \oplus b_{i+1}, \qquad 0 \le i \le n-2$$

$$g_{n-1} = b_{n-1},$$

where the symbol $\oplus$ denotes modulo 2 addition, (exclusive OR) which is defined as follows:

$$0 \oplus 0 = 0, \quad 1 \oplus 0 = 1, \quad 0 \oplus 1 = 1, \quad 1 \oplus 1 = 0$$

For example, the Gray code corresponding to the binary number **101101** is obtained as illustrated below.

b_5	b_4	b_3	b_2	b_1	b_0
1	0	1	1	0	1
	$\oplus$	$\oplus$	$\oplus$	$\oplus$	$\oplus$
1	1	1	0	1	1
g_5	g_4	g_3	g_2	g_1	g_0

Gray code-to-binary conversion. To convert from Gray code to binary, we start with the left most digit and move to the right, making $b_i = g_i$ if the number of 1's preceding g_i is even, and making $b_i = \bar{g}_i$ (bar denoting complement) if the number of 1's preceding g_i is odd. During this process, zero 1's is treated as an even number of 1's. For example, the binary number which corresponds to the Gray code word **1001011** is found to be **1110010** in the following manner:

g_6	g_5	g_4	g_3	g_2	g_1	g_0
1	0	0	1	0	1	1
↓			↓		↓	↓
1	1	1	0	0	1	0
b_6	b_5	b_4	b_3	b_2	b_1	b_0

References

1. Ahmed, N., Schreiber, H., and Lopresti, P.: On Notation and Definition of Terms Related to a Class of Complete Orthogonal Functions. *IEEE Trans. Electromagnetic Compatability* EMC-15 (1973) 75–80.
2. Harmuth, H.: *Transmission of Information by Orthogonal Functions*. 2nd ed. New York, Heidelberg, Berlin: Springer, 1972, Chapter 1.
3. Pichler, F.: Das System der sal- und cal-Funktionen als Erweiterung des Systems der Walsh-Funktionen und die Theorie der sal- und cal-Fourier Transformation. Thesis, Dept. of Mathematics, Innsbruck Univ., Austria 1967.

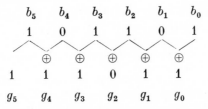

4. Rademacher, H.: Einige Sätze von allgemeinen Orthogonalfunktionen. *Math. Annalen* 87 (1922) 122–138.
5. Corrington, M. S.: Advanced Analytical and Signal Processing Techniques. (1962) ASTIA Document No. AD 277–942.
6. Haar, A.: Zur Theorie der Orthogonalen Funktionensysteme. *Math. Ann.* 69 (1910) 331–371; 71 (1912) 38–53.
7. Nagy, B. S.: *Introduction to Real Functions and Orthogonal Expansions.* New York: Oxford University Press, 1965.
8. Walsh, J. L.: A Closed Set of Orthogonal Functions. *Amer. J. of Mathematics* 45 (1923) 5–24.
9. Lackey, R. B., and Meltzer, D.: A Simplified Definition of Walsh Functions. *IEEE Trans. Computers* C-20 (1971) 211–213.
10. Pratt, W. K., and Andrews, H. C.: Hadamard Transform Image Coding. *Proc. IEEE* 57 (1969) 58–68.
11. Paley, R. E. A. C.: A Remarkable Series of Orthogonal Functions. *Proc. London Math. Soc.* (2) 34 (1932) 241–279.
12. Yuen, C.: Walsh Functions and Gray Code. *Proc. 1971 Walsh Functions Symposium,* 68–73, National Technical Information Service, Springfield, Va. 22151, order no. AD-707431.
13. Ahmed, N., and Rao, K. R.: Walsh Functions and Hadamard Transform. *Proc. 1972 Walsh Functions Symposium,* 8–13, National Technical Information Service, Springfield, Va. 22151, order no. AD-744650.
14. Lackey, R. B.: The Wonderful World of Walsh Functions. Ibid., 2–7.
15. Shore, J. E.: On the Application of Haar Functions. Naval Research Laboratory, Washington, D. C. NRL Report 7467, January 1973.
16. Shore, J. E., and Berkowitz, R. L.: Convergence Properties of Haar Series. Ibid., NRL Report 7470, January 1973.
17. Shore, J. E.: On the Application of Haar Functions. *IEEE Trans. Communications* COM-21 (1973) 209–216.
18. Corrington, M. S.: Solution of Differential and Integral Equations with Walsh Functions. *IEEE Trans. Circuit Theory* CT-20 (1973) 470–476.
19. Lackey, R. B.: So What's a Walsh Function. *IEEE Fall Electronics Conference,* Oct. 18–20, 1971, Chicago, Ill., 368–371.
20. Ahmed, N., and Rao, K. R.: Transform Properties of Walsh Functions. Ibid., 378–382.

Problems

5-1 Let the binary and Gray code forms of i be given by

$$i_{\text{binary}} = b_n\, b_{n-1} \cdots b_2 b_1$$

$$i_{\text{Gray}} = g_n\, g_{n-1} \cdots g_2 g_1$$

Then the Walsh functions $\text{wal}_w(i, t)$ and $\text{Wal}_w(i, t)$ can be generated using the Rademacher functions as follows [9, 13, 14].

$$\text{wal}_w(i, t) = \prod_{k=1}^{n} [\text{rad}(k, t)]^{g_k}$$

and

$$\text{Wal}_w(i, t) = \prod_{k=1}^{n} [\text{Rad}(k, t)]^{g_k}.$$

From Fig. 5.3 it is apparent that

$$\begin{aligned}
\text{Rad}(1, t) &= + + + + + + + + - - - - - - - - \\
\text{Rad}(2, t) &= + + + + - - - - + + + + - - - - \\
\text{Rad}(3, t) &= + + - - + + - - + + - - + + - - \\
\text{Rad}(4, t) &= + - + - + - + - + - + - + - + -
\end{aligned}$$

where "+" and "−" abbreviate +1 and −1 respectively.

Use the above information to show that

$$\begin{aligned}
\text{Wal}_w(13, t) &= \text{Rad}(1, t)\,\text{Rad}(2, t)\,\text{Rad}(4, t) \\
&= + - + - - + - + - + - + + - + - \\
\text{Wal}_w(9, t) &= \text{Rad}(1, t)\,\text{Rad}(3, t)\,\text{Rad}(4, t) \\
&= + - - + + - - + - + + - - + + - \\
\text{Wal}_w(8, t) &= \text{Rad}(3, t)\,\text{Rad}(4, t) \\
&= + - - + + - - + + - - + + - - +
\end{aligned}$$

5-2 It can be shown that the product of two Walsh functions yields another Walsh function [2]:

$$\begin{aligned}
\text{wal}_w(h, t)\ \text{wal}_w(k, t) &= \text{wal}_w(h \oplus k, t) \\
\text{Wal}_w(h, t)\ \text{Wal}_w(k, t) &= \text{Wal}_w(h \oplus k, t) \qquad \text{(P5-2-1)}
\end{aligned}$$

where $\oplus$ denotes modulo 2 addition.

From Fig. 5.5b it can be seen that:

$$\begin{aligned}
\text{Wal}_w(1, t) &= + + + + - - - - \\
\text{Wal}_w(2, t) &= + + - - - - + + \\
\text{Wal}_w(3, t) &= + + - - + + - - \\
\text{Wal}_w(4, t) &= + - - + + - - + \\
\text{Wal}_w(5, t) &= + - - + - + + - \\
\text{Wal}_w(6, t) &= + - + - - + - + \\
\text{Wal}_w(7, t) &= + - + - + - + -
\end{aligned}$$

Apply Eq. (P5-2-1) to the above set of Walsh functions to verify that

$$\begin{aligned}
\text{Wal}_w(4, t)\ \text{Wal}_w(7, t) &= \text{Wal}_w(3, t) \\
\text{Wal}_w(3, t)\ \text{Wal}_w(6, t) &= \text{Wal}_w(5, t) \\
\text{Wal}_w(1, t)\ \text{Wal}_w(5, t) &= \text{Wal}_w(4, t) \\
\text{Wal}_w(2, t)\ \text{Wal}_w(4, t) &= \text{Wal}_w(6, t)
\end{aligned}$$

5-3 Consider the following identities:

(i) $2i \oplus 2k = 2(i \oplus k)$

(ii) $(2i - 1) \oplus (2k - 1) = 2[(i - 1) \oplus (k - 1)]$

(iii) $(2i - 1) \oplus 2k = 2[k \oplus (i - 1) + 1] - 1$

(iv) $(2i) \oplus (2k - 1) = 2[i \oplus (k - 1) + 1] - 1$

Substitute the relations $\text{Wal}_w(2i, t) = \text{Cal}(i, t)$ and $\text{Wal}_w(2i - 1, t) = \text{Sal}(i, t)$ for $i = 1, 2, \ldots$ in $\text{Wal}_w(h, t)\,\text{Wal}_w(k, t) = \text{Wal}_w(h \oplus k, t)$, and show that [2]

$$\begin{aligned}
\text{Cal}(i, t)\,\text{Cal}(k, t) &= \text{Cal}(i \oplus k, t)\\
\text{Sal}(i, t)\,\text{Cal}(k, t) &= \text{Sal}\{[k \oplus (i - 1)] + 1, t\}\\
\text{Cal}(i, t)\,\text{Sal}(k, t) &= \text{Sal}\{[i \oplus (k - 1)] + 1, t\}\\
\text{Sal}(i, t)\,\text{Sal}(k, t) &= \text{Cal}[(i - 1) \oplus (k - 1), t]\\
\text{Cal}(0, t) &= \text{Wal}_w(0, t)
\end{aligned}$$

5-4 If u and v take the values $0, 1, \ldots, 7$, one can obtain Table P5-4-1, where $r_0(u)$, $r_1(u)$ and $r_2(u)$ are defined in Eq. (5.4-4).

Table P5-4-1

u	u_2	u_1	u_0	u_0	u_1	u_2	$r_0(u)$	$r_1(u)$	$r_2(u)$	v	v_2	v_1	v_0	v_0	v_1	v_2
0	0	0	0	0	0	0	0	0	0	0	0	0	0	0	0	0
1	0	0	1	1	0	0	0	0	1	1	0	0	1	1	0	0
2	0	1	0	0	1	0	0	1	1	2	0	1	0	0	1	0
3	0	1	1	1	1	0	0	1	0	3	0	1	1	1	1	0
4	1	0	0	0	0	1	1	1	0	4	1	0	0	0	0	1
5	1	0	1	1	0	1	1	1	1	5	1	0	1	1	0	1
6	1	1	0	0	1	1	1	0	1	6	1	1	0	0	1	1
7	1	1	1	1	1	1	1	0	0	7	1	1	1	1	1	1

Array =

u \ v	0	1	2	3	4	5	6	7
0	+	+	+	+	+	+	+	+
1	+							
2	+							
3	+							
4	+							
5	+							
6	+							
7	+							

Using Table P5-4-1 and Eqs. (5.4-4), (5.4-7) and (5.4-11), construct three arrays, (say Array m, $m = 1, 2, 3$), of the type shown alongside.

Verify that the resulting Arrays 1, 2, and 3 above are identical to $\mathbf{H}_w(3)$, $\mathbf{H}_p(3)$ and $\mathbf{H}_h(3)$ described in Figs. 5.5b, 5.6b, and 5.7b respectively.

5-5 In Fig. 5.4a the first eight Haar functions are shown. Using Eq. (5.3-2) draw the next eight Haar functions and obtain $\mathbf{H}^*(4)$.

5-6 As described in problem 5-1, Walsh functions can be expressed as products of Rademacher functions [9, 13, 14]. In Fig. 5.5a, the first eight Walsh-ordered Walsh functions are shown. Draw the next eight Walsh functions [2, 9] and obtain $\mathbf{H}_w(4)$. Identify the sal and cal functions and their sequencies.

5-7 Develop tables similar to Tables 5.4-1 and 5.4-2 for $N = 16$. Rearrange the Walsh-ordered Walsh functions of Prob. 5-6 to obtain corresponding Paley-ordered and Hadamard-ordered Walsh functions. Identify the sal and cal functions and their sequencies. Using the first sixteen Walsh functions, obtain $\mathbf{H}_p(4)$ and $\mathbf{H}_h(4)$.

Chapter Six

Walsh-Hadamard Transform

This chapter is devoted to the study of the Walsh-Hadamard transform (WHT), which is perhaps the most well-known of the nonsinusoidal orthogonal transforms. The WHT has gained prominence in various digital signal processing applications, since it can essentially be computed using additions and subtractions only. Consequently its hardware implementation is also simpler.

Fast algorithms to compute the WHT are developed and the notion of Walsh spectra is introduced. Properties of Walsh spectra are studied and physical interpretations are provided. Throughout this chapter, the analogy between the Walsh-Hadamard and discrete Fourier transforms is explored.

6.1 Walsh Series Representation

Before considering the development of Walsh-Hadamard transforms[1], it is instructive to study some aspects of representing a given continuous signal $x(t)$ in the form of a Walsh series. For the purposes of discussion, we will assume that $x(t)$ is defined on the half-open unit interval $t \in [0, 1)$.

It has been shown [1] that the set of Walsh functions $\{\mathrm{wal}_\omega(i, t)\}$ is *closed*. Thus, every signal $x(t)$ which is absolutely integrable[2] in $t \in [0, 1)$ can be expanded in a series of the form

$$x(t) = \sum_{k=0}^{\infty} d_k \, \mathrm{wal}_\omega(k, t) \tag{6.1-1}$$

Since the set of functions $\{\mathrm{wal}_\omega(k, t)\}$ forms an orthonormal system in the closed interval $t \in [0, 1]$, the coefficients d_k are given by

$$d_k = \int_0^1 x(t) \, \mathrm{wal}_\omega(k, t) \, dt, \qquad k = 0, 1, 2, \ldots \tag{6.1-2}$$

We recall that [see Eq. (5.4-3)]

$$\begin{aligned} \mathrm{wal}_\omega(k, t) &= \mathrm{cal}(s_k, t), \qquad && k \text{ even} \\ \mathrm{wal}_\omega(k, t) &= \mathrm{sal}(s_k, t), \qquad && k \text{ odd} \end{aligned}$$

[1] That is, the different forms of the WHT.

[2] That is $\int_0^1 |x(t)| \, dt$ is finite.

where s_k is the sequency of $\text{wal}_\omega(k, t)$, which is defined as [see Eq. (5.4-2)]

$$s_k = \begin{cases} 0, & k = 0 \\ k/2, & k \text{ even} \\ (k+1)/2, & k \text{ odd} \end{cases}$$

With $\text{wal}_\omega(k, t)$ expressed in terms of its sal and cal components, Eq. (1) becomes

$$x(t) = a_0 \, \text{wal}_\omega(0, t) + \sum_{k=1}^{\infty} [a_k \, \text{cal}\,(k, t) + b_k \, \text{sal}\,(k, t)] \tag{6.1-3}$$

where

$$a_0 = d_0$$
$$a_k = d_{2k},$$
$$b_k = d_{2k-1},$$

In order to obtain a finite series expansion which consists of N terms, where N is of the form $N = 2^n$, we truncate the above series to obtain

$$x(t) \simeq a_0 \, \text{wal}_\omega(0, t) + \sum_{k=1}^{N/2-1} [a_k \, \text{cal}\,(k, t) + b_k \, \text{sal}\,(k, t)] + b_{N/2} \, \text{sal}\,(N/2, t) \tag{6.1-4}$$

The following convergence conditions for the series in Eq. (4) were given by Walsh [1], Paley [2], and Fine [3]:

(i) If $x(t)$ is continuous in $t \in [0, 1)$, the series converges *uniformly* to the value of $x(t)$; that is

$$\lim_{k \to \infty} \{a_k, b_k\} = 0 \tag{6.1-5}$$

Thus there exists some $k = N_0$ such that all a_k and b_k for which $k > N_0$, are smaller than any predetermined $\varepsilon > 0$. If we omit such coefficients in the Walsh representation of $x(t)$, the information lost is likely to be insignificant.

(ii) At points where $x(t)$ is discontinuous in $t \in [0, 1)$, there is convergence in the mean.

From the above discussion it is apparent that the Walsh representation of signals is analogous to the Fourier representation, which was considered in Section 2.1. This is what one would expect intuitively, because of the strong resemblance between Fourier sinusoids and Walsh functions, as illustrated in Fig. 6.1 for the case $N = 8$.

In the section that follows, we shall develop Walsh-Hadamard transforms which are analogs of the discrete Fourier transform (DFT). Walsh-Hadamard transforms are used for the Walsh representation of the data

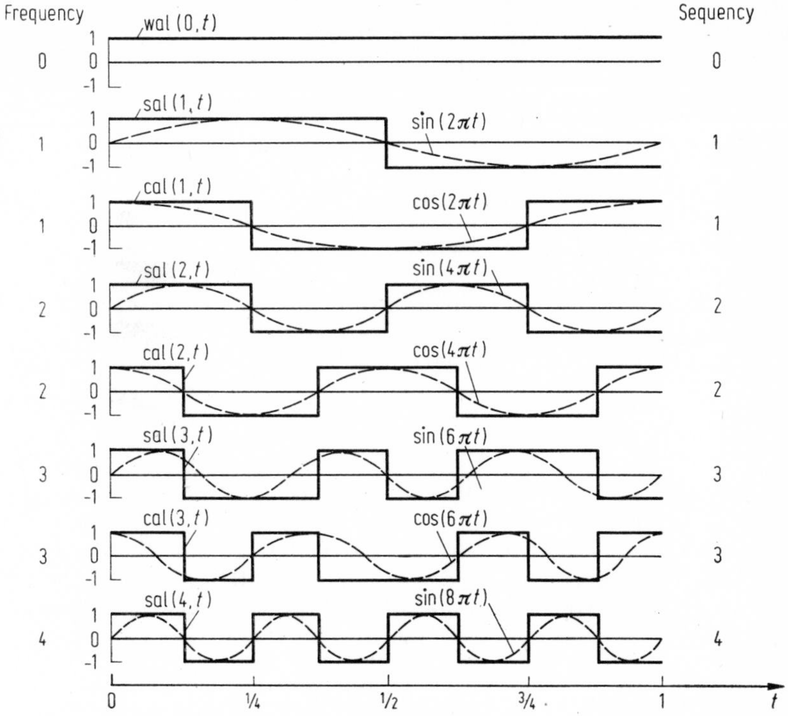

Fig. 6.1 Walsh functions and Fourier harmonics

sequences. Their basis functions are sampled Walsh functions which can be expressed in terms of the Hadamard matrices $\mathbf{H}_h(n)$. These matrices can be generated using the following recurrence relation [see Eq. (5.4-10)]:

$$\mathbf{H}_h(k) = \begin{bmatrix} \mathbf{H}_h(k-1) & \mathbf{H}_h(k-1) \\ \mathbf{H}_h(k-1) & -\mathbf{H}_h(k-1) \end{bmatrix}, \qquad k = 1, 2, \ldots, n \qquad (6.1\text{-}6)$$

where $\mathbf{H}_h(0) = 1$ and $n = \log_2 N$.

For example, with $k = 1$, and $k = 2$, Eq. (6) yields

$$\mathbf{H}_h(1) = \begin{bmatrix} 1 & 1 \\ 1 & -1 \end{bmatrix}, \quad \mathbf{H}_h(2) = \begin{bmatrix} 1 & 1 & 1 & 1 \\ 1 & -1 & 1 & -1 \\ 1 & 1 & -1 & -1 \\ 1 & -1 & -1 & 1 \end{bmatrix}$$

It is straightforward to show that the matrices $\mathbf{H}_h(k)$ have the following properties:

(i) $\mathbf{H}_h(k)$ is a symmetric matrix, i.e.

$$\mathbf{H}_h(k)' = \mathbf{H}_h(k), \tag{6.1-7}$$

prime denoting transpose.

(ii) $\mathbf{H}_h(k)$ is orthogonal, i.e.

$$\mathbf{H}_h(k)'\,\mathbf{H}_h(k) = 2^k\,\mathbf{I}(k) \tag{6.1-8}$$

where $\mathbf{I}(k)$ is the $(2^k \times 2^k)$ identity matrix.

(iii) The inverse of $\mathbf{H}_h(k)$ is proportional to itself; i.e.

$$[\mathbf{H}_h(k)]^{-1} = \frac{1}{2^k}\,\mathbf{H}_h(k) \tag{6.1-9}$$

where $[\mathbf{H}_h(k)]^{-1}$ is the inverse of $\mathbf{H}_h(k)$.

6.2 Hadamard-ordered Walsh-Hadamard Transform $(\mathrm{WHT})_h$

This transform is also referred to as the BIFORE (*BI*nary *FO*urier *RE*presentation) transform (BT). The notion of BIFORE was introduced by Ohnsorg [4]. The $(\mathrm{WHT})_h$ can be defined using matrix or exponential notation.

Matrix definition. Let $\{X(m)\}$ denote an N-periodic sequence $X(m)$, $m = 0, 1, \ldots, N-1$, of finite valued real[1] numbers; then,

$$\{X(m)\} = \{X(0)X(1)\cdots X(N-1)\} \tag{6.2-1}$$

$\{X(m)\}$ is represented by means of an N-vector $\boldsymbol{X}(n)$ to obtain

$$\boldsymbol{X}(n)' = [X(0)X(1)\cdots X(N-1)] \tag{6.2-2}$$

where $n = \log_2 N$, and $\boldsymbol{X}(n)'$ is the transpose of $\boldsymbol{X}(n)$.

The $(\mathrm{WHT})_h$ of $\{X(m)\}$ is defined as

$$\boldsymbol{B}_x(n) = \frac{1}{N}\,\mathbf{H}_h(n)\,\boldsymbol{X}(n) \tag{6.2-3}$$

where $B_x(k)$ is the k-th $(\mathrm{WHT})_h$ coefficient and $\boldsymbol{B}_x(n)' = [B_x(0)\;B_x(1)\cdots B_x(N-1)]$.

From Eqs. (6.1-9) and (3) it follows that the inverse $(\mathrm{WHT})_h$ transform $(\mathrm{IWHT})_h$ is defined as

$$\boldsymbol{X}(n) = \mathbf{H}_h(n)\,\boldsymbol{B}_x(n) \tag{6.2-4}$$

Since Eqs. (3) and (4) constitute a transform pair, the $(\mathrm{WHT})_h$ representation of $\{X(m)\}$ is unique. A simple example is considered next.

[1] Most of the results we shall develop can be extended to complex number sequences.

Example 6.2-1

Let $\{X(m)\} = \{1\ 2\ -1\ 3\}$. Then, $n = 2$ and Eq. (3) yields

$$\boldsymbol{B}_x(2) = \frac{1}{4}\mathbf{H}_h(2)\,\boldsymbol{X}(2) \tag{6.2-5}$$

Substituting for $\mathbf{H}_h(2)$ we obtain

$$\begin{bmatrix} B_x(0) \\ B_x(1) \\ B_x(2) \\ B_x(3) \end{bmatrix} = \frac{1}{4}\begin{bmatrix} 1 & 1 & 1 & 1 \\ 1 & -1 & 1 & -1 \\ 1 & 1 & -1 & -1 \\ 1 & -1 & -1 & 1 \end{bmatrix}\begin{bmatrix} 1 \\ 2 \\ -1 \\ 3 \end{bmatrix} \tag{6.2-6}$$

Evaluation of Eq. (6) leads to the $(\mathrm{WHT})_h$ coefficients

$$\boldsymbol{B}_x(2)' = \begin{bmatrix} \frac{5}{4} & \frac{-5}{4} & \frac{1}{4} & \frac{3}{4} \end{bmatrix}$$

To verify that the transformation in Eq. (6) is unique, we substitute $B_x(k)$, $k = 0, 1, 2, 3$ in Eq. (4) to obtain

$$\begin{bmatrix} X(0) \\ X(1) \\ X(2) \\ X(3) \end{bmatrix} = \begin{bmatrix} 1 & 1 & 1 & 1 \\ 1 & -1 & 1 & -1 \\ 1 & 1 & -1 & -1 \\ 1 & -1 & -1 & 1 \end{bmatrix}\begin{bmatrix} 5/4 \\ -5/4 \\ 1/4 \\ 3/4 \end{bmatrix} \tag{6.2-7}$$

Evaluation of Eq. (7) leads to recovery of the original sequence $\{X(m)\} = \{1\ 2\ -1\ 3\}$.

Exponential definition. The $(\mathrm{WHT})_h$ can alternately be defined as

$$B_x(u) = \frac{1}{N}\sum_{m=0}^{N-1} X(m)\,(-1)^{\langle m,\, u\rangle}, \qquad u = 0, 1, \ldots, N-1 \tag{6.2-8}$$

where

$$\langle m, u\rangle = \sum_{s=0}^{n-1} u_s m_s; \qquad n = \log_2 N$$

The terms u_s and m_s in Eq. (8) are the coefficients of the binary representations of u and m rexpectively. That is, for each decimal value of $0 \le u \le N-1$, we have

$$u = u_{n-1}2^{n-1} + u_{n-2}2^{n-2} + \cdots + u_1 2^1 + u_0 2^0 \tag{6.2-9}$$

where

$$u_v = 0 \text{ or } 1, \qquad v = 0, 1, 2, \ldots, n-1$$

Similarly each decimal value of m, $0 \leq m \leq N - 1$ is expressed as

$$m = m_{n-1} 2^{n-1} + m_{n-2} 2^{n-2} + \cdots + m_1 2^1 + m_0 2^0 \qquad (6.2\text{-}10)$$

where

$$m_v = 0 \text{ or } 1, \qquad v = 0, 1, \ldots, n-1$$

It is instructive to demonstrate that the matrix and exponential definitions are equivalent by considering the case $N = 4$. Then Eq. (8) yields

$$\begin{aligned}
4B_x(0) &= X(0)(-1)^{\langle 0,0\rangle} + X(1)(-1)^{\langle 1,0\rangle} + X(2)(-1)^{\langle 2,0\rangle} + X(3)(-1)^{\langle 3,0\rangle}\\
4B_x(1) &= X(0)(-1)^{\langle 0,1\rangle} + X(1)(-1)^{\langle 1,1\rangle} + X(2)(-1)^{\langle 2,1\rangle} + X(3)(-1)^{\langle 3,1\rangle}\\
4B_x(2) &= X(0)(-1)^{\langle 0,2\rangle} + X(1)(-1)^{\langle 1,2\rangle} + X(2)(-1)^{\langle 2,2\rangle} + X(3)(-1)^{\langle 3,2\rangle}\\
4B_x(3) &= X(0)(-1)^{\langle 0,3\rangle} + X(1)(-1)^{\langle 1,3\rangle} + X(2)(-1)^{\langle 2,3\rangle} + X(3)(-1)^{\langle 3,3\rangle}
\end{aligned} \qquad (6.2\text{-}11)$$

To evaluate the exponents of (-1), we generate a matrix $[\langle m, u\rangle]$, $m, u = 0, 1, 2, 3$ as indicated below.

m	m_1	m_0
0	0	0
1	0	1
2	1	0
3	1	1

u	u_1	u_0
0	0	0
1	0	1
2	1	0
3	1	1

$$[\langle m, u\rangle] = \begin{array}{c} \\ \\ 0\\1\\2\\3 \end{array}\!\!\begin{array}{c} u\rightarrow \\ m\;\; 0 \;\; 1 \;\; 2 \;\; 3 \\ \downarrow \\ \begin{bmatrix} 0 & 0 & 0 & 0\\ 0 & 1 & 0 & 1\\ 0 & 0 & 1 & 1\\ 0 & 1 & 1 & 2 \end{bmatrix}\end{array}$$

From the elements of the matrix $[\langle m, u\rangle]$ and Eq. (11), it follows that

$$\begin{aligned}
4B_x(0) &= X(0) + X(1) + X(2) + X(3)\\
4B_x(1) &= X(0) - X(1) + X(2) - X(3)\\
4B_x(2) &= X(0) + X(1) - X(2) - X(3)\\
4B_x(3) &= X(0) - X(1) - X(2) + X(3)
\end{aligned}$$

In matrix form, the above equations can be expressed as

$$\boldsymbol{B}_x(2) = \frac{1}{4}\,\mathbf{H}_h(2)\,\boldsymbol{X}(2)$$

which is the $(\text{WHT})_h$ matrix definition for $N = 4$.

Again, the $(\text{IWHT})_h$ corresponding to the $(\text{WHT})_h$ in Eq. (8) is defined as

$$X(m) = \sum_{u=0}^{N-1} B_x(u)(-1)^{\langle m, u\rangle}, \qquad m = 0, 1, \ldots, N-1 \qquad (6.2\text{-}12)$$

6.3 Fast Hadamard-ordered Walsh-Hadamard Transform $(\mathrm{FWHT})_h$

We recall that the FFT is an algorithm to compute the DFT efficiently. Similarly, the $(\mathrm{FWHT})_h$ is an algorithm to compute the $(\mathrm{WHT})_h$ efficiently. The $(\mathrm{FWHT})_h$ can be derived using matrix factoring [5, 6] or matrix partitioning techniques [7]. We illustrate the matrix partitioning technique for the case $N = 8$. With $N = 8$, Eq. (6.2-3) yields

$$\boldsymbol{B}_x(3) = \frac{1}{8}\,\mathbf{H}_h(3)\,\boldsymbol{X}(3) \tag{6.3-1}$$

Using Eq. (6.1-6), $\mathbf{H}_h(3)$ is expressed in terms of $\mathbf{H}_h(2)$ to obtain

$$\begin{bmatrix} B_x(0) \\ B_x(1) \\ B_x(2) \\ B_x(3) \\ \hdashline B_x(4) \\ B_x(5) \\ B_x(6) \\ B_x(7) \end{bmatrix} = \frac{1}{8}\begin{bmatrix} \mathbf{H}_h(2) & \mathbf{H}_h(2) \\ \mathbf{H}_h(2) & -\mathbf{H}_h(2) \end{bmatrix} \begin{bmatrix} X(0) \\ X(1) \\ X(2) \\ X(3) \\ \hdashline X(4) \\ X(5) \\ X(6) \\ X(7) \end{bmatrix} \tag{6.3-2}$$

From the matrix partitioning indicated in Eq. (2) it follows that

$$\begin{bmatrix} B_x(0) \\ B_x(1) \\ B_x(2) \\ B_x(3) \end{bmatrix} = \frac{1}{8}\,\mathbf{H}_h(2) \begin{bmatrix} X_1(0) \\ X_1(1) \\ X_1(2) \\ X_1(3) \end{bmatrix} \tag{6.3-3a}$$

and

$$\begin{bmatrix} B_x(4) \\ B_x(5) \\ B_x(6) \\ B_x(7) \end{bmatrix} = \frac{1}{8}\,\mathbf{H}_h(2) \begin{bmatrix} X_1(4) \\ X_1(5) \\ X_1(6) \\ X_1(7) \end{bmatrix} \tag{6.3-3b}$$

where

$$X_1(l) = X(l) + X(4 + l), \qquad l = 0, 1, 2, 3$$
$$X_1(l) = X(l - 4) - X(l), \qquad l = 4, 5, 6, 7$$

The sequence of additions and subtractions in Eqs. (3a) and (3b) are shown in the signal flow graph in Fig. 6.2, and are indicated by iteration #1.

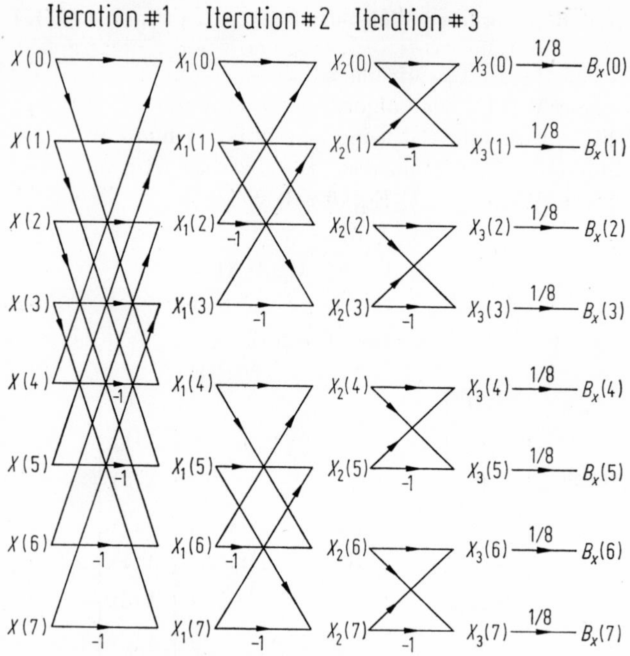

Fig. 6.2 (FWHT)$_h$ signal flow graph, $N = 8$

Again, application of Eq. (6.1-6) to Eqs. (3a) and (3b) results in

$$\begin{bmatrix} B_x(0) \\ B_x(1) \\ \hdashline B_x(2) \\ B_x(3) \end{bmatrix} = \frac{1}{8} \begin{bmatrix} \mathbf{H}_h(1) & \mathbf{H}_h(1) \\ \mathbf{H}_h(1) & -\mathbf{H}_h(1) \end{bmatrix} \begin{bmatrix} X_1(0) \\ X_1(1) \\ \hdashline X_1(2) \\ X_1(3) \end{bmatrix} \tag{6.3-4a}$$

and

$$\begin{bmatrix} B_x(4) \\ B_x(5) \\ \hdashline B_x(6) \\ B_x(7) \end{bmatrix} = \frac{1}{8} \begin{bmatrix} \mathbf{H}_h(1) & \mathbf{H}_h(1) \\ \mathbf{H}_h(1) & -\mathbf{H}_h(1) \end{bmatrix} \begin{bmatrix} X_1(4) \\ X_1(5) \\ \hdashline X_1(6) \\ X_1(7) \end{bmatrix} \tag{6.3-4b}$$

From the matrix partitioning indicated in Eqs. (4a) and (4b), we obtain the following set of equations:

$$\begin{bmatrix} B_x(0) \\ B_x(1) \end{bmatrix} = \frac{1}{8} \mathbf{H}_h(1) \begin{bmatrix} X_1(0) + X_1(2) \\ X_1(1) + X_1(3) \end{bmatrix} = \frac{1}{8} \mathbf{H}_h(1) \begin{bmatrix} X_2(0) \\ X_2(1) \end{bmatrix} \tag{6.3-5a}$$

$$\begin{bmatrix} B_x(2) \\ B_x(3) \end{bmatrix} = \frac{1}{8} \mathbf{H}_h(1) \begin{bmatrix} X_1(0) - X_1(2) \\ X_1(1) - X_1(3) \end{bmatrix} = \frac{1}{8} \mathbf{H}_h(1) \begin{bmatrix} X_2(2) \\ X_2(3) \end{bmatrix} \tag{6.3-5b}$$

$$\begin{bmatrix} B_x(4) \\ B_x(5) \end{bmatrix} = \frac{1}{8} \mathbf{H}_h(1) \begin{bmatrix} X_1(4) + X_1(6) \\ X_1(5) + X_1(7) \end{bmatrix} = \frac{1}{8} \mathbf{H}_h(1) \begin{bmatrix} X_2(4) \\ X_2(5) \end{bmatrix} \tag{6.3-5c}$$

$$\begin{bmatrix} B_x(6) \\ B_x(7) \end{bmatrix} = \frac{1}{8} \mathbf{H}_h(1) \begin{bmatrix} X_1(4) - X_1(6) \\ X_1(5) - X_1(7) \end{bmatrix} = \frac{1}{8} \mathbf{H}_h(1) \begin{bmatrix} X_2(6) \\ X_2(7) \end{bmatrix} \tag{6.3-5d}$$

The sequence of additions and subtractions in Eqs. (5a) through (5d) are indicated by iteration #2 in Fig. 6.2. Since

$$\mathbf{H}_h(1) = \begin{bmatrix} 1 & 1 \\ 1 & -1 \end{bmatrix},$$

these equations reduce to

$$\begin{aligned} 8B_x(0) &= X_2(0) + X_2(1) = X_3(0) \\ 8B_x(1) &= X_2(0) - X_2(1) = X_3(1) \\ 8B_x(2) &= X_2(2) + X_2(3) = X_3(2) \\ 8B_x(3) &= X_2(2) - X_2(3) = X_3(3) \\ 8B_x(4) &= X_2(4) + X_2(5) = X_3(4) \\ 8B_x(5) &= X_2(4) - X_2(5) = X_3(5) \\ 8B_x(6) &= X_2(6) + X_2(7) = X_3(6) \\ 8B_x(7) &= X_2(6) - X_2(7) = X_3(7) \end{aligned} \tag{6.3-6}$$

The sequence of additions and subtractions in Eq. (6) are indicated by iteration #3 in Fig. 6.2. From the $(\mathrm{FWHT})_h$ signal flow graph, it is apparent that apart from the 1/8 multiplier, only additions and subtractions are required. The number of additions and subtractions required to compute the eight $(\mathrm{WHT})_h$ coefficients is $8 \times \log_2 8 = 24$.

Generalizations. The generalizations pertaining to the $(\mathrm{FWHT})_h$ are straightforward. The overall structure of the signal flow graph for any $N = 2^n$ is similar to that in Fig. 6.2. The following remarks can be made for the general case, $N = 2^n$.

1. The total number of iterations is given by $n = \log_2 N$. If r is an iteration index, then $r = 1, 2, \ldots, n$.

2. The r-th iteration results in 2^{r-1} groups with $N/2^{r-1}$ members in each group. Half the members in each group are associated with an addition operation while the remaining half are associated with a subtraction operation.

3. The total number of arithmetic operations to compute all the transform coefficients is approximately $N \log_2 N$, compared to N^2 as implied by Eq. (6.2-3).

4. The algorithm can also be used to compute the $(\text{IWHT})_h$ in Eq. (6.2-4).

Example 6.3-1

Given the data sequence, $\{X(m)\} = \{1\ 2\ 1\ 1\ 3\ 2\ 1\ 2\}$. Use the $(\text{FWHT})_h$ to compute the $(\text{WHT})_h$ coefficients $B_x(k)$, $k = 0, 1, \ldots, 7$.

Solution: Following the sequence of computations in Fig. 6.2, we obtain:

$$
\begin{array}{lllllll}
X(0) = 1 & X_1(0) = 4 & X_2(0) = 6 & X_3(0) = 13 & \xrightarrow{1/8} & B_x(0) \\
X(1) = 2 & X_1(1) = 4 & X_2(1) = 7 & X_3(1) = -1 & \xrightarrow{1/8} & B_x(1) \\
X(2) = 1 & X_1(2) = 2 & X_2(2) = 2 & X_3(2) = 3 & \xrightarrow{1/8} & B_x(2) \\
X(3) = 1 & X_1(3) = 3 & X_2(3) = 1 & X_3(3) = 1 & \xrightarrow{1/8} & B_x(3) \\
X(4) = 3 & X_1(4) = -2 & X_2(4) = -2 & X_3(4) = -3 & \xrightarrow{1/8} & B_x(4) \\
X(5) = 2 & X_1(5) = 0 & X_2(5) = -1 & X_3(5) = -1 & \xrightarrow{1/8} & B_x(5) \\
X(6) = 1 & X_1(6) = 0 & X_2(6) = -2 & X_3(6) = -1 & \xrightarrow{1/8} & B_x(6) \\
X(7) = 2 & X_1(7) = -1 & X_2(7) = 1 & X_3(7) = -3 & \xrightarrow{1/8} & B_x(7)
\end{array}
$$

Hence, $\{B_x(k)\} = \left\{\frac{13}{8} \quad \frac{-1}{8} \quad \frac{3}{8} \quad \frac{1}{8} \quad \frac{-3}{8} \quad \frac{-1}{8} \quad \frac{-1}{8} \quad \frac{-3}{8}\right\}$

Sequencies associated with $(\text{WHT})_h$ coefficients. Using a bit-reversal calculation, one can obtain the sequency associated with a given $B_x(k)$. We start with Eq. (5.4-9), from which it follows that

$$\text{Wal}_h(k, t) = \text{Wal}_w[b(\langle k\rangle), t] \tag{6.3-7}$$

where $\langle k\rangle$ is obtained by the bit-reversal of k, $b(\langle k\rangle)$ is the Gray code-to-binary conversion of $\langle k\rangle$ and $\text{Wal}_h(k, t)$, $\text{Wal}_w(k, t)$ are sampled Hadamard-ordered and Walsh-ordered Walsh functions respectively.

Combining Eqs. (5.4-2) and (7) we obtain

$$s_{B_x(k)} = \text{sequency of } \text{Wal}_h(k, t) = \begin{cases} 0, & k = 0 \\ b(\langle k\rangle)/2, & b(\langle k\rangle) \text{ even} \\ \dfrac{b(\langle k\rangle) + 1}{2}, & b(\langle k\rangle) \text{ odd} \end{cases} \tag{6.3-8}$$

where $s_{B_x(k)}$ is the sequency associated with the $(\text{WHT})_h$ coefficient $B_x(k)$.

For the purposes of illustration, Eq. (8) is evaluated for $N = 8$ and the results are summarized in Table 6.3-1.

Table 6.3-1 Sequencies associated with $(\mathrm{WHT})_h$ coefficients

k	k_{binary}	$\langle k \rangle_{\text{binary}}$	$b(\langle k \rangle)_{\text{binary}}$	$b(\langle k \rangle)_{\text{decimal}}$	$(\mathrm{WHT})_h$ coefficient	$s_{B_x(k)}$
0	000	000	000	0	$B_x(0)$	0
1	001	100	111	7	$B_x(1)$	4
2	010	010	011	3	$B_x(2)$	2
3	011	110	100	4	$B_x(3)$	2
4	100	001	001	1	$B_x(4)$	1
5	101	101	110	6	$B_x(5)$	3
6	110	011	010	2	$B_x(6)$	1
7	111	111	101	5	$B_x(7)$	3

6.4 Walsh-ordered Walsh-Hadamard Transform $(\mathrm{WHT})_w$

In Table 6.3-1 it is observed that the $s_{B_x(k)}$ are *not* in natural increasing order; that is, they do not have the following sequency ordering:

Transform coefficient #:	0 1 2 3 4 5 6 7
Associated sequency:	0 1 1 2 2 3 3 4

In certain applications it may be desirable to have the sequencies associated with the transform coefficients in natural increasing order. An orthogonal transformation which has this property is the $(\mathrm{WHT})_w$. Hence the $(\mathrm{WHT})_w$ is also known as the sequency-ordered WHT.

Matrix definition. The $(\mathrm{WHT})_w$ of the data sequence $\{X(m)\} = \{X(0)\ X(1) \cdots X(N-1)\}$ is defined as

$$\boldsymbol{W}_x(n) = \frac{1}{N}\, \mathbf{H}_w(n)\, \boldsymbol{X}(n) \tag{6.4-1}$$

where $W_x(k)$ is the k-th $(\mathrm{WHT})_w$ coefficient, and

$$\boldsymbol{W}_x(n)' = [W_x(0)\ W_x(1) \cdots W_x(N-1)]$$

$\mathbf{H}_w(n)$ is the $(N \times N)$ Walsh-ordered Hadamard matrix. Since $\mathbf{H}_w(n)$ is orthogonal and symmetric, the inverse transform $(\mathrm{IWHT})_w$ is given by

$$\boldsymbol{X}(n) = \mathbf{H}_w(n)\, \boldsymbol{W}_x(n) \tag{6.4-2}$$

Example 6.4-1

Let $\{X(m)\} = \{1\ 2\ 1\ 1\ 3\ 2\ 1\ 2\}$. Evaluate $W_x(k)$, $k = 0, 1, \ldots, 7$.

Solution: Substituting $\mathbf{H}_w(3)$ given in Fig. 5.5b into Eq. (1), we obtain

$$\begin{bmatrix} W_x(0) \\ W_x(1) \\ W_x(2) \\ W_x(3) \\ W_x(4) \\ W_x(5) \\ W_x(6) \\ W_x(7) \end{bmatrix} = \frac{1}{8} \begin{bmatrix} 1 & 1 & 1 & 1 & 1 & 1 & 1 & 1 \\ 1 & 1 & 1 & 1 & -1 & -1 & -1 & -1 \\ 1 & 1 & -1 & -1 & -1 & -1 & 1 & 1 \\ 1 & 1 & -1 & -1 & 1 & 1 & -1 & -1 \\ 1 & -1 & -1 & 1 & 1 & -1 & -1 & 1 \\ 1 & -1 & -1 & 1 & -1 & 1 & 1 & -1 \\ 1 & -1 & 1 & -1 & -1 & 1 & -1 & 1 \\ 1 & -1 & 1 & -1 & 1 & -1 & 1 & -1 \end{bmatrix} \begin{bmatrix} 1 \\ 2 \\ 1 \\ 1 \\ 3 \\ 2 \\ 1 \\ 2 \end{bmatrix}$$

Evaluation of the above matrix equation leads to

$$\{W_x(k)\} = \left\{\frac{13}{8} \quad \frac{-3}{8} \quad \frac{-1}{8} \quad \frac{3}{8} \quad \frac{1}{8} \quad \frac{-3}{8} \quad \frac{-1}{8} \quad \frac{-1}{8}\right\}$$

From the above example it is evident that N^2 additions and subtractions are required to compute the $(\mathrm{WHT})_w$ coefficients $W_x(k)$, $k = 0, 1, \ldots, N-1$. In Section 6.5 we will develop an algorithm which yields the $W_x(k)$, $k = 0, 1, \ldots, N-1$ in $N \log_2 N$ additions and subtractions.

Exponential definition. Since the elements of $\mathbf{H}_w(n)$ can be generated using Eq. (5.4-4), it follows that the $(\mathrm{WHT})_w$ and $(\mathrm{IWHT})_w$ can alternately be defined as

$$W_x(u) = \frac{1}{N} \sum_{m=0}^{N-1} X(m)\,(-1)^{\langle m,\, r(u)\rangle}$$

and

$$X(m) = \sum_{u=0}^{N-1} W_x(u)\,(-1)^{\langle u,\, r(m)\rangle}, \qquad u, m = 0, 1, \ldots, N-1 \qquad (6.4\text{-}3)$$

where

$$\langle m, r(u)\rangle = \sum_{s=0}^{n-1} r_s(u)\, m_s\,, \qquad n = \log_2 N,$$

and

$$r_0(u) = u_{n-1}, \quad r_1(u) = u_{n-1} + u_{n-2}, \quad r_2(u) = u_{n-2} + u_{n-3}, \quad \ldots, \quad r_{n-1}(u) = u_1 + u_0$$

The terms u_s and m_s are the coefficients of the binary representations of u and m and are defined in Eqs. (6.2-9) and (6.2-10) respectively.

The relation between the $(\mathrm{WHT})_w$ and $(\mathrm{WHT})_h$ coefficients can be established in a straightforward manner. From the definitions of these transforms and from Eq. (6.3-7), it follows that

$$B_x(k) = W_x[b(\langle k\rangle)], \qquad k = 0, 1, \ldots, N-1 \qquad (6.4\text{-}4)$$

where $\langle k \rangle$ denotes the bit-reversal of k, and $b(\langle k \rangle)$ is the Gray code-to-binary conversion of $\langle k \rangle$.

For example, when $N = 8$, Eq. (4) yields (see Table **6.3-1**).

$$B_x(0) = W_x(0), \quad B_x(1) = W_x(7), \quad B_x(2) = W_x(3), \quad B_x(3) = W_x(4),$$
$$B_x(4) = W_x(1), \quad B_x(5) = W_x(6), \quad B_x(6) = W_x(2), \quad B_x(7) = W_x(5)$$

6.5 Fast Walsh-ordered Walsh-Hadamard Transform $(\text{FWHT})_w$

The $(\text{FWHT})_w$ is an algorithm that enables one to compute the $(\text{WHT})_w$ using additions and subtractions, ignoring multiplications by the $1/N$ factor in Eq. (6.4-1). One approach is to first compute the $(\text{WHT})_h$ coefficients $B_x(k)$, $k = 0, 1, \ldots, N-1$. Subsequently, the $(\text{WHT})_w$ coefficients can be obtained using Eq. (6.4-4). However, since this approach involves a Gray code-to-binary conversion in addition to a bit-reversal, it is not a very efficient procedure. Andrews and Pratt have developed another algorithm [8] which does not have a Cooley-Tukey type flow graph; it can be implemented in $\log_2 N$ additions and subtractions.

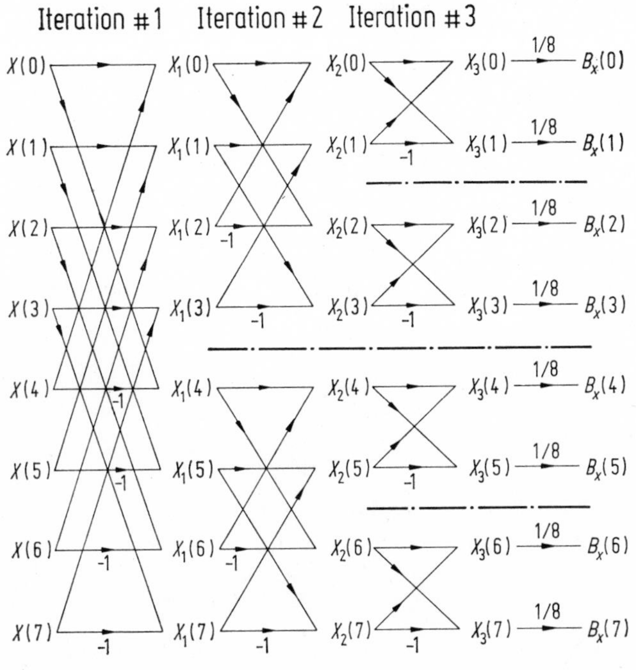

Fig. 6.3 $(\text{FWHT})_h$ signal flow graph, $N = 8$. The heavy broken lines separate blocks

In this section we shall consider a version of the $(\mathrm{FWHT})_w$ which was introduced by Manz [9]. This algorithm uses a Cooley-Tukey type signal flow graph and is implemented in $N \log_2 N$ additions and subtractions. It is essentially a simple modification of the $(\mathrm{FWHT})_h$, and is best illustrated for the case $N = 8$. For convenience, the $(\mathrm{FWHT})_h$ signal flow graph in Fig. 6.2 has been repeated in Fig. 6.3.

The first step is to bit-reverse the input and order it in ascending index order. If $\{X(m)\} = \{X(0)\, X(1) \cdots X(N-1)\}$ is the given data sequence, then we denote the bit-reversed sequence in ascending index order by $\{\hat{X}(m)\} = \{\hat{X}(0)\, \hat{X}(1) \cdots \hat{X}(N-1)\}$.

The second step is to define a "reversal", which is best illustrated by the simple case considered in Fig. 6.4.

Normally (i.e., without a reversal), we would have

$$\left.\begin{aligned} \hat{X}_{k+1}(s) &= \hat{X}_k(s) + \hat{X}_k(s+2) \\ \hat{X}_{k+1}(s+1) &= \hat{X}_k(s+1) + \hat{X}_k(s+3) \end{aligned}\right\} \text{additions}$$

and

$$\left.\begin{aligned} \hat{X}_{k+1}(s+2) &= \hat{X}_k(s) - \hat{X}_k(s+2) \\ \hat{X}_{k+1}(s+3) &= \hat{X}_k(s+1) - \hat{X}_k(s+3) \end{aligned}\right\} \text{subtractions} \qquad (6.5\text{-}1)$$

However, with a reversal, we have

$$\left.\begin{aligned} \hat{X}_{k+1}(s) &= \hat{X}_k(s) - \hat{X}_k(s+2) \\ \hat{X}_{k+1}(s+1) &= \hat{X}_k(s+1) - \hat{X}_k(s+3) \end{aligned}\right\} \text{subtractions}$$

and

$$\left.\begin{aligned} \hat{X}_{k+1}(s+2) &= \hat{X}_k(s) + \hat{X}_k(s+2) \\ \hat{X}_{k+1}(s+3) &= \hat{X}_k(s+1) + \hat{X}_k(s+3) \end{aligned}\right\} \text{additions} \qquad (6.5\text{-}2)$$

From Eqs. (1) and (2), it is apparent that a reversal causes the additions in an iteration of the signal flow graph to be replaced by subtractions, and vice versa.

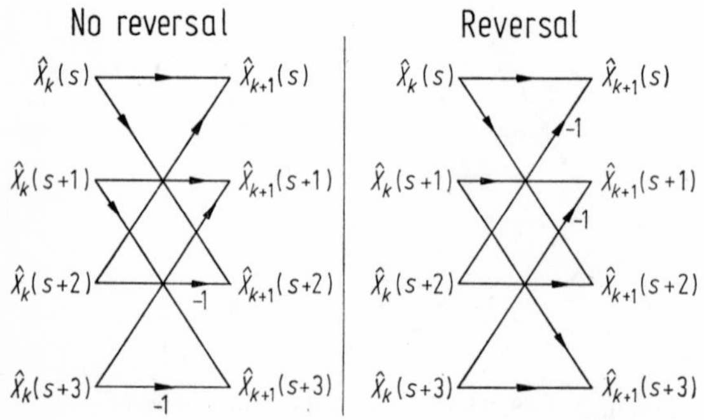

Fig. 6.4 Illustration of the effect of a reversal

The third step is to define a "block". A block is defined as a group of additions/subtractions which are disconnected from its neighbors, above or below. The blocks in Fig. 6.3 are separated by heavy broken lines. For example, iteration #1 has $2^0 = 1$ block; iteration #2 has 2 blocks, etc. In general, iteration #k has 2^{k-1} blocks.

Finally, we list the rules for locating the blocks where reversals occur.

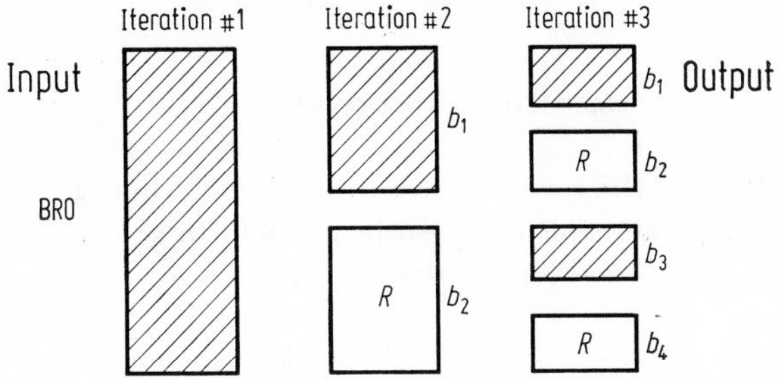

Fig. 6.5 Illustration of the positions of blocks in the $(\mathrm{FWHT})_w$ flow graph for $N = 8$. R indicates a block which has reversal; BRO means bit-reversed order

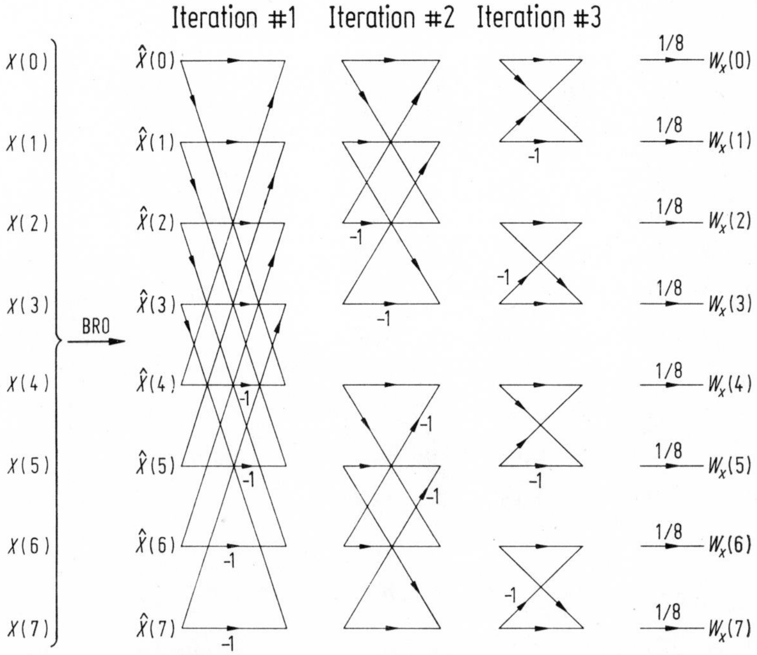

Fig. 6.6 $(\mathrm{FWHT})_w$ signal flow graph, $N = 8$

Rule 1: No reversals occur in iteration #1.

Rule 2: If b_m, $m = 1, 2, \ldots, 2^{k-1}$ denote the blocks in the k-th iteration, then the blocks with reversals are obtained as illustratd below, where R represents reversal.

$$\begin{array}{ccccccc} b_1 & b_2 & b_3 & b_4 & b_5 & b_6 \cdots & b_{2^{k-1}} \\ & \downarrow & & \downarrow & & \downarrow & \downarrow \\ & R & & R & & R & R \end{array}$$

In other words, *every other* block experiences a reversal, starting with block b_2.

Figure 6.5 shows the reversal pattern for $N = 8$ while the corresponding $(\mathrm{FWHT})_w$ signal flow graph is shown in Fig. 6.6. In general, the $(\mathrm{FWHT})_w$ signal flow graph has $n = \log_2 N$ iterations.

Example 6.5-1

Use the $(\mathrm{FWHT})_w$ to find the $(\mathrm{WHT})_w$ of $\{X(m)\} = \{1\ 2\ 1\ 1\ 3\ 2\ 1\ 2\}$.

Solution: If $\{\hat{X}(m)\} = \{\hat{X}(0)\ \hat{X}(1)\ \hat{X}(2)\ \hat{X}(3)\ \hat{X}(4)\ \hat{X}(5)\ \hat{X}(6)\ \hat{X}(7)\}$ denotes the sequence obtained by the bit-reversal of $\{X(m)\}$, then it follows that

$$\hat{X}(0) = X(0), \quad \hat{X}(1) = X(4), \quad \hat{X}(2) = X(2), \quad \hat{X}(3) = X(6),$$
$$\hat{X}(4) = X(1), \quad \hat{X}(5) = X(5), \quad \hat{X}(6) = X(3), \quad \hat{X}(7) = X(7)$$

Substituting for $\{X(m)\}$ and $\{\hat{X}(m)\}$ in Fig. 6.6, we obtain:

$$\begin{array}{llllllr} X(0) = 1 & \hat{X}(0) = 1 & X_1(0) = 3 & X_2(0) = 5 & X_3(0) = 13 & \xrightarrow{1/8} & W_x(0) \\ X(1) = 2 & \hat{X}(1) = 3 & X_1(1) = 5 & X_2(1) = 8 & X_3(1) = -3 & \xrightarrow{1/8} & W_x(1) \\ X(2) = 1 & \hat{X}(2) = 1 & X_1(2) = 2 & X_2(2) = 1 & X_3(2) = -1 & \xrightarrow{1/8} & W_x(2) \\ X(3) = 1 & \hat{X}(3) = 1 & X_1(3) = 3 & X_2(3) = 2 & X_3(3) = 3 & \xrightarrow{1/8} & W_x(3) \\ X(4) = 3 & \hat{X}(4) = 2 & X_1(4) = -1 & X_2(4) = -1 & X_3(4) = 1 & \xrightarrow{1/8} & W_x(4) \\ X(5) = 2 & \hat{X}(5) = 2 & X_1(5) = 1 & X_2(5) = 2 & X_3(5) = -3 & \xrightarrow{1/8} & W_x(5) \\ X(6) = 1 & \hat{X}(6) = 1 & X_1(6) = 0 & X_2(6) = -1 & X_3(6) = -1 & \xrightarrow{1/8} & W_x(6) \\ X(7) = 2 & \hat{X}(7) = 2 & X_1(7) = -1 & X_2(7) = 0 & X_3(7) = -1 & \xrightarrow{1/8} & W_x(7) \end{array}$$

Thus
$$\{W_x(k)\} = \left\{\frac{13}{8}\ \frac{-3}{8}\ \frac{-1}{8}\ \frac{3}{8}\ \frac{1}{8}\ \frac{-3}{8}\ \frac{-1}{8}\ \frac{-1}{8}\right\}$$

These values of $W_x(k)$ are the same as those obtained by evaluating a matrix product in Example 6.4-1.

6.6 Cyclic and Dyadic Shifts

Cyclic shift. Let $\{X(m)\}$ be a real-valued N-periodic sequence. Consider a sequence $\{Z(m)\}_l$ such that

$$\{Z(m)\}_l = \{Z(0)\ Z(1) \cdots Z(N-1)\} \tag{6.6-1}$$

where [see Eq. (3.2-3)]

$$Z(m) = X(m+l), \qquad m = 0, 1, \ldots, N-1$$

For example

$$\{Z(m)\}_1 = \{X(1) \quad X(2) \cdots X(N-2) \quad X(N-1) \quad X(0)\},$$
$$\{Z(m)\}_2 = \{X(2) \quad X(3) \cdots X(N-1) \quad X(0) \quad X(1)\}, \text{ etc.}$$

Since the arguments of $\{X(m)\}$ and $\{Z(m)\}_l$ are interpreted as modulo N, $\{Z(m)\}_l$ is sometimes referred to as being obtained by means of a *cyclic* (or *circular*) shift of $\{X(m)\}$. The size of the shift is l.

Now, if $C_{z,l}(k)$ denotes the k-th DFT coefficient of $\{Z(m)\}_l$, then from the DFT shift theorem [see Eq. (3.2-4)], we have

$$C_{z,l}(k) = W^{-kl}\, C_x(k), \qquad k = 0, 1, \ldots, N-1$$

That is,

$$|C_{z,l}(k)|^2 = |C_x(k)|^2, \qquad k = 0, 1, \ldots, N-1 \tag{6.6-2}$$

Equation (2) implies that $|C_x(k)|^2$ is independent of l, and hence *invariant* to cyclic shifts of $\{X(m)\}$.

Dyadic shift. If $\{\tilde{Z}(m)\}_l$ denotes the sequence obtained by subjecting $\{X(m)\}$ to a *dyadic* shift of size l, then $\{\tilde{Z}(m)\}_l$ is defined as

$$\{\tilde{Z}(m)\}_l = \{\tilde{Z}(0) \quad \tilde{Z}(1) \cdots \tilde{Z}(N-1)\} \tag{6.6-3}$$

where $\tilde{Z}(m) = X(m \oplus l)$, $m = 0, 1, \ldots, N-1$, and $\oplus$ denotes modulo 2 addition.

As an illustrative example, the modulo 2 addition in Eq. (3) is evaluated for $N = 8$, and the results are summarized in Table 6.6-1.

Table 6.6-1

($\rightarrow m$, $\downarrow l$)

$m \oplus l$	0	1	2	3	4	5	6	7
0	0	1	2	3	4	5	6	7
1	1	0	3	2	5	4	7	6
2	2	3	0	1	6	7	4	5
3	3	2	1	0	7	6	5	4
4	4	5	6	7	0	1	2	3
5	5	4	7	6	1	0	3	2
6	6	7	4	5	2	3	0	1
7	7	6	5	4	3	2	1	0

For example, using this table we obtain

$$\begin{aligned}
\{\tilde{Z}(m)\}_1 &= \{X(1) \quad X(0) \quad X(3) \quad X(2) \quad X(5) \quad X(4) \quad X(7) \quad X(6)\}\,,\\
\{\tilde{Z}(m)\}_2 &= \{X(2) \quad X(3) \quad X(0) \quad X(1) \quad X(6) \quad X(7) \quad X(4) \quad X(5)\}\,,\\
\{\tilde{Z}(m)\}_3 &= \{X(3) \quad X(2) \quad X(1) \quad X(0) \quad X(7) \quad X(6) \quad X(5) \quad X(4)\}\,,\\
\{\tilde{Z}(m)\}_4 &= \{X(4) \quad X(5) \quad X(6) \quad X(7) \quad X(0) \quad X(1) \quad X(2) \quad X(3)\}\,, \text{ etc.}
\end{aligned} \tag{6.6-4}$$

Corresponding to Eq. (2) it can be shown that

$$B^2_{\tilde{z},l}(k) = B^2_x(k)$$

and

$$W^2_{\tilde{z},l}(k) = W^2_x(k), \qquad k = 0, 1, \ldots, N-1 \tag{6.6-5}$$

where $B_{\tilde{z},l}(k)$ and $W_{\tilde{z},l}(k)$ are the k-th $(\mathrm{WHT})_h$ and $(\mathrm{WHT})_w$ coefficients respectively.

Equation (5) states that $B^2_x(k)$ and $W^2_x(k)$ are independent of l, and hence *invariant* to dyadic shifts of $\{X(m)\}$. For the purposes of illustration, we show that Eq. (5) is valid for $N = 8$ and $l = 1$.

The $(\mathrm{WHT})_h$ of $\{\tilde{Z}(m)\}_1$ is given by

$$\boldsymbol{B}_{\tilde{z},1}(3) = \frac{1}{8}\mathbf{H}_h(3)\,\tilde{\boldsymbol{Z}}(3)_1 \tag{6.6-6}$$

where

$$\boldsymbol{B}_{\tilde{z},1}(3)' = [B_{\tilde{z},1}(0)\; B_{\tilde{z},1}(1) \cdots B_{\tilde{z},1}(7)]$$

and

$$\tilde{\boldsymbol{Z}}(3)_1' = [X(1) \quad X(0) \quad X(3) \quad X(2) \quad X(5) \quad X(4) \quad X(7) \quad X(6)]$$

From Eq. (4) it follows that

$$\tilde{\boldsymbol{Z}}(3)_1 = \tilde{\mathbf{M}}(3)\,\boldsymbol{X}(3) \tag{6.6-7}$$

where

$$\tilde{\mathbf{M}}(3) = \left[\begin{array}{cccc:cccc}
0 & 1 & 0 & 0 & 0 & 0 & 0 & 0\\
1 & 0 & 0 & 0 & 0 & 0 & 0 & 0\\
0 & 0 & 0 & 1 & 0 & 0 & 0 & 0\\
0 & 0 & 1 & 0 & 0 & 0 & 0 & 0\\
\hdashline
0 & 0 & 0 & 0 & 0 & 1 & 0 & 0\\
0 & 0 & 0 & 0 & 1 & 0 & 0 & 0\\
0 & 0 & 0 & 0 & 0 & 0 & 0 & 1\\
0 & 0 & 0 & 0 & 0 & 0 & 1 & 0
\end{array}\right]$$

Substituting Eq. (7) in Eq. (6) and using the relation

$$\boldsymbol{X}(3) = \mathbf{H}_h(3)\,\boldsymbol{B}_x(3)$$

we obtain

$$\boldsymbol{B}_{\tilde{z},1}(3) = \frac{1}{8}\mathbf{H}_h(3)\tilde{\boldsymbol{M}}(3)\mathbf{H}_h(3)\boldsymbol{B}_x(3) \tag{6.6-8}$$

The matrix product $\tilde{\mathbf{A}}(3) = \frac{1}{8}\mathbf{H}_h(3)\ \tilde{\mathbf{M}}(3)\ \mathbf{H}_h(3)$ is called a *similarity transformation* corresponding to the (WHT)$_h$. Substituting for $\mathbf{H}_h(3)$ (see Fig. 5.7b) and subsequently evaluating $\tilde{\mathbf{A}}(3)$, one obtains

$$\tilde{\mathbf{A}}(3) = \begin{bmatrix} 1 & & & & & & & 0 \\ & -1 & & & & & & \\ & & 1 & & & & & \\ & & & -1 & & & & \\ & & & & 1 & & & \\ & & & & & -1 & & \\ & & & & & & 1 & \\ 0 & & & & & & & -1 \end{bmatrix}$$

Clearly, since $\tilde{\mathbf{A}}(3)$ is a *diagonal* matrix, Eq. (8) yields

$$B_{\tilde{z},1}(k) = (-1)^k B_x(k), \qquad k = 0, 1, \ldots, 7,$$

which implies that

$$B^2_{\tilde{z},1}(k) = B^2_x(k), \qquad k = 0, 1, \ldots, 7 \tag{6.6-9}$$

Thus $B^2_x(k)$ is independent of l, and hence is invariant to dyadic shifts of $\{X(m)\}$. The diagonal form of the similarity transformation is also valid for other values of l.

In conclusion, we observe that Eq. (9) also implies that $W^2_x(k)$ is invariant to dyadic shifts of $\{X(m)\}$, since $B_x(k)$ and $W_x(k)$ are uniquely related by virtue of Eq. (6.4-4).

6.7 (WHT)$_w$ Spectra [10–12]

(WHT)$_w$ power spectrum. We recall that the DFT power spectrum of a real-valued data sequence $\{X(m)\}$ is defined as

$$P(k) = |C_x(k)|^2, \qquad k = 0, 1, \ldots, N/2 \tag{6.7-1}$$

where $C_x(k)$ is the k-th DFT coefficient. Again, by definition,

$$C_x(k) = R_x(k) + iI_x(k) \tag{6.7-2}$$

where $R_x(k)$ and $I_x(k)$ are the real and imaginary parts of $C_x(k)$. It is observed that $R_x(k)$ and $I_x(k)$ correspond to the cosine and sine coefficients in a Fourier representation. Thus by analogy, a $(\mathrm{WHT})_w$ power spectrum is defined as the sum of the squares of the cal and sal coefficients in the Walsh representation. For example, when $N = 8$, it follows that [see Fig. 6.1] the $(\mathrm{WHT})_w$ power spectrum is given by

$$P_w(0) = W_x^2(0)$$
$$P_w(1) = W_x^2(1) + W_x^2(2)$$
$$P_w(2) = W_x^2(3) + W_x^2(4)$$
$$P_w(3) = W_x^2(5) + W_x^2(6)$$

and

$$P_w(4) = W_x^2(7)$$

Thus the general form of the above spectrum is as follows:

$$P_w(0) = W_x^2(0)$$
$$P_w(s) = W_x^2(2s-1) + W_x^2(2s), \qquad s = 1, 2, \ldots, N/2 - 1$$

and

$$P_w(N/2) = W_x^2(N-1) \tag{6.7-3}$$

where $P_w(s)$ is the $(\mathrm{WHT})_w$ power spectral point associated with sequency s. Clearly, this spectrum has $(N/2 + 1)$ points.

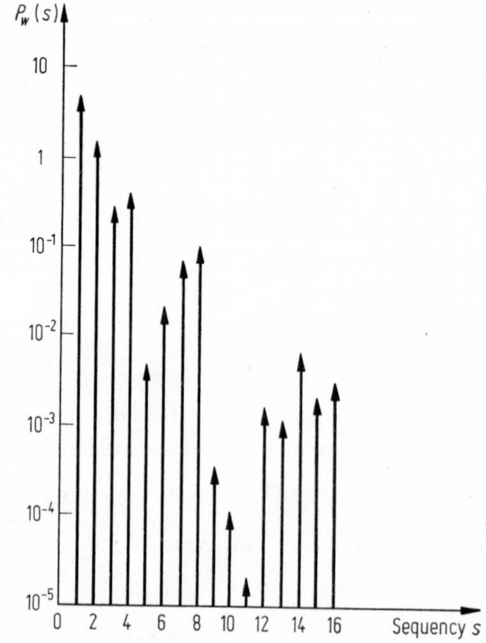

Fig. 6.7 $(\mathrm{WHT})_w$ power spectrum of an exponentially decaying sequence

From Eqs. (6.6-5) and (3) it follows that the $(\text{WHT})_w$ power spectrum is invariant to dyadic shift of $\{X(m)\}$. As an illustrative example, the $(\text{WHT})_w$ power spectrum of the exponentially decaying sequence

$$X(m) = 10e^{-4m/31}, \qquad m = 0, 1, \ldots, 31$$

is shown in Fig. 6.7.

$(\text{WHT})_w$ phase spectrum. By analogy with respect to the DFT phase spectrum, the $(\text{WHT})_w$ phase spectrum is defined as

$$\psi_w(0) = 0, \pi$$

$$\psi_w\left(\frac{N}{2}\right) = 2k\pi \pm \frac{\pi}{2}, \qquad k = 0, 1, 2, \ldots$$

and

$$\psi_w(s) = \tan^{-1}\left\{\frac{W_x(2s-1)}{W_x(2s)}\right\}, \qquad s = 1, 2, \ldots, N/2 - 1 \tag{6.7-4}$$

6.8 (WHT)$_h$ Spectra [4, 13]

$(\text{WHT})_h$ power spectrum. While the $(\text{WHT})_w$ power spectrum is invariant to dyadic shifts of the data sequence $\{X(m)\}$, one can develop a $(\text{WHT})_h$ power spectrum which is invariant to its cyclic shifts. This development is best illustrated for $N = 8$. Let $\{Z(m)\}_l$ denote $\{X(m)\}$ shifted cyclically to the left by l positions. That is,

$$\{Z(m)\}_l = \{X(l)\ X(l+1) \cdots X(l-2)\ X(l-1)\}, \qquad l = 1, 2, \ldots, 7$$

With $l = 1$, we have

$$\boldsymbol{Z}(3)_1 = \mathbf{M}(3)\,\boldsymbol{X}(3) \tag{6.8-1}$$

where

$$\mathbf{M}(3) = \begin{bmatrix} 0 & 1 & 0 & 0 & 0 & 0 & 0 & 0 \\ 0 & 0 & 1 & 0 & 0 & 0 & 0 & 0 \\ 0 & 0 & 0 & 1 & 0 & 0 & 0 & 0 \\ 0 & 0 & 0 & 0 & 1 & 0 & 0 & 0 \\ 0 & 0 & 0 & 0 & 0 & 1 & 0 & 0 \\ 0 & 0 & 0 & 0 & 0 & 0 & 1 & 0 \\ 0 & 0 & 0 & 0 & 0 & 0 & 0 & 1 \\ 1 & 0 & 0 & 0 & 0 & 0 & 0 & 0 \end{bmatrix}$$

The $(\text{WHT})_h$ of $\{Z(m)\}_1$ is given by

$$\boldsymbol{B}_{z,1}(3) = \frac{1}{8}\,\mathbf{H}_h(3)\,\mathbf{M}(3)\,\boldsymbol{X}(3) \tag{6.8-2}$$

That is

$$\boldsymbol{B}_{z,1}(3) = \mathbf{A}(3)\,\boldsymbol{B}_x(3) \tag{6.8-3}$$

where $\mathbf{A}(3)$ is a similarity transformation given by

$$\mathbf{A}(3) = \frac{1}{8}\,\mathbf{H}_h(3)\,\mathbf{M}(3)\,\mathbf{H}_h(3) \tag{6.8-4}$$

Evaluation of $\mathbf{A}(3)$ in Eq. (4) results in

$$\mathbf{A}(3) = \frac{1}{2}\left[\begin{array}{cc:cc:cccc} 2 & 0 & \multicolumn{2}{c:}{\mathbf{0}_2} & & & & \\ 0 & -2 & & & \multicolumn{4}{c}{\mathbf{0}_4} \\ \hdashline & & 0 & -2 & & & & \\ \multicolumn{2}{c:}{\mathbf{0}_2} & 2 & 0 & & & & \\ \hdashline & & & & 1 & -1 & -1 & -1 \\ & & & & 1 & -1 & 1 & 1 \\ \multicolumn{4}{c:}{\mathbf{0}_4} & 1 & 1 & 1 & -1 \\ & & & & -1 & -1 & 1 & -1 \end{array}\right]$$

Repetitive application of Eq. (4) yields

$$\boldsymbol{B}_{z,l}(3) = \mathbf{A}(3)^l\,\boldsymbol{B}_x(3), \qquad l = 1, 2, \ldots, 7 \tag{6.8-5}$$

We observe that $\mathbf{A}(3)$ is made up of square matrices of increasing order along the diagonal. From this "block diagonal" structure of $\mathbf{A}(3)$ and Eq. (5), the following set of equations is obtained:

$$B_{z,l}(0) = B_x(0)$$

$$B_{z,l}(1) = (-1)^l\,B_x(1)$$

$$\begin{bmatrix} B_{z,l}(2) \\ B_{z,l}(3) \end{bmatrix} = \mathbf{D}(1)^l \begin{bmatrix} B_x(2) \\ B_x(3) \end{bmatrix}$$

and

$$\begin{bmatrix} B_{z,l}(4) \\ B_{z,l}(5) \\ B_{z,l}(6) \\ B_{z,l}(7) \end{bmatrix} = \mathbf{D}(2)^l \begin{bmatrix} B_x(4) \\ B_x(5) \\ B_x(6) \\ B_x(7) \end{bmatrix}$$

where $\mathbf{D}(1) = \begin{bmatrix} 0 & -1 \\ 1 & 0 \end{bmatrix}$, and $\mathbf{D}(2) = \frac{1}{2}\begin{bmatrix} 1 & -1 & -1 & -1 \\ 1 & -1 & 1 & 1 \\ 1 & 1 & 1 & -1 \\ -1 & -1 & 1 & -1 \end{bmatrix}$ (6.8-6)

are orthogonal. That is, $\mathbf{D}(1)'\,\mathbf{D}(1) = \mathbf{I}(1)$, and $\mathbf{D}(2)'\,\mathbf{D}(2) = \mathbf{I}(2)$.

Equation (6) implies that

$$B_{z,l}^2(k) = B_x^2(k), \qquad k = 0, 1;$$

$$\sum_{k=2}^{3} B_{z,l}^2(k) = \sum_{k=2}^{3} B_x^2(k),$$

and

$$\sum_{k=4}^{7} B_{z,l}^2(k) = \sum_{k=4}^{7} B_x^2(k), \qquad l = 1, 2, \ldots, 7 \tag{6.8-7}$$

Clearly, the quantities on the right hand side of Eq. (7) are independent of l, and hence invariant to cyclic shifts of $\{X(m)\}$. In other words, they constitute the (WHT)$_h$ power spectrum for $N = 8$. In general, we denote this spectrum as follows:

$$P_h(0) = B_x^2(0)$$

and

$$P_h(r) = \sum_{k=2^{r-1}}^{2^r-1} B_x^2(k), \qquad r = 1, 2, \ldots, n; \quad n = \log_2 N \tag{6.8-8}$$

From Eq. (8) it follows that the (WHT)$_h$ power spectrum consists of $n + 1$ points.

Computational considerations. By suitably modifying the (FWHT)$_h$, the power spectrum can be computed without having to actually compute *all* the coefficients $B_x(k)$. The modification is illustrated for the case when $N = 8$. From the (FWHT)$_h$ signal flow graph in Fig. 6.2, it follows that

$$B_x(0) = \frac{1}{8} X_3(0)$$

$$B_x(1) = \frac{1}{8} X_3(1)$$

$$\begin{bmatrix} B_x(2) \\ B_x(3) \end{bmatrix} = \frac{1}{8} \mathbf{H}_h(1) \begin{bmatrix} X_2(2) \\ X_2(3) \end{bmatrix}$$

and

$$\begin{bmatrix} B_x(4) \\ B_x(5) \\ B_x(6) \\ B_x(7) \end{bmatrix} = \frac{1}{8} \mathbf{H}_h(2) \begin{bmatrix} X_1(4) \\ X_1(5) \\ X_1(6) \\ X_1(7) \end{bmatrix} \tag{6.8-9}$$

Since $\mathbf{H}_h(k)$ is orthogonal, Eq. (9) yields

$$\sum_{k=2}^{3} B_x^2(k) = \frac{2}{8^2} \sum_{m=2}^{3} X_2^2(m)$$

and

$$\sum_{k=4}^{7} B_x^2(k) = \frac{4}{8^2} \sum_{m=4}^{7} X_1^2(m)$$

The power spectrum then becomes

$$P_h(0) = \frac{1}{8^2} X_3^2(0)$$

$$P_h(1) = \frac{1}{8^2} X_3^2(1)$$

$$P_h(2) = \frac{2}{8^2} \sum_{m=2}^{3} X_2^2(m)$$

and

$$P_h(3) = \frac{2^2}{8^2} \sum_{m=4}^{7} X_1^2(m) \tag{6.8-10}$$

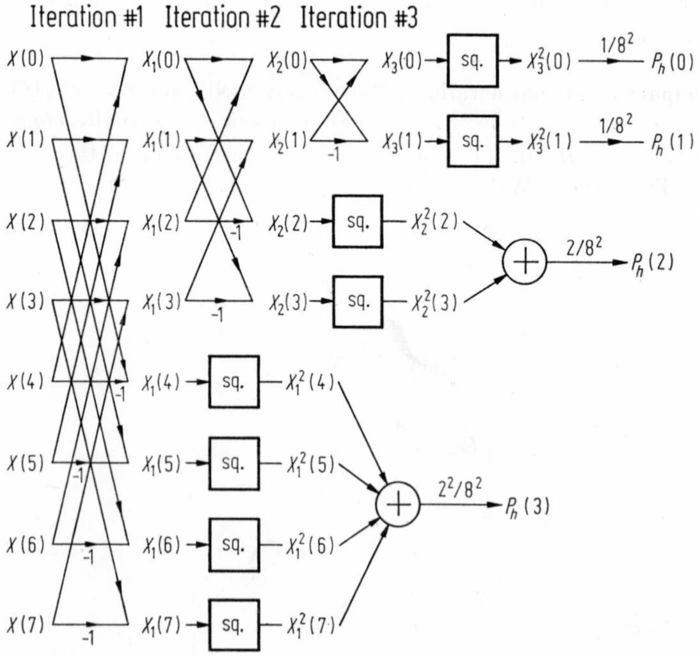

Fig. 6.8 $(\text{WHT})_h$ power spectrum signal flow graph, $N = 8$

A generalization of the above $(\mathrm{FWHT})_h$ modification leads to the following expression for the $(\mathrm{WHT})_h$ power spectrum.

$$P_h(0) = \frac{X_n^2(0)}{N^2}$$

and

$$P_h(r) = \frac{2^{r-1}}{N^2} \sum_{m=2^{r-1}}^{2^r-1} X_{n+1-r}^2(m), \qquad r = 1, 2, \ldots, n \qquad (6.8\text{-}11)$$

The signal flow graph corresponding to Eq. (11) for $N = 8$ is shown in Fig. 6.8.

6.9 Physical Interpretations for the $(\mathrm{WHT})_h$ Power Spectrum [13]

The $(\mathrm{WHT})_h$ power spectrum has the following interesting physical interpretations:

(1) The spectral points $P_h(r)$ represent the average powers in a set of $(n + 1)$ mutually orthogonal subsequences.

(2) Each $P_h(r)$ represents the power content of a *group* of sequencies, rather than that of a *single* sequency, as is the case with a $(\mathrm{WHT})_w$ spectral point.

To show the first interpretation, we use the result that an N-periodic sequence $\{X(m)\}$ can be decomposed into an $N/2$-periodic sequence $\{F_1(m)\}$, and an $N/2$-antiperiodic[1] sequence $\{G_1(m)\}$ as follows:

$$\{X(m)\} = \{F_1(m)\} + \{G_1(m)\}$$

where

$$F_1(m) = \frac{1}{2}\left\{X(m) + X\left(m + \frac{N}{2}\right)\right\}$$

and

$$G_1(m) = \frac{1}{2}\left\{X(m) - X\left(m + \frac{N}{2}\right)\right\}, \qquad m = 0, 1, \ldots, N/2 - 1$$

Now, $\{F_1(m)\}$ can be further decomposed into an $N/4$-periodic sequence $\{F_2(m)\}$ and an $N/4$-antiperiodic sequence $\{G_2(m)\}$; i.e.,

$$\{F_1(m)\} = \{F_2(m)\} + \{G_2(m)\}$$

where

$$F_2(m) = \frac{1}{2}\left\{F_1(m) + F_1\left(m + \frac{N}{4}\right)\right\}$$

and

$$G_2(m) = \frac{1}{2}\left\{F_1(m) - F_1\left(m + \frac{N}{4}\right)\right), \qquad m = 0, 1, \ldots, N/4 - 1$$

Continuing this process, it follows that $\{X(m)\}$ can be decomposed into the following $n + 1$ subsequences:

$$\{X(m)\} = \{F_n(m)\} + \{G_n(m)\} + \{G_{n-1}(m)\} + \cdots + \{G_1(m)\} \qquad (6.9\text{-}1)$$

[1] A sequence is said to be M-antiperiodic if $X(m) = -X(M + m)$.

where $\{F_n(m)\}$ is a 1-periodic sequence and $\{G_{n-r}(m)\}$ is a 2^r-antiperiodic sequence, $r = 0, 1, \ldots, n - 1$.

The above decomposition process is illustrated in Fig. 6.9 for $N = 8$. $\{X(m)\}$ is decomposed in the form

$$\{X(m)\} = \{F_3(m)\} + \{G_3(m)\} + \{G_2(m)\} + \{G_1(m)\}$$

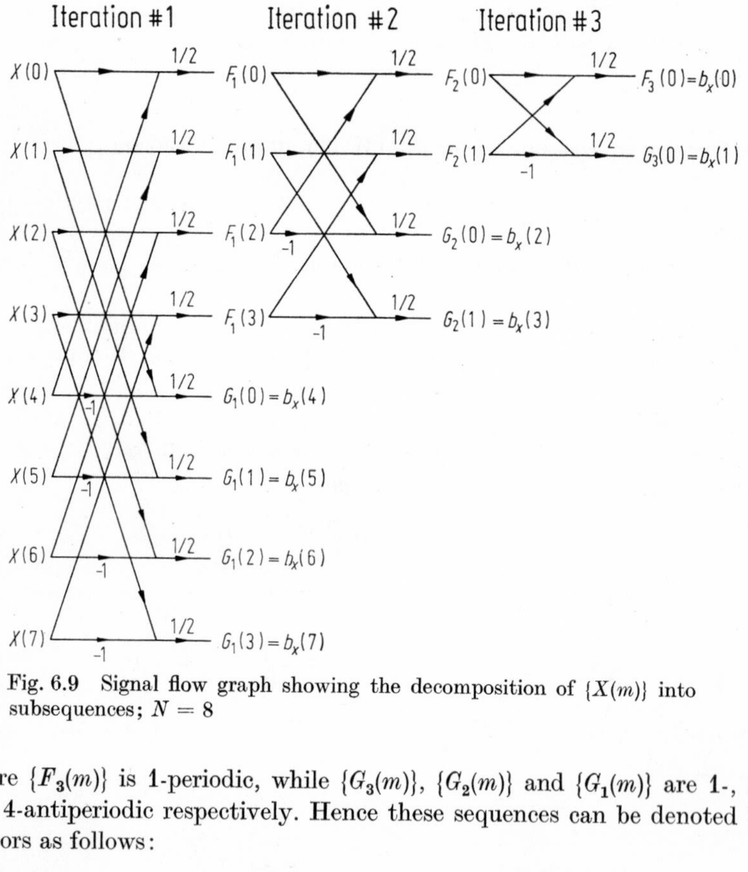

Fig. 6.9 Signal flow graph showing the decomposition of $\{X(m)\}$ into subsequences; $N = 8$

where $\{F_3(m)\}$ is 1-periodic, while $\{G_3(m)\}$, $\{G_2(m)\}$ and $\{G_1(m)\}$ are 1-, 2-, and 4-antiperiodic respectively. Hence these sequences can be denoted by vectors as follows:

$$\begin{aligned}
\boldsymbol{F}_3(3)' &= [b_x(0) \quad b_x(0) \quad b_x(0) \quad b_x(0) \quad b_x(0) \quad b_x(0) \quad b_x(0) \quad b_x(0)] \\
\boldsymbol{G}_3(3)' &= [b_x(1) \quad -b_x(1) \quad b_x(1) \quad -b_x(1) \quad b_x(1) \quad -b_x(1) \quad b_x(1) \quad -b_x(1)] \\
\boldsymbol{G}_2(3)' &= [b_x(2) \quad b_x(3) \quad -b_x(2) \quad -b_x(3) \quad b_x(2) \quad b_x(3) \quad -b_x(2) \quad -b_x(3)] \\
\boldsymbol{G}_1(3)' &= [b_x(4) \quad b_x(5) \quad b_x(6) \quad b_x(7) \quad -b_x(4) \quad -b_x(5) \quad -b_x(6) \quad -b_x(7)]
\end{aligned} \tag{6.9-2}$$

where the coefficients $b_x(k)$ will be expressed in terms of the $(\mathrm{WHT})_w$ coefficients at a later stage.

Inspection of the vectors in Eq. (2) shows that they are all *mutually orthogonal* and hence it follows that

$$\|\boldsymbol{X}(3)\|^2 = \|\boldsymbol{F}_3(3)\|^2 + \|\boldsymbol{G}_3(3)\|^2 + \|\boldsymbol{G}_2(3)\|^2 + \|\boldsymbol{G}_1(3)\|^2 \qquad (6.9\text{-}3)$$

where $\|\cdot\|$ denotes the norm of the vector enclosed. Also

$$\|\boldsymbol{F}_3(3)\|^2 = 8b_x^2(0), \qquad \|\boldsymbol{G}_3(3)\|^2 = 8b_x^2(1)$$

$$\|\boldsymbol{G}_2(3)\|^2 = 4\sum_{k=2}^{3} b_x^2(k)$$

and

$$\|\boldsymbol{G}_1(3)\|^2 = 2\sum_{k=4}^{7} b_x^2(k) \qquad (6.9\text{-}4)$$

Now, we recall that the average power of $\{X(m)\}$ is given by

$$P_{av} = \frac{1}{8}\sum_{m=0}^{7} X^2(m) = \frac{1}{8}\|\boldsymbol{X}(3)\|^2$$

Substitution of Eq. (4) in Eq. (3) leads to

$$P_{av} = b_x^2(0) + b_x^2(1) + \frac{1}{2}\sum_{k=2}^{3} b_x^2(k) + \frac{1}{4}\sum_{k=4}^{7} b_x^2(k) \qquad (6.9\text{-}5)$$

To relate the coefficients $b_x(k)$ to the corresponding coefficients $B_x(k)$, the $(\mathrm{WHT})_h$ of $\{X(m)\}$ is obtained, where $\{X(m)\}$ is expressed in terms of its decomposed subsequences. This results in the matrix equation

$$\begin{bmatrix} B_x(0) \\ B_x(1) \\ B_x(2) \\ B_x(3) \\ B_x(4) \\ B_x(5) \\ B_x(6) \\ B_x(7) \end{bmatrix} = \frac{1}{8}\left[\begin{array}{cc:cc:cc:cc} 1 & 1 & 1 & 1 & 1 & 1 & 1 & 1 \\ 1 & -1 & 1 & -1 & 1 & -1 & 1 & -1 \\ \hdashline \multicolumn{2}{c:}{\mathbf{H}_h(1)} & \multicolumn{2}{c:}{-\mathbf{H}_h(1)} & \multicolumn{2}{c:}{\mathbf{H}_h(1)} & \multicolumn{2}{c}{-\mathbf{H}_h(1)} \\ \hdashline \multicolumn{4}{c:}{\mathbf{H}_h(2)} & \multicolumn{4}{c}{-\mathbf{H}_h(2)} \end{array}\right]$$

$$\times \left[\begin{bmatrix} b_x(0) \\ b_x(0) \\ b_x(0) \\ b_x(0) \\ b_x(0) \\ b_x(0) \\ b_x(0) \\ b_x(0) \end{bmatrix} + \begin{bmatrix} b_x(1) \\ -b_x(1) \\ b_x(1) \\ -b_x(1) \\ b_x(1) \\ -b_x(1) \\ b_x(1) \\ -b_x(1) \end{bmatrix} + \begin{bmatrix} b_x(2) \\ b_x(3) \\ -b_x(2) \\ -b_x(3) \\ b_x(2) \\ b_x(3) \\ -b_x(2) \\ -b_x(3) \end{bmatrix} + \begin{bmatrix} b_x(4) \\ b_x(5) \\ b_x(6) \\ b_x(7) \\ -b_x(4) \\ -b_x(5) \\ -b_x(6) \\ -b_x(7) \end{bmatrix}\right] \qquad (6.9\text{-}6)$$

From Eq. (6), it follows that

$$B_x(0) = b_x(0), \qquad B_x(1) = b_x(1)$$

$$\begin{bmatrix} B_x(2) \\ B_x(3) \end{bmatrix} = \frac{1}{2}\mathbf{H}_h(1)\begin{bmatrix} b_x(2) \\ b_x(3) \end{bmatrix}$$

$$\begin{bmatrix} B_x(4) \\ B_x(5) \\ B_x(6) \\ B_x(7) \end{bmatrix} = \frac{1}{4}\mathbf{H}_h(2)\begin{bmatrix} b_x(4) \\ b_x(5) \\ b_x(6) \\ b_x(7) \end{bmatrix} \tag{6.9-7}$$

Since $\mathbf{H}_h(1)$ and $\mathbf{H}_h(2)$ are orthogonal, Eq. (7) yields

$$\sum_{k=2}^{3} b_x^2(k) = 2\sum_{k=2}^{3} B_x^2(k)$$

and

$$\sum_{k=4}^{7} b_x^2(k) = 4\sum_{k=4}^{7} B_x^2(k)$$

Thus

$$\begin{aligned} P_{av} &= B_x^2(0) + B_x^2(1) + \sum_{k=2}^{3} B_x^2(k) + \sum_{k=4}^{7} B_x^2(k) \\ &= P_h(0) + P_h(1) + P_h(2) + P_h(3) \end{aligned} \tag{6.9-8}$$

Clearly, $P_h(0)$ represents the average power in the 1-periodic sequence $\{F_3(m)\}$, while $P_h(1)$, $P_h(2)$ and $P_h(3)$ represent the average powers in $\{G_3(m)\}$, $\{G_2(m)\}$ and $\{G_1(m)\}$, which are 1-, 2-, and 4-antiperiodic respectively.

A generalization of the above analysis leads to the following conclusions:

(i) $P_h(0)$ represents the average power in the 1-periodic sequence $\{F_n(m)\}$.
(ii) $P_h(r)$ represents the average power in the 2^r-antiperiodic sequence $\{G_{n+1-r}(m)\}$, $r = 1, 2, \ldots, n$.

To illustrate the second of the above physical interpretations, we examine the sequency content of the $(\mathrm{WHT})_h$ power spectrum for $N = 8$. From Eq. (6.8-8) we obtain

$$\begin{aligned} P_h(0) &= B_x^2(0) \\ P_h(1) &= B_x^2(1) \\ P_h(2) &= B_x^2(2) + B_x^2(3) \\ P_h(3) &= B_x^2(4) + B_x^2(5) + B_x^2(6) + B_x^2(7) \end{aligned} \tag{6.9-9}$$

Now, from Table 6.3-1 we have

$$\begin{aligned}
s_{B_x(0)} &= 0\\
s_{B_x(1)} &= 4\\
s_{B_x(2)} &= s_{B_x(3)} = 2\\
s_{B_x(4)} &= s_{B_x(6)} = 1\\
s_{B_x(5)} &= s_{B_x(7)} = 3
\end{aligned} \tag{6.9-10}$$

If $F[P_h(r)]$ denotes the sequency content of the spectral point $P_h(r)$, then from Eqs. (9) and (10) it follows that

$$\begin{aligned}
F[P_h(0)] &= 0\\
F[P_h(3)] &= 1, 3\\
F[P_h(2)] &= 2\\
F[P_h(1)] &= 4
\end{aligned} \tag{6.9-11}$$

It can be shown that the general form of Eq. (11) is given by

$$\begin{aligned}
F[P_h(0)] &= 0\\
F[P_h(n)] &= 1, 3, 5, \ldots, N/2 - 1\\
F[P_h(n-1)] &= 2, 6, 10, \ldots, N/2 - 2\\
F[P_h(n-2)] &= 4, 12, 20, \ldots, N/2 - 4\\
&\vdots\\
F[P_h(n-k)] &= 2^k, 3 \cdot 2^k, 5 \cdot 2^k, \ldots, N/2 - 2^k\\
&\vdots\\
F[P_h(1)] &= N/2
\end{aligned} \tag{6.9-12}$$

Clearly, each spectral point $P_h(k)$ represents the power content in a group of sequencies rather than that in a single sequency, as is the case with the (WHT)$_w$ spectrum. This sequency grouping, however, is not arbitrary. Each group consists of a fundamental and the set of all odd sequencies relative to that fundamental.

Relation to the DFT spectrum. A natural sequel to the above discussion is the relationship between the (WHT)$_h$ and DFT power spectra. We consider the case $N = 8$. Then the (WHT)$_h$ power spectrum and FFT signal flow graphs are as shown in Fig. 6.10.

From the FFT flow graph, it follows that

$$\begin{bmatrix} C_x(1)\\ C_x(5)\\ C_x(3)\\ C_x(7)\end{bmatrix} = \frac{1}{8}\left[\begin{array}{cc:cc} 1 & W & \multicolumn{2}{c}{\multirow{2}{*}{0}}\\ 1 & -W & & \\ \hdashline \multicolumn{2}{c:}{\multirow{2}{*}{0}} & 1 & W^3\\ & & 1 & -W^3\end{array}\right]\begin{bmatrix} 1 & 0 & W^2 & 0\\ 0 & 1 & 0 & W^2\\ 1 & 0 & -W^2 & 0\\ 0 & 1 & 0 & -W^2\end{bmatrix}\begin{bmatrix} X_1(4)\\ X_1(5)\\ X_1(6)\\ X_1(7)\end{bmatrix}$$

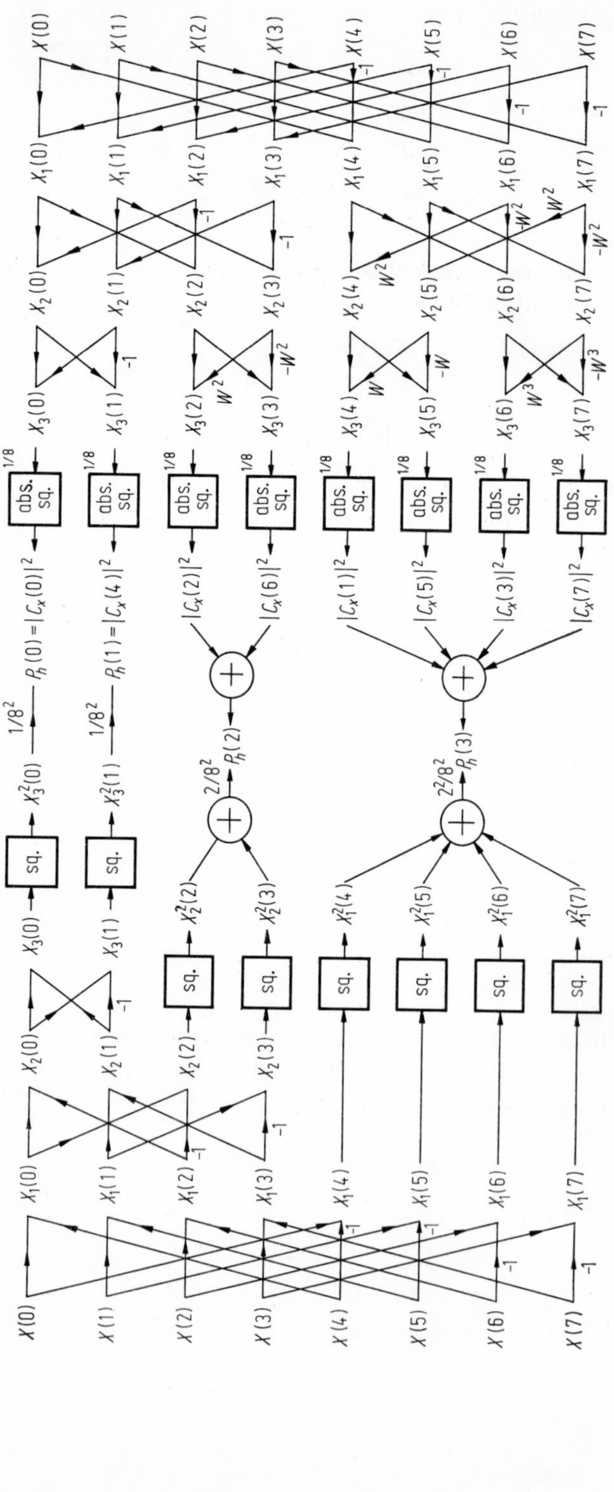

Fig. 6.10 Signal flow graphs showing the relationship between the $(WHT)_h$ and DFT power spectra

That is

$$\begin{bmatrix} C_x(1) \\ C_x(5) \\ C_x(3) \\ C_x(7) \end{bmatrix} = \frac{1}{8} \begin{bmatrix} 1 & W & W^2 & W^3 \\ 1 & -W & W^2 & -W^3 \\ 1 & W^3 & -W^2 & -W^5 \\ 1 & -W^3 & -W^2 & W^5 \end{bmatrix} \begin{bmatrix} X_1(4) \\ X_1(5) \\ X_1(6) \\ X_1(7) \end{bmatrix} \tag{6.9-13}$$

Since the (4×4) matrix in Eq. (13) is orthogonal, it follows that

$$|C_x(1)|^2 + |C_x(5)|^2 + |C_x(3)|^2 + |C_x(7)|^2 = \frac{2^2}{8^2} \sum_{m=4}^{7} X_1^2(m) \tag{6.9-14}$$

Inspection of Eq. (6.8-10) shows that the right side of Eq. (14) is $P_h(3)$. Thus,

$$P_h(3) = |C_x(1)|^2 + |C_x(5)|^2 + |C_x(3)|^2 + |C_x(7)|^2$$

Similarly, we obtain

$$P_h(2) = |C_x(2)|^2 + |C_x(6)|^2$$

$$P_h(1) = |C_x(4)|^2$$

and

$$P_h(0) = C_x^2(0)$$

A generalization of the above analysis leads to the following relationship between the (WHT)$_h$ and DFT power spectra.

$$P_h(0) = C_x^2(0)$$

and

$$P_h(r) = \sum_{m=2^{r-1}}^{2^r-1} |C_x(\langle m \rangle)|^2, \qquad r = 1, 2, \ldots, n \tag{6.9-15}$$

where $\langle m \rangle$ denotes the decimal number obtained from the bit-reversal of an n-bit binary representation of m.

(WHT)$_h$ phase spectrum. Corresponding to the (WHT)$_h$ power spectrum, a phase or "position" spectrum can be developed. Using the concepts of average power and phase angle, the phase spectrum is defined with respect to a multidimensional space in terms of a reference vector. It can be shown that [14]

$$\psi_h(0) = \frac{B_x(0)}{\left|\sqrt{P_h(0)}\right|}$$

and

$$\psi_h(r) = \frac{\sum_{k=2^{r-1}}^{2^r-1} B_x(k)}{2^{(r-1)/2}\left|\sqrt{P_h(r)}\right|}, \qquad r = 1, 2, \ldots, n \tag{6.9-16}$$

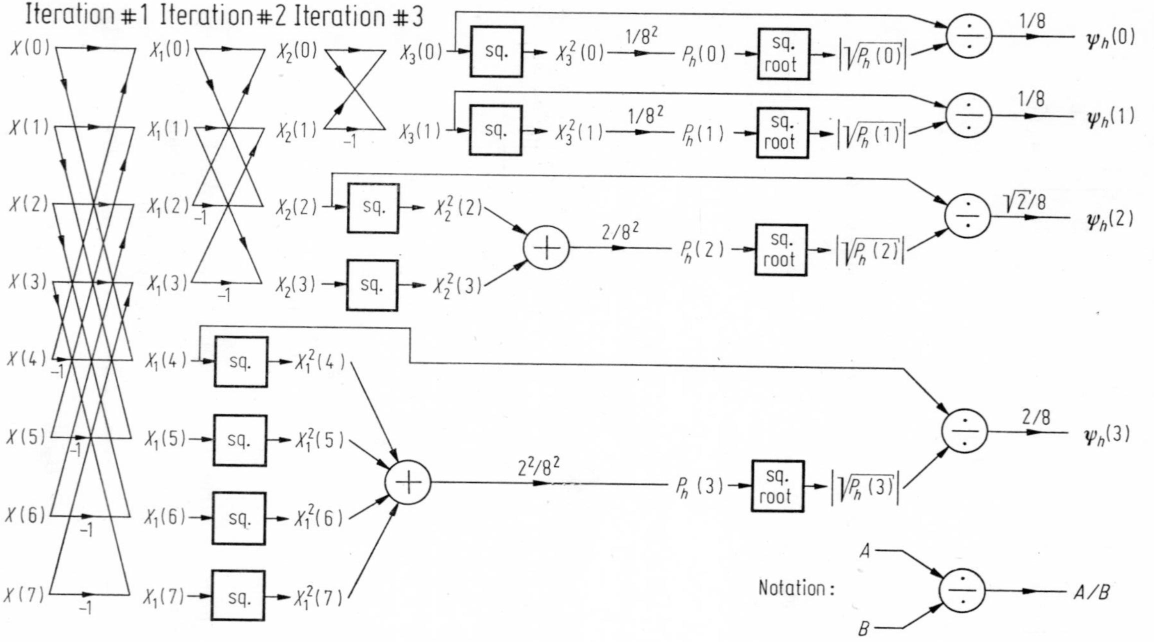

Fig. 6.11 $(\mathrm{WHT})_h$ power and phase spectra signal flow graph, $N = 8$

where $\psi_h(r)$ is the r-th $(\text{WHT})_h$ phase spectral point. For $N = 8$, Eq. (16) yields

$$\psi_h(0) = \frac{B_x(0)}{\left|\sqrt{P_h(0)}\right|}, \qquad \psi_h(1) = \frac{B_x(1)}{\left|\sqrt{P_h(1)}\right|}, \qquad \psi_h(2) = \frac{\sum_{k=2}^{3} B_x(k)}{\left|\sqrt{2P_h(2)}\right|},$$

and $\psi_h(3) = \dfrac{\sum_{k=4}^{7} B_x(k)}{\left|\sqrt{4P_x(3)}\right|}$

The above phase spectrum along with the power spectrum can be evaluated rapidly using the signal flow graph shown in Fig. 6.11.

It can be shown that the $(\text{WHT})_h$ phase spectrum is *invariant* to multiplication of the data sequence $\{X(m)\}$ by a real number. On the other hand, it changes in a specific manner [see Prob. 6-7] when $\{X(m)\}$ is shifted *cyclically*. Hence, the alternate name, "position" spectrum. Clearly, these properties are analogous to those of the DFT phase spectrum. However, while each spectral point in the DFT spectrum is defined with respect to a single frequency, the concept of phase for the $(\text{WHT})_h$ is defined for groups of sequencies. The composition of these sequencies is the same as that of the $(\text{WHT})_h$ power spectrum. Because of this sequency grouping and consequent data compression, the data sequence $\{X(m)\}$ *cannot* be recovered when the corresponding $(\text{WHT})_h$ power and phase spectra are specified.

6.10 Modified Walsh-Hadamard Transform (MWHT)

The MWHT [15, 16] is obtained by a simple modification of the $(\text{WHT})_h$. It is an orthogonal transform which is defined as

$$\boldsymbol{F}(n) = \frac{1}{N}\,\hat{\mathbf{H}}(n)\,\boldsymbol{X}(n) \tag{6.10-1}$$

where

$$\boldsymbol{F}(n)' = [F(0)\;F(1)\;\cdots\;F(N-1)]$$

is the MWHT vector,

$$\boldsymbol{X}(n)' = [X(0)\;X(1)\;\cdots\;X(N-1)],$$

and $\hat{\mathbf{H}}(n)$ is the $(N \times N)$ "modified" Hadamard matrix.

The matrices $\hat{\mathbf{H}}(k)$ are defined by the recurrence relation

$$\hat{\mathbf{H}}(k+1) = \begin{bmatrix} \hat{\mathbf{H}}(k) & \hat{\mathbf{H}}(k) \\ 2^{k/2}\,\mathbf{I}(k) & -2^{k/2}\,\mathbf{I}(k) \end{bmatrix}, \qquad k = 0, 1, \ldots, n \tag{6.10-2}$$

where $\hat{\mathbf{H}}(0) = 1$ and $\mathbf{I}(k)$ is the $(2^k \times 2^k)$ identity matrix. For example, with $N = 8$, Eq. (2) yields

$$\hat{\mathbf{H}}(3) = \left[\begin{array}{cccc:cccc} 1 & 1 & 1 & 1 & 1 & 1 & 1 & 1 \\ 1 & -1 & 1 & -1 & 1 & -1 & 1 & -1 \\ \sqrt{2} & 0 & -\sqrt{2} & 0 & \sqrt{2} & 0 & -\sqrt{2} & 0 \\ 0 & \sqrt{2} & 0 & -\sqrt{2} & 0 & \sqrt{2} & 0 & -\sqrt{2} \\ \hdashline 2 & 0 & 0 & 0 & -2 & 0 & 0 & 0 \\ 0 & 2 & 0 & 0 & 0 & -2 & 0 & 0 \\ 0 & 0 & 2 & 0 & 0 & 0 & -2 & 0 \\ 0 & 0 & 0 & 2 & 0 & 0 & 0 & -2 \end{array}\right] \tag{6.10-3}$$

Since the matrices $\hat{\mathbf{H}}(k)$ are orthogonal, we have

$$\hat{\mathbf{H}}(k)' \cdot \hat{\mathbf{H}}(k) = 2^k \, \mathbf{I}(k), \qquad k = 0, 1, \ldots, n \tag{6.10-4}$$

From Eqs. (1) and (4) it follows that the inverse modified Walsh-Hadamard transform (IMWHT) is given by

$$\boldsymbol{X}(n) = \hat{\mathbf{H}}(n)' \, \boldsymbol{F}(n) \tag{6.10-5}$$

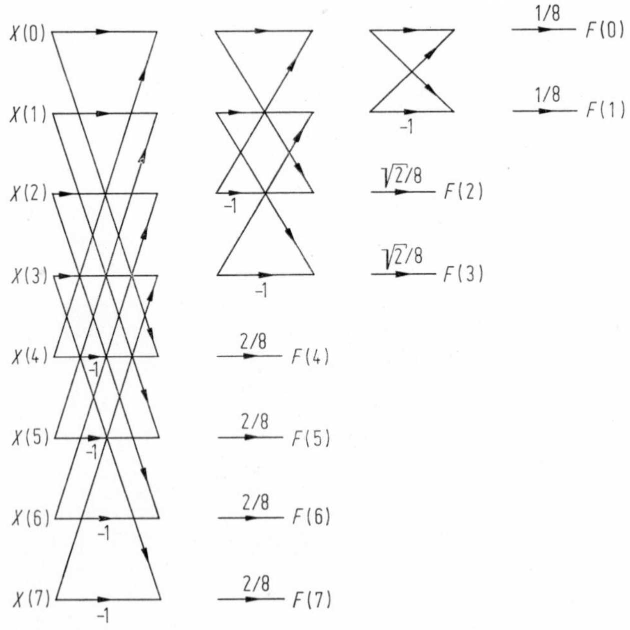

Fig. 6.12 MWHT signal flow graph, $N = 8$

We observe that unlike $\mathbf{H}_h(n)$, the modified Hadamard matrix $\hat{\mathbf{H}}(n)$ is *non-symmetric*. Consequently, the signal flow graphs associated with the algorithms to compute the MWHT and IMWHT have different structures. For the purposes of illustration, the MWHT and IMWHT signal flow graphs for $N = 8$ are shown in Figs. 6.12 and 6.13 respectively.

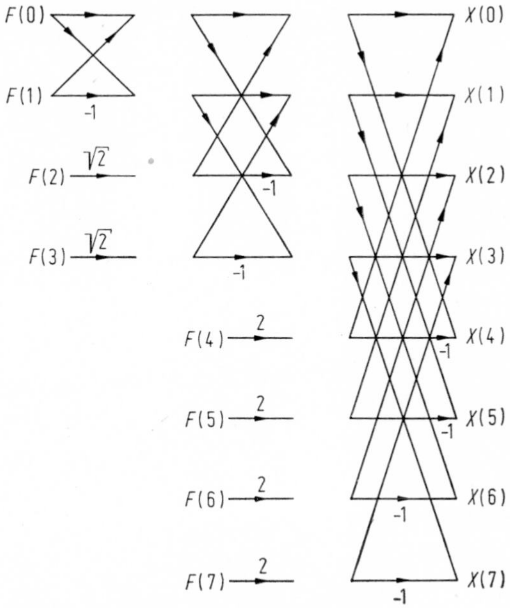

Fig. 6.13 IMWHT signal flow graph, $N = 8$

Comparing Fig. 6.12 with Fig. 6.8 it follows that the $(\mathrm{WHT})_h$ power spectrum can be expressed in terms of the MWHT coefficients as follows:

$$P_h(0) = F^2(0)$$

$$P_h(1) = F^2(1)$$

$$P_h(2) = \sum_{k=2}^{3} F^2(k)$$

and

$$P_h(3) = \sum_{k=4}^{7} F^2(k) \tag{6.10-6}$$

The general form of Eq. (6) is given by

$$P_h(0) = F^2(0)$$

and

$$P_h(r) = \sum_{k=2^{r-1}}^{2^r-1} F^2(k)\,, \qquad r = 1, 2, \ldots, n \tag{6.10-7}$$

A class of cyclic shift invariants[1]. We recall that the $(\text{WHT})_h$ power spectrum has $\log_2 N + 1$ spectral points which are invariant to cyclic shifts of $\{X(m)\}$. We now show that the $(\text{WHT})_h$ spectral points are a subset of a set of $(N/2 + 1)$ shift invariants.

Consider the case $N = 8$ and let $\{Z(m)\}_l$ denote the sequence obtained by a cyclic shift of $\{Z(m)\}$, l places to the left. With $l = 1$ we have

$$\boldsymbol{Z}(3)_1' = [X(1)\ X(2) \cdots X(7)\ X(0)]$$

and hence

$$\boldsymbol{Z}(3)_1 = \mathbf{M}(3)\,\boldsymbol{X}(3) \tag{6.10-8}$$

where $\mathbf{M}(3)$ is defined in Eq. (6.8-1).

If $\boldsymbol{F}_{z,1}(3)' = [F_{z,1}(0)\ F_{z,1}(1) \cdots F_{z,1}(7)]$ denotes the MWHT of $\boldsymbol{Z}(3)_1$, then

$$\boldsymbol{F}_{z,1}(3) = \frac{1}{8}\,\hat{\mathbf{H}}(3)\,\mathbf{M}(3)\,\boldsymbol{X}(3) \tag{6.10-9}$$

Substitution of Eq. (5) in Eq. (9) leads to the similarity transform

$$\boldsymbol{F}_{z,1}(3) = \frac{1}{8}\,\hat{\mathbf{H}}(3)\,\mathbf{M}(3)\,\hat{\mathbf{H}}(3)'\,\boldsymbol{F}_x(3) \tag{6.10-10}$$

Evaluating Eq. (10) we obtain

$$\begin{bmatrix} F_{z,1}(0) \\ F_{z,1}(1) \\ F_{z,1}(2) \\ F_{z,1}(3) \\ F_{z,1}(4) \\ F_{z,1}(5) \\ F_{z,1}(6) \\ F_{z,1}(7) \end{bmatrix} = \begin{bmatrix} 1 & & & & & & & 0 \\ & -1 & & & & & & \\ & & 0 & 1 & & & & \\ & & -1 & 0 & & & & \\ & & & & 0 & 1 & 0 & 0 \\ & & & & 0 & 0 & 1 & 0 \\ & & & & 0 & 0 & 0 & 1 \\ 0 & & & & -1 & 0 & 0 & 0 \end{bmatrix} \begin{bmatrix} F_x(0) \\ F_x(1) \\ F_x(2) \\ F_x(3) \\ F_x(4) \\ F_x(5) \\ F_x(6) \\ F_x(7) \end{bmatrix} \tag{6.10-11}$$

which implies that

$$F_{z,1}(0) = F_x(0)$$

$$F_{z,1}(1) = (-1)\,F_x(1)$$

$$\begin{bmatrix} F_{z,1}(2) \\ F_{z,1}(3) \end{bmatrix} = \begin{bmatrix} 0 & 1 \\ -1 & 0 \end{bmatrix} \begin{bmatrix} F_x(2) \\ F_x(3) \end{bmatrix}$$

[1] This material may be omitted without loss of continuity.

and

$$\begin{bmatrix} F_{z,1}(4) \\ F_{z,1}(5) \\ F_{z,1}(6) \\ F_{z,1}(7) \end{bmatrix} = \begin{bmatrix} 0 & 1 & 0 & 0 \\ 0 & 0 & 1 & 0 \\ 0 & 0 & 0 & 1 \\ -1 & 0 & 0 & 0 \end{bmatrix} \begin{bmatrix} F_x(4) \\ F_x(5) \\ F(_x6) \\ F_x(7) \end{bmatrix} \qquad (6.10\text{-}12)$$

Since each of the square matrices in Eq. (12) is orthonormal, it follows that the following quantities denoted by Q_0 and $Q_{i,0}$ are invariant to cyclic shifts of the data sequence $\{X(m)\}$:

$$Q_0 = F_x^2(0)$$

$$Q_{1,0} = F_x^2(1)$$

$$Q_{2,0} = F_x^2(2) + F_x^2(3)$$

and

$$Q_{3,0} = F_x^2(4) + F_x^2(5) + F_x^2(6) + F_x^2(7) \qquad (6.10\text{-}13)$$

Comparing Eqs. (13) and (6) one obtains

$$Q_0 = P_h(0)$$

$$Q_{1,0} = P_h(1)$$

$$Q_{2,0} = P_h(2)$$

and

$$Q_{3,0} = P_h(3) \qquad (6.10\text{-}14)$$

It can be shown that [17] the general form of Eq. (14) is given by

$$Q_0 = P_h(0)$$

and

$$Q_{r,0} = P_h(r), \qquad r = 1, 2, \ldots, n \qquad (6.10\text{-}15)$$

That is, the $(\text{WHT})_h$ power spectrum is equivalent to the set of invariants $\{Q_0, Q_{i,0}\}$.

Again, from Eq. (12) it follows that

$$\begin{bmatrix} F_{z,l}(2) \\ F_{z,l}(3) \end{bmatrix} = \begin{bmatrix} 0 & 1 \\ -1 & 0 \end{bmatrix}^l \begin{bmatrix} F_x(2) \\ F_x(3) \end{bmatrix}, \qquad l = 1, 2$$

and

$$\begin{bmatrix} F_{z,l}(4) \\ F_{z,l}(5) \\ F_{z,l}(6) \\ F_{z,l}(7) \end{bmatrix} = \begin{bmatrix} 0 & 1 & 0 & 0 \\ 0 & 0 & 1 & 0 \\ 0 & 0 & 0 & 1 \\ -1 & 0 & 0 & 0 \end{bmatrix}^l \begin{bmatrix} F_x(4) \\ F_x(5) \\ F_x(6) \\ F_x(7) \end{bmatrix} \qquad (6.10\text{-}16)$$

Evaluating Eq. (16) and noting that powers of orthonormal matrices are also orthonormal, it follows that additional shift invariants can be obtained by evaluating the following dot products, where "·" denotes the dot product operation.

$$Q_{2,1} = \begin{matrix} \{ & F_x(2) & F_x(3)\} \\ \cdot\{ & F_x(3) & -F_x(2)\} \end{matrix} = 0$$

$$Q_{2,2} = \begin{matrix} \{ & F_x(2) & F_x(3)\} \\ \cdot\{ & -F_x(2) & -F_x(3)\} \end{matrix} = -Q_{2,0}$$

$$Q_{3,1} = \begin{matrix} \{ & F_x(4) & F_x(5) & F_x(6) & F_x(7)\} \\ \cdot\{ & F_x(5) & F_x(6) & F_x(7) & -F_x(4)\} \end{matrix}$$

$$Q_{3,2} = \begin{matrix} \{ & F_x(4) & F_x(5) & F_x(6) & F_x(7)\} \\ \cdot\{ & F_x(6) & F_x(7) & -F_x(4) & -F_x(5)\} \end{matrix} = 0$$

and

$$Q_{3,3} = \begin{matrix} \{ & F_x(4) & F_x(5) & F_x(6) & F_x(7)\} \\ \cdot\{ & F_x(7) & -F_x(4) & -F_x(5) & -F_x(6)\} \end{matrix} = -Q_{3,1} \qquad (6.10\text{-}17)$$

An examination of Eqs. (13) and (17) results in the following observations:

(i) There are $(8/2 + 1)$ independent shift invariants, namely Q_0, $Q_{1,0}$, $Q_{2,0}$, $Q_{3,0}$, and $Q_{3,1}$.
(ii) Computation of the $Q_{i,j}$ involves successive cyclic shifts, sign changes, and dot products, and hence can be done rapidly [18, 19].

For the general case, it can be shown that there are $(N/2 + 1)$ shift invariants in the set $\{Q_0, Q_{i,j}\}$, which is referred to as the WHT quadratic spectrum or the Q-spectrum. A compact expression for the Q-spectrum is as follows [18]:

$$Q_0 = F_x^2(0)$$

$$Q_{m,0} = \sum_{k=2^{m-1}}^{2^m-1} F_x^2(k), \qquad m = 1, 2, \ldots, n$$

and

$$Q_{m,q} = \sum_{k=2^{m-1}}^{2^m-1-q} F_x(k)\, F_x(q+k) - \sum_{j=0}^{q-1} F_x(2^m + j - q)\, F_x(2^{m-1} + j)$$

for

$$k = 2^{m-1},\quad 2^{m-1}+1, \ldots,\quad 2^m - 1, \quad q = 1, 2, \ldots, 2^{m-1}, \qquad m = 1, 2, \ldots, n. \qquad (6.10\text{-}18)$$

The Q-spectrum will be used in the next section in connection with an autocorrelation theorem.

6.11 Cyclic and Dyadic Correlation/Convolution

Let $\{X(m)\}$ and $\{Y(m)\}$ be two real-valued N-periodic data sequences. One can derive several properties pertaining to dyadic correlation/convolution for the $(\text{WHT})_w$ which parallel those of the DFT, for cyclic correlation/convolution [20–22]. In the interest of summarizing, we present a comparison of such properties in Table 6.11-1.

Table 6.11-1 DFT and $(\text{WHT})_w$ Properties Pertaining to Correlation/Convolution

DFT	$(\text{WHT})_w$
1. Correlation	
Cyclic	Dyadic
If $\hat{Z}(m) = \frac{1}{N} \sum_{h=0}^{N-1} X(h)\, Y(m+h)$,	If $\hat{Z}(m) = \frac{1}{N} \sum_{h=0}^{N-1} X(h)\, Y(m \oplus h)$,
$m = 0, 1, \ldots, N-1$, then	$m = 0, 1, \ldots, N-1$, then
$C_{\hat{z}}(k) = C_x(k)\, \bar{C}_y(k), \quad k = 0, 1, \ldots, N-1$	$W_{\hat{z}}(k) = W_x(k) W_y(k), \quad k = 0, 1, \ldots, N-1$
2. Autocorrelation	
Cyclic	Dyadic
If $\hat{Z}(m) = \frac{1}{N} \sum_{h=0}^{N-1} X(h)\, X(m+h)$,	If $\hat{Z}(m) = \frac{1}{N} \sum_{h=0}^{N-1} X(h)\, X(m \oplus h)$,
$m = 0, 1, \ldots, N-1$, then	$m = 0, 1, \ldots, N-1$, then
$C_{\hat{z}}(k) = \lvert C_x(k) \rvert^2, \quad k = 0, 1, \ldots, N/2$	$W_{\hat{z}}(k) = W_x^2(k), \quad k = 0, 1, \ldots, N/2$
3. Convolution	
Cyclic	Dyadic
If $Z(m) = \frac{1}{N} \sum_{h=0}^{N-1} X(h)\, Y(m-h)$,	If $Z(m) = \frac{1}{N} \sum_{h=0}^{N-1} X(h) Y(m \ominus h)$,
$m = 0, 1, \ldots, N-1$, then	$m = 0, 1, \ldots, N-1$, where $\ominus$ denotes modulo 2 subtraction, then
$C_z(k) = C_x(k)\, C_y(k), \quad k = 0, 1, \ldots, N-1$	$W_z(k) = W_x(k)\, W_y(k), \quad k = 0, 1, \ldots, N-1$

Comment: Since modulo 2 addition and subtractions are identical operations, there is no difference between dyadic correlation and convolution.

Cyclic autocorrelation theorem for $(\text{WHT})_h$[1]. In Table 6.11-1 we consider the cyclic autocorrelation property of the DFT,

$$C_{\hat{z}}(k) = \lvert C_x(k) \rvert^2, \qquad k = 0, 1, \ldots, N/2 \tag{6.11-1}$$

Since $\lvert C_x(k) \rvert^2$ denotes the k-th DFT spectral point, Eq. (1) states that the DFT of the cyclic autocorrelation sequence $\{\hat{Z}(m)\}$ yields a set of $(N/2 + 1)$ cyclic shift invariants. It can be shown that the $(\text{WHT})_h$ has an analogous

[1] This material may be omitted without loss of continuity.

property which involves the Q-spectrum defined in Eq. (6.10-18). A detailed proof of this theorem is presented elsewhere [19, 23].

The $(\text{WHT})_h$ cyclic autocorrelation theorem can be expressed as follows:

$$\hat{\boldsymbol{Q}}(n) = \frac{1}{N}\hat{\mathbf{H}}(n)\,\hat{\boldsymbol{Z}}(n) \tag{6.11-2}$$

where

$$\hat{\boldsymbol{Q}}(n)' = [\hat{Q}_0\,\hat{Q}_{1,0}\,\hat{Q}_{2,0}\,\hat{Q}_{2,1}\cdots\hat{Q}_{n,2^{n-1}-1}],$$

with

$$\hat{Q}_0 = Q_0,\quad \hat{Q}_{1,0} = Q_{1,0}$$

$$\hat{Q}_{m,l} = 2^{-(m-1)/2}\,Q_{m,l} \quad \text{for} \quad m = 2, 3, \ldots, n, \quad l = 0, 1, \ldots, 2^{m-1}-1$$

and

$$\hat{\boldsymbol{Z}}(m)' = [\hat{Z}(0)\,\hat{Z}(1)\cdots\hat{Z}(N-1)]$$

is the vector representation of the cyclic autocorrelation sequence.

For example, when $N = 8$, Eq. (2) yields

$$\begin{bmatrix} Q_0 \\ Q_{1,0} \\ \frac{1}{\sqrt{2}}Q_{2,0} \\ \frac{1}{\sqrt{2}}Q_{2,1} \\ \frac{1}{2}Q_{3,0} \\ \frac{1}{2}Q_{3,1} \\ \frac{1}{2}Q_{3,2} \\ \frac{1}{2}Q_{3,3} \end{bmatrix} = \frac{1}{8}\,\hat{\mathbf{H}}(3) \begin{bmatrix} \hat{Z}(0) \\ \hat{Z}(1) \\ \hat{Z}(2) \\ \hat{Z}(3) \\ \hat{Z}(4) \\ \hat{Z}(5) \\ \hat{Z}(6) \\ \hat{Z}(7) \end{bmatrix}$$

where $\hat{\mathbf{H}}(3)$ is defined in Eq. (6.10-3).

In words, Eq. (2) states that the MWHT of the cyclic autocorrelation sequence $\{\hat{Z}(m)\}$ yields a set of $(N/2 + 1)$ independent Q-spectral points which are invariant to cyclic shifts of $\{X(m)\}$, as is the case with the DFT power spectral points $|C_x(k)|^2$. However, the Q-spectrum can not be interpreted as a power spectrum since some of the $Q_{i,j}$ can be negative. For a physical interpretation of the Q-spectrum, the interested reader may refer [19] or [23].

6.12 Multidimensional $(\mathrm{WHT})_h$ and $(\mathrm{WHT})_w$

We shall restrict our attention to the 2-dimensional case which is used for image processing. The extension to the r-dimensional case is straightforward.

The definition of the 2-dimensional $(\mathrm{WHT})_h$ follows directly from Eq. (6.2-8):

$$B_{xx}(u_1, u_2) = \frac{1}{N_1 N_2} \sum_{m_2=0}^{N_2-1} \sum_{m_1=0}^{N_1-1} X(m_1, m_2)\,(-1)^{\langle m_1, u_1\rangle + \langle m_2, u_2\rangle} \tag{6.12-1}$$

where $X(m_1, m_2)$ is an input data point, $B_{xx}(u_1, u_2)$ is a transform coefficient, and

$$\langle m_i, u_i\rangle = \sum_{s=0}^{n_i-1} m_i(s)\, u_i(s), \qquad i = 1, 2$$

The terms $m_i(s)$ and $u_i(s)$ are the binary representations of m_i and u_i respectively; i.e.,

$$[u_i]_{\mathrm{decimal}} = [u_i(n_i-1),\quad u_i(n_i-2), \ldots, u_i(1),\quad u_i(0)]_{\mathrm{binary}}$$

$X(m_1, m_2)$ can be recovered uniquely from the inverse 2-dimensional $(\mathrm{WHT})_h$ which is given by

$$X(m_1, m_2) = \sum_{u_2=0}^{N_2-1} \sum_{u_1=0}^{N_1-1} B_{xx}(u_1, u_2)\,(-1)^{\langle m_1, u_1\rangle + \langle m_2, u_2\rangle} \tag{6.12-2}$$

The data can be expressed in the form of an $(N_1 \times N_2)$ matrix $[X(m_1, m_2)]$, where

$$[X(m_1, m_2)] = \begin{bmatrix} X(0,0) & X(0,1) & \cdots & X(0, N_2-1) \\ X(1,0) & X(1,1) & \cdots & X(1, N_2-1) \\ \cdot & \cdot & \cdot & \cdot \\ X(N_1-1, 0) & X(N_1-1, 1) & \cdots & X(N_1-1, N_2-1) \end{bmatrix} \tag{6.12-3}$$

In Eq. (1) we consider the inner summation which is given by

$$\frac{1}{N_1} \sum_{m_1=0}^{N_1-1} X(m_1, m_2)\,(-1)^{\langle m_1, u_1\rangle} =$$
$$\frac{1}{N_1}\{X(0, m_2)(-1)^{\langle 0, u_1\rangle} + X(1, m_2)(-1)^{\langle 1, u_1\rangle} + \cdots + X(N_1-1, m_2)(-1)^{\langle N_1-1, u_1\rangle}\} \tag{6.12-4}$$

Inspection of Eq. (4) shows that its right-hand side represents the $(\mathrm{WHT})_h$ of each *column* of the data matrix $[X(m_1, m_2)]$. Thus we introduce the notation

$$\frac{1}{N_1} \sum_{m_1=0}^{N_1-1} X(m_1, m_2)\,(-1)^{\langle m_1, u_1\rangle} = B_x(u_1, m_2) \tag{6.12-5}$$

The coefficients $B_x(u_1, m_2)$ are written in matrix form to obtain

$$[B_x(u_1, m_2)] = \begin{bmatrix} B_x(0,0) & B_x(0,1) & \cdots B_x(0, N_2-1) \\ B_x(1,0) & B_x(1,1) & \cdots B_x(1, N_2-1) \\ \cdots & \cdots & \cdots \\ B_x(N_1-1, 0) & B_x(N_1-1, 1) & \cdots B_x(N_1-1, N_2-1) \end{bmatrix} \tag{6.12-6}$$

Substitution of Eq. (5) in Eq. (1) leads to

$$B_{xx}(u_1, u_2) = \frac{1}{N_2} \sum_{m_2=0}^{N_2-1} B_x(u_1, m_2)(-1)^{\langle m_2, u_2 \rangle}$$

which yields

$$B_{xx}(u_1, u_2) =$$
$$\frac{1}{N_2}\{B_x(u_1, 0)(-1)^{\langle 0, u_2 \rangle} + B_x(u_1, 1)(-1)^{\langle 1, u_2 \rangle} + \cdots + B_x(u_1, N_2-1)(-1)^{\langle N_2-1, u_2 \rangle}\} \tag{6.12-7}$$

Equation (7) implies that the coefficients $B_{xx}(u_1, u_2)$ can be obtained by taking the $(\text{WHT})_h$ of each *row* of $[B_x(u_1, m_2)]$ in Eq. (6). This results in a set of $N_1 N_2$ coefficients whose matrix form is given by

$$[B_{xx}(u_1, u_2)] = \begin{bmatrix} B_{xx}(0,0) & B_{xx}(0,1) & \cdots B_{xx}(0, N_2-1) \\ B_{xx}(1,0) & B_{xx}(1,1) & \cdots B_{xx}(1, N_2-1) \\ \cdots & \cdots & \cdots \\ B_{xx}(N_1-1, 0) & B_{xx}(N_1-1, 1) & \cdots B_{xx}(N_1-1, N_2-1) \end{bmatrix} \tag{6.12-8}$$

The above discussion shows that the 2-dimensional $(\text{WHT})_h$ can be computed using the 1-dimensional $(\text{FWHT})_h$, a total of $N_1 N_2$ times, as follows:

(i) With $N = N_1$, the $(\text{FWHT})_h$ of each of the N_2 columns of $[X(m_1, m_2)]$ is taken to obtain $[B_x(u_1, m_2)]$ in Eq. (6).

(ii) Subsequently with $N = N_2$, the $(\text{FWHT})_h$ of each of the N_1 rows of $[B_x(u_1, m_2)]$ is taken to obtain $[B_{xx}(u_1, u_2)]$ in Eq. (8).

Alternately, the 2-dimensional $(\text{WHT})_h$ can be computed as a 1-dimensional $(\text{WHT})_h$, where the data matrix is rearranged in the form of a $(N_1 N_2)$-vector [see Prob. 6-8]. A discussion of the r-dimensional $(\text{WHT})_h$, $r \geq 2$, along with a development of the corresponding power spectrum is available in [24].

In conclusion, we observe that as a consequence of (i) and (ii) above, the 2-dimensional $(\text{WHT})_h$ and its inverse can be equivalently defined in matrix form as follows:

$$[B_{xx}(u_1, u_2)] = \frac{1}{N_1 N_2} \mathbf{H}_h(n_1)[X(m_1, m_2)]\mathbf{H}_h(n_2)$$

and

$$[X(m_1, m_2)] = \mathbf{H}_h(n_1)\,\mathbf{B}_{xx}(u_1, u_2)]\,\mathbf{H}_h(n_2) \qquad (6.12\text{-}9)$$

Similarly, it can be shown that the matrix form definitions of the 2-dimensional $(\mathrm{WHT})_w$ and its inverse are given by

$$[W_{xx}(u_1, u_2)] = \frac{1}{N_1 N_2}\mathbf{H}_w(n_1)\,[X(m_1, m_2)]\,\mathbf{H}_w(n_2)$$

and

$$[X(m_1, m_2)] = \mathbf{H}_w(n_1)\,[W_{xx}(u_1, u_2)]\,\mathbf{H}_w(n_2) \qquad (6.12\text{-}10)$$

where

$$[W_{xx}(u_1, u_2)] = \begin{bmatrix} W_{xx}(0,0) & W_{xx}(0,1) & \cdots & W_{xx}(0, N_2-1) \\ W_{xx}(1,0) & W_{xx}(1,1) & \cdots & W_{xx}(1, N_2-1) \\ \cdots & \cdots & \cdots & \cdots \\ W_{xx}(N_1-1, 0) & W_{xx}(N_1-1, 1) & \cdots & W_{xx}(N_1-1, N_2-1) \end{bmatrix}$$

From Eq. (10) it follows that the 2-dimensional transform or its inverse can be computed using the 1-dimensional $(\mathrm{FWHT})_w$, a total of $N_1 N_2$ times, as was the case with the $(\mathrm{WHT})_h$.

6.13 Summary

The Walsh-Hadamard transforms $(\mathrm{WHT})_w$ and $(\mathrm{WHT})_h$ were introduced, and algorithms to compute them were developed. Corresponding power and phase spectra were defined. It was shown that the $(\mathrm{WHT})_w$ and $(\mathrm{WHT})_h$ power spectra are respectively invariant to dyadic and cyclic shifts of the data sequence. Physical interpretations for the $(\mathrm{WHT})_w$ and $(\mathrm{WHT})_h$ power spectra were provided, and a cyclic autocorrelation theorem for the $(\mathrm{WHT})_h$ was presented.

While discussing the topics cited above, we emphasized the analogy between the Walsh-Hadamard and the discrete Fourier transforms. The $(\mathrm{WHT})_w$, and $(\mathrm{WHT})_h$, like the DFT, were extended to two dimensions. It was shown that the 2-dimensional $(\mathrm{WHT})_w/(\mathrm{WHT})_h$ of an $(N_1 \times N_2)$ data matrix can be obtained by $N_1 N_2$ applications of the 1-dimensional $(\mathrm{FWHT})_w/(\mathrm{FWHT})_h$.

Although our attention was restricted to the $(\mathrm{WHT})_w$ and $(\mathrm{WHT})_h$, we could have alternately considered the Paley-ordered Walsh-Hadamard transform $(\mathrm{WHT})_p$. Some aspects of the $(\mathrm{WHT})_p$ are discussed in [26].

Appendix 6.1

This appendix provides a listing of a subroutine for computing the $(\mathrm{WHT})_h$ and $(\mathrm{WHT})_w$ using the corresponding fast algorithms developed in Sections 6.3 and 6.5 respectively.

$(\mathrm{WHT})_h$ Computer Program

```
      SUBROUTINE WHT (NUM,X,II)
C
C     II=0  HADAMARD-ORDERED WHT
C     II=1  INVERSE HADAMARD-ORDERED WHT
C     II=2  WALSH-ORDERED WHT
C     II=3  INVERSE WALSH-ORDERED WHT
C
      DIMENSION IPOWER(10),X(32),Y(32)
C
C     THIS ROUTINE CALCULATES THE FAST WALSH-HADAMARD
C     TRANSFORMS FOR ANY GIVEN NUMBER  WHICH
C     IS A POWER OF TWO
C
C     NUM  NUMBER OF POINTS
C
      IF(II.LE.1) GO TO 14
C
C     BIT REVERSE THE INPUT
C
      DO 11 I = 1,NUM
      IB = I - 1
      IL = 1
    9 IBD = IB/2
      IPOWER(IL) = 1
      IF (IB.EQ.(IBD*2)) IPOWER(IL) = 0
      IF (IBD.EQ.0) GO TO 10
      IB = IBD
      IL = IL + 1
      GO TO 9
   10 CONTINUE
      IP = 1
      IFAC = NUM
      DO 12 I1 = 1,IL
      IFAC = IFAC/2
   12 IP = IP + IFAC * IPOWER(I1)
   11 Y(IP) = X(I)
      DO 13 I=1,NUM
   13 X(I) = Y(I)
   14 CONTINUE
C
C     CALCULATE NUMBER OF ITERATIONS.
C
   65 ITER=0
      IREM=NUM
    1 IREM=IREM/2
      IF (IREM.EQ.0) GO TO 2
      ITER=ITER+1
      GO TO 1
    2 CONTINUE
C
C     BEGIN A LOOP FOR (LOG TO BASE TWO OF NUM) ITERATIONS.
C
      DO 50 M = 1,ITER
C
C     CALCULATE NUMBER OF PARTITIONS
C
      IF (M.EQ.1) NUMP = 1
      IF (M.NE.1) NUMP = NUMP * 2
      MNUM =NUM/NUMP
```

```
      MNUM2 = MNUM/2
C
C     BEGIN A LOOP FOR THE NUMBER OF PARTITIONS.
C
      ALPH = 1.
      DO 49 MP = 1,NUMP
      IB = (MP-1) * MNUM
C
C     BEGIN A LOOP THROUGH THIS PARTITION.
C
      DO 48 MP2 = 1,MNUM2
      MNUM21 = MNUM2 + MP2 + IB
      IBA = IB + MP2
      Y(IBA) = X(IBA) + ALPH * X(MNUM21)
      Y(MNUM21) = X(IBA) - ALPH * X(MNUM21)
   48 CONTINUE
      IF (II.GE.2)ALPH=-ALPH
   49 CONTINUE
C
C
      DO 7 I = 1,NUM
    7 X(1) = Y(I)
   50 CONTINUE
      IF(II.EQ.1.OR.II.EQ.3) RETURN
      R=1./NUM
      DO 15 I=1,NUM
   15 X(I)=X(I)*R
      RETURN
      END
```

References

1. Walsh, J. L.: A Closed Set of Orthogonal Functions. *Amer. J. of Mathematics* 45 (1923) 5–24.
2. Paley, R. E. A. C.: A Remarkable Series of Orthogonal Functions. *Proc. London Math. Soc.* (2) 34 (1932) 241–279.
3. Fine, N. J.: On the Walsh Functions. *Trans. Amer. Math. Soc.* 69 (1950) 66–77.
4. Ohnsorg, F.: Binary Fourier Representation. Presented at the Spectrum Analysis Techniques Symp., Honeywell Research Center, Hopkins, Minn., 20–21 Sept. 1966.
5. Andrews, H. C., and Caspari, K. L.: A Generalized Technique For Spectral Analysis. *IEEE Trans. Computers* C-19 (1970) 16–25.
6. Glassman, J. A.: A Generalization of the Fast Fourier Transform. *IEEE Trans. Computers* C-19 (1970) 105–116.
7. Ahmed, N., and Cheng, S. M.: On Matrix Partitioning and a Class of Algorithms. *IEEE Trans. Education* E-13 (1970) 103–105.
8. Pratt, W. K., Andrews, H. C., and Kane, J.: Hadamard Transform Image Coding. *Proc. IEEE* 57 (1969) 58–68.
9. Manz, J. W.: A Sequency-Ordered Fast Walsh Transform. *IEEE Trans. Audio and Electroacoustics* AU-20 (1972) 204–205.
10. Boesswetter, C.: Analog Sequency Analysis and Synthesis of Voice Signals. *Proc. 1970 Walsh Functions Symposium*, 220–229.

11. Robinson, G. S., and Campanella, S. J.: Digital Sequency Decomposition of Voice Signals. *Proc. 1970 Walsh Functions Symposium*, 230–237.
12. Harmuth, H. F.: *Transmission of Information by Orthogonal Functions*. 1st. ed. New York, Heidelberg, Berlin: Springer-Verlag, 1969.
13. Ahmed, N., Rao, K. R., and Abdussattar, A. L.: BIFORE or Hadamard Transform. *IEEE Trans. Audio and Electroacoustics* AU-19 (1971) 225–234.
14. Ahmed, N., and Rao, K. R.: A Phase Spectrum For Binary Fourier Representation. *Intern. J. Computer Math.* Section B, (3) (1971) 85–101.
15. Ahmed, N., and Schultz, R. B.: Position Spectrum Considerations. *IEEE Trans. Audio and Electroacoustics* AU-19 (1971) 326–327.
16. Ahmed, N., Schultz, R. B., and Rao, K. R.: On the Characterization of the Amplitude and Shift of N-Periodic Sequences. *Information and Control* 20 (1972) 9–19.
17. Ohnsorg, F. R.: Spectral Modes of the Walsh-Hamard Transform. *Proc. 1971 Walsh Functions Symposium*, 55–59.
18. Ahmed, N., Abdussattar, A. L., and Rao, K. R.: An Algorithm to Compute the Walsh-Hadamard Transform Spectral Modes. *Proc. 1972 Walsh Functions Symposium*, 276–280.
19. Abdussattar, A. L.: Spectral Modes of the Walsh-Hadamard Transform. Ph. D. dissertation, Kansas State University, Manhattan, Kansas, U.S.A., 1972.
20. Corrington, M. S., and Adams, R. N.: Advanced Analytical and Signal Processing Techniques. Applications of Walsh Functions to Nonlinear Analysis (1962). AD 277–942.
21. Pichler, F.: Walsh Functions and Optimal Linear Systems. Proc. *1970 Walsh Functions Symposium*, 17–22.
22. Frank, H. F.: Implementation of Dynamic Correlation. *Proc. 1971 Walsh Functions Symposium*, 111–117.
23. Ahmed, N., Rao, K. R., and Abdussattar, A. L.: On Cyclic Autocorrelation and the Walsh-Hadamard Transform. *IEEE Trans. Electromagnetic Compatability* EMC-15 (1973) 141–146.
24. Bates, R. M.: Multi-dimensional BIFORE Transform. Ph. D. Dissertation, Kansas State University, Manhattan, Kansas, U.S.A., 1971.
25. Ahmed, N., and Bates, R. M.: A Power Spectrum and Related Physical Interpretation for the Multi-dimensional BIFORE Transform. *Proc. 1971 Walsh Functions Symposium*, 47–50.
26. Yuen, C.: Walsh Functions and Gray Code. *Proc. 1971 Walsh Functions Symposium*, 68–73.
27. Ohnsorg, F. R.: Quantization of the Fourier Sine-Cosine Functions. Honeywell Memo No. U-RD 6299, Systems and Research Center, St. Paul, Minn. August 1963.
28. Robinson, G. S.: Logical Convolution and Discrete Walsh and Power Spectra. *IEEE Trans. Audio and Electroacoustics* AU-20 (1972) 271–279.
29. Gulamhusein, M. N., and Fallside, F.: Short-Time Spectral and Autocorrelation Analysis in the Walsh Domain. *IEEE Trans. Info. Theory* IT-19 (1973) 615–623.
30. Bhagavan, B. K., and Polge, R. J.: Sequencing the Hadamard Transform. *IEEE Trans. Audio and Electroacoustics* AU-21 (1973) 472–473.
31. Mar, H. Y. L., and Sheng, C. L.: Fast Hadamard Transform using the H Diagram. *IEEE Trans. Computers* C-22 (1973) 957–959.
32. Ahmed, N., and Natarajan, T.: On Logical and Arithmetic Autocorrelation Functions. *IEEE Trans. Electromagnetic Compatability* EMC-16 (1974) 177–183.

Problems

6-1 Use the signal flow graphs in Figs. 6.2 and 6.6 respectively to show that $\mathbf{H}_h(3)$ and $\mathbf{H}_w(3)$ can be factored as follows:

$$\mathbf{H}_h(3) = \left[\begin{array}{cc:cc:cc:cc} 1 & 1 & & & & & & \mathbf{0} \\ 1 & -1 & & & & & & \\ \hdashline & & 1 & 1 & & & & \\ & & 1 & -1 & & & & \\ \hdashline & & & & 1 & 1 & & \\ & & & & 1 & -1 & & \\ \hdashline & & & & & & 1 & 1 \\ \mathbf{0} & & & & & & 1 & -1 \end{array}\right] \left[\begin{array}{cc:cc:cc:cc} 1 & 0 & 1 & 0 & & & & \\ 0 & 1 & 0 & 1 & & \mathbf{0}_4 & & \\ \hdashline 1 & 0 & -1 & 0 & & & & \\ 0 & 1 & 0 & -1 & & & & \\ \hdashline & & & & 1 & 0 & 1 & 0 \\ & \mathbf{0}_4 & & & 0 & 1 & 0 & 1 \\ \hdashline & & & & 1 & 0 & -1 & 0 \\ & & & & 0 & 1 & 0 & -1 \end{array}\right]$$

$$\times \left[\begin{array}{cccc:cccc} 1 & 0 & 0 & 0 & 1 & 0 & 0 & 0 \\ 0 & 1 & 0 & 0 & 0 & 1 & 0 & 0 \\ 0 & 0 & 1 & 0 & 0 & 0 & 1 & 0 \\ 0 & 0 & 0 & 1 & 0 & 0 & 0 & 1 \\ \hdashline 1 & 0 & 0 & 0 & -1 & 0 & 0 & 0 \\ 0 & 1 & 0 & 0 & 0 & -1 & 0 & 0 \\ 0 & 0 & 1 & 0 & 0 & 0 & -1 & 0 \\ 0 & 0 & 0 & 1 & 0 & 0 & 0 & -1 \end{array}\right]$$

and

$$\mathbf{H}_w(3) = \left[\begin{array}{cc:cc:cc:cc} 1 & 1 & & & & & & \mathbf{0} \\ 1 & -1 & & & & & & \\ \hdashline & & 1 & -1 & & & & \\ & & 1 & 1 & & & & \\ \hdashline & & & & 1 & 1 & & \\ & & & & 1 & -1 & & \\ \hdashline & & & & & & 1 & -1 \\ \mathbf{0} & & & & & & 1 & 1 \end{array}\right] \left[\begin{array}{cc:cc:cc:cc} 1 & 0 & 1 & 0 & & & & \\ 0 & 1 & 0 & 1 & & \mathbf{0}_4 & & \\ \hdashline 1 & 0 & -1 & 0 & & & & \\ 0 & 1 & 0 & -1 & & & & \\ \hdashline & & & & 1 & 0 & -1 & 0 \\ & \mathbf{0}_4 & & & 0 & 1 & 0 & -1 \\ \hdashline & & & & 1 & 0 & 1 & 0 \\ & & & & 0 & 1 & 0 & 1 \end{array}\right]$$

$$\times \left[\begin{array}{cccc:cccc} 1 & 0 & 0 & 0 & 1 & 0 & 0 & 0 \\ 0 & 1 & 0 & 0 & 0 & 1 & 0 & 0 \\ 0 & 0 & 1 & 0 & 0 & 0 & 1 & 0 \\ 0 & 0 & 0 & 1 & 0 & 0 & 0 & 1 \\ \hdashline 1 & 0 & 0 & 0 & -1 & 0 & 0 & 0 \\ 0 & 1 & 0 & 0 & 0 & -1 & 0 & 0 \\ 0 & 0 & 1 & 0 & 0 & 0 & -1 & 0 \\ 0 & 0 & 0 & 1 & 0 & 0 & 0 & -1 \end{array}\right] \left[\begin{array}{cccc:cccc} 1 & 0 & 0 & 0 & 0 & 0 & 0 & 0 \\ 0 & 0 & 0 & 0 & 1 & 0 & 0 & 0 \\ 0 & 0 & 1 & 0 & 0 & 0 & 0 & 0 \\ 0 & 0 & 0 & 0 & 0 & 0 & 1 & 0 \\ \hdashline 0 & 1 & 0 & 0 & 0 & 0 & 0 & 0 \\ 0 & 0 & 0 & 0 & 0 & 1 & 0 & 0 \\ 0 & 0 & 0 & 1 & 0 & 0 & 0 & 0 \\ 0 & 0 & 0 & 0 & 0 & 0 & 0 & 1 \end{array}\right]$$

6-2 If $\{X(m)\} = \{1\ 2\ 1\ 1\ 3\ 2\ 1\ 2\}$, then from Example 6.5-1, we have

$$\{W_x(m)\} = \left\{\frac{13}{8}\ \ \frac{-3}{8}\ \ \frac{-1}{8}\ \ \frac{3}{8}\ \ \frac{1}{8}\ \ \frac{-3}{8}\ \ \frac{-1}{8}\ \ \frac{-1}{8}\right\}$$

(a) Show that the $(\mathrm{WHT})_w$ power and phase spectra of $\{X(m)\}$ are as follows:

$$P_w(0) = \frac{169}{64}, \quad P_w(1) = \frac{10}{64}, \quad P_w(2) = \frac{10}{64},$$

$$P_w(3) = \frac{10}{64}, \quad \text{and} \quad P_w(4) = \frac{1}{64}$$

$$\psi_w(0) = 0, \quad \psi_w(1) = 251.58°, \quad \psi_w(2) = 71.58°,$$

$$\psi_w(3) = 251.58° \quad \text{and} \quad \psi_w(4) = 90°$$

(b) Using the above power and phase spectra, show how $\{X(m)\} = \{1\ 2\ 1\ 1\ 3\ 2\ 1\ 2\}$ can be recovered.

6-3 A general recursive formula for the similarity transform corresponding to $\mathbf{A}(3)$ in Eq. (6.8-4) was developed by Ohnsorg [27]:

$$\mathbf{A}(k+1) = \begin{bmatrix} \mathbf{A}(k) & \mathbf{0} \\ \mathbf{0} & \mathbf{D}(k) \end{bmatrix}, \qquad k = 1, 2, \ldots, n-1, \quad n = \log_2 N$$

where

$$\mathbf{A}(1) = \begin{bmatrix} 1 & 0 \\ 0 & -1 \end{bmatrix}$$

$$\mathbf{D}(k) = \mathbf{A}(k) - 2^{-(k-1)}[\mathbf{J}(k)]$$

$$\mathbf{J}(k) = [\boldsymbol{H}_h(k)\ \boldsymbol{H}_h(k) \cdots \boldsymbol{H}_h(k)]$$

and $\boldsymbol{H}_h(k)$ is a 2^k-vector which is identical to the *last* column of $\mathbf{H}_h(k)$.
Use the above recurrence formula to show that

$$\mathbf{A}(2) = \begin{bmatrix} 1 & 0 & & \\ 0 & -1 & \multicolumn{2}{c}{\mathbf{0}_2} \\ & & 0 & -1 \\ \multicolumn{2}{c}{\mathbf{0}_2} & 1 & 0 \end{bmatrix}$$

and

$$\mathbf{A}(3) = \frac{1}{2} \begin{bmatrix} 2 & 0 & & & & & & \\ 0 & -2 & & & & & & \\ & & 0 & -2 & & & & \\ & & 2 & 0 & & & & \\ & & & & 1 & -1 & -1 & -1 \\ & & & & 1 & -1 & 1 & 1 \\ & & & & 1 & 1 & 1 & -1 \\ & & & & -1 & -1 & 1 & -1 \end{bmatrix}$$

6-4 Consider the data sequence

$$\{X(m)\} = \{1\ 2\ 3\ 1\ 5\ 3\ 4\ 1\}$$

(a) Use the $(\mathrm{FWHT})_w$ to show that $\{W_x(m)\} = \left\{\frac{20}{8}\ \ \frac{-6}{8}\ \ \frac{-4}{8}\ \ \frac{2}{8}\ \ \frac{-4}{8}\ \ \frac{-2}{8}\ \ \frac{-4}{8}\ \ \frac{6}{8}\right\}$

(b) Find the $(\mathrm{WHT})_w$ power spectrum of $\{X(m)\}$

Answer:

$$P_w(0) = \frac{400}{64}, \quad P_w(1) = \frac{52}{64}, \quad P_w(2) = \frac{20}{64},$$

$$P_w(3) = \frac{20}{64} \quad \text{and} \quad P_w(4) = \frac{36}{64}$$

(c) Let $\{Z(m)\}_3$ denote the sequence obtained by subjecting $\{X(m)\}$ to a dyadic shift of size 3. Then [see Table 6.6-1]

$$\{Z(m)\}_3 = \{1\ 3\ 2\ 1\ 1\ 4\ 3\ 5\}$$

Use the $(\mathrm{FWHT})_w$ to show that the $(\mathrm{WHT})_w$ of $\{Z(m)\}_3$ is

$$\{W_{z,3}(m)\} = \left\{\frac{20}{8}\ \frac{-6}{8}\ \frac{4}{8}\ \frac{-2}{8}\ \frac{-4}{8}\ \frac{-2}{8}\ \frac{4}{8}\ \frac{-6}{8}\right\}$$

(d) Verify that the $(\mathrm{WHT})_w$ power spectrum of $\{Z(m)\}_3$ is the same as that of $\{X(m)\}$.

6-5 (a) Decompose the data sequence

$$\{X(m)\} = \{1\ 2\ 1\ 1\ 3\ 2\ 1\ 2\}$$

into 4 mutually orthogonal subsequences such that

$$\{X(m)\} = \{F_3(m)\} + \{G_3(m)\} + \{G_2(m)\} + \{G_1(m)\},$$

where $\{F_3(m)\}$ is 1-periodic and $\{G_3(m)\}$, $\{G_2(m)\}$, $\{G_1(m)\}$ are respectively 1-, 2-, and 4-antiperiodic.

Answer:

$$\begin{aligned}
\{F_3(m)\} &= \{1.625\ \ 1.625\ \ 1.625\ \ 1.625\ \ 1.625\ \ 1.625\ \ 1.625\ \ 1.625\} \\
\{G_3(m)\} &= \{-0.125\ \ 0.125\ \ -0.125\ \ 0.125\ \ -0.125\ \ 0.125\ \ -0.125\ \ 0.125\} \\
\{G_2(m)\} &= \{0.5\ \ 0.25\ \ -0.5\ \ -0.25\ \ 0.5\ \ 0.25\ \ -0.5\ \ -0.25\} \\
\{G_1(m)\} &= \{-1\ \ 0\ \ 0\ \ -0.5\ \ 1\ \ 0\ \ 0\ \ 0.5\}
\end{aligned}$$

(b) From Example 6.3-1 we know that

$$\{B_x(k)\} = \left\{\frac{13}{8}\ \frac{-1}{8}\ \frac{3}{8}\ \frac{1}{8}\ \frac{-3}{8}\ \frac{-1}{8}\ \frac{-1}{8}\ \frac{-3}{8}\right\}$$

Use this information to evaluate the $(\mathrm{WHT})_h$ power spectrum $P_h(0)$, $P_h(1)$, $P_h(2)$, and $P_h(3)$.

(c) Verify that

$$P_h(0) = \text{average power in } \{F_3(m)\}$$

and

$$P_h(r) = \text{average power in } \{G_{4-r}(m)\}, \qquad r = 1, 2, 3$$

6-6 (a) $\mathbf{D}(k)$ is an orthogonal matrix such that

$$\mathbf{D}(k)'\,\mathbf{D}(k) = c\,\mathbf{I}(k)$$

where c is a constant. If $\mathbf{D}^m(k)$ denotes the m-th power of $\mathbf{D}(k)$, show that $\mathbf{D}^m(k)$ is also orthogonal for any m which is a finite positive integer.

(b) Verify that Eq. (6.9-12) can be expressed in the following compact form.

$$F[P_h(0)] = 0$$

$$F[P_h(1)] = N/2$$

and

$$F[P_h(r)] = 2^{n-r}(2k+1), \qquad r = 2, 3, \ldots, n; \quad k = 0, 1, \ldots, 2^{r-2} - 1,$$

where $n = \log_2 N$.

6-7 In this problem, an interesting property of the $(\text{WHT})_h$ phase spectrum is illustrated for the case $N = 8$. Consider the 8-periodic data sequence

$$\{X(m)\} = \{1\ 2\ 1\ 1\ 3\ 2\ 1\ 2\}$$

Let $\{Z(m)\}_l$ denote the sequence obtained by subjecting $\{X(m)\}$ to the left cyclic shift of size l. That is,

$$\{Z(m)\}_l = \{X(l)\ X(l+1) \ldots X(l-2)\ X(l-1)\}, \qquad l = 0, 1, 2, \ldots, 7$$

where $\{Z(m)\}_0 = \{X(m)\}$.

Compute the $(\text{WHT})_h$ phase spectra of $\{Z(m)\}_l$, $l = 0, 1, \ldots, 7$ and show that it varies as shown in the following table.

$\psi_h(l)$ ↓ $l \rightarrow$	0	1	2	3	4	5	6	7
$\psi_h(0)$	1	1	1	1	1	1	1	1
$\psi_h(1)$	-1	1	-1	1	-1	1	-1	1
$\psi_h(2)$	$\frac{2}{\sqrt{5}}$	$\frac{1}{\sqrt{5}}$	$\frac{-2}{\sqrt{5}}$	$\frac{-1}{\sqrt{5}}$	$\frac{2}{\sqrt{5}}$	$\frac{1}{\sqrt{5}}$	$\frac{-2}{\sqrt{5}}$	$\frac{-1}{\sqrt{5}}$
$\psi_h(3)$	$\frac{-2}{\sqrt{5}}$	0	0	$\frac{-1}{\sqrt{5}}$	$\frac{2}{\sqrt{5}}$	0	0	$\frac{1}{\sqrt{5}}$

Observations. (i) $\psi_h(0)$ does not change with respect to $l = 1, 2, \ldots, 7$.
(ii) $\psi_h(r)$ is 2^{r-1}-antiperiodic $r = 1, 2, 3$, as l is varied from 0 through 7.

Comment. For any $N = 2^n$, $\psi_h(r)$ is 2^{r-1}-antiperiodic $r = 1, 2, \ldots, n$, as l is varied from 0 through $N - 1$.

6-8 Given the (2×4) data matrix

$$[X(m_1, m_2)] = \begin{bmatrix} 1 & 1 & 3 & 1 \\ 2 & 1 & 2 & 2 \end{bmatrix}$$

(a) Show that the 2-dimensional $(\text{WHT})_h$ of $[X(m_1, m_2)]$ is given by

$$[B_{xx}(u_1, u_2)] = \frac{1}{8}\begin{bmatrix} 13 & 3 & -3 & -1 \\ -1 & 1 & -1 & -3 \end{bmatrix}$$

(b) Arrange the columns of $[X(m_1, m_2)]$ in the form of a 8-vector to obtain

$$\mathbf{Y}(3) = \begin{bmatrix} 1 \\ 2 \\ \hdashline 1 \\ 1 \\ \hdashline 3 \\ 2 \\ \hdashline 1 \\ 2 \end{bmatrix}$$

Use the 1-dimensional $(\text{FWHT})_h$ to compute $B_y(k)$, $k = 0, 1, \ldots, 7$. Hence show that

$$\begin{bmatrix} B_y(0) \\ B_y(1) \\ B_y(2) \\ B_y(3) \\ B_y(4) \\ B_y(5) \\ B_y(6) \\ B_y(7) \end{bmatrix} = \begin{bmatrix} B_{xx}(0, 0) \\ B_{xx}(1, 0) \\ B_{xx}(0, 1) \\ B_{xx}(1, 1) \\ B_{xx}(0, 2) \\ B_{xx}(1, 2) \\ B_{xx}(0, 3) \\ B_{xx}(1, 3) \end{bmatrix} \begin{matrix} \left.\begin{matrix} \\ \end{matrix}\right\} \text{column 1 of } [B_{xx}(u_1, u_2)] \\ \left.\begin{matrix} \\ \end{matrix}\right\} \text{column 2 of } [B_{xx}(u_1, u_2)] \\ \left.\begin{matrix} \\ \end{matrix}\right\} \text{column 3 of } [B_{xx}(u_1, u_2)] \\ \left.\begin{matrix} \\ \end{matrix}\right\} \text{column 4 of } [B_{xx}(u_1, u_2)] \end{matrix}$$

6-9 Fig. P6-9-2 shows the functions $X(t \oplus \tau_i)$, $i = 1, 2, \ldots, 6$, which are obtained by subjecting $x(t)$ in Fig. P6.9-1 to dyadic shifts. What are the values of τ_i, $i = 1, 2, \ldots, 6$? *Hint:* Use Table 6.6-1.

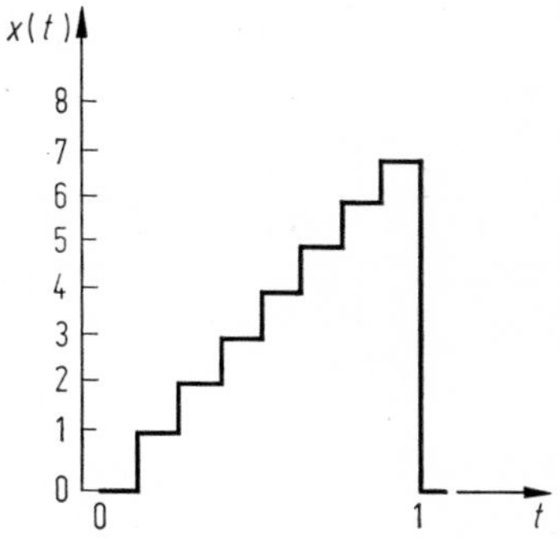

Fig. P6-9-1 A staircase function

6-10 Consider the exponentially decaying sequence

$$X(m) = 10e^{-4m/31}, \qquad m = 0, 1, \ldots, 31$$

Use the $(\text{FWHT})_h$ program in Appendix 6.1 to compute the $(\text{WHT})_h$ power spectrum of the above sequence.

Answer: $P_h(0) = 6.8046$, $P_h(1) = 0.0271$, $P_h(2) = 0.1044$,
$P_h(3) = 0.4208$, $P_h(4) = 1.5697$, and $P_h(5) = 5.1749$

6-11 Extend the 1-dimensional $(\text{WHT})_w$ and $(\text{WHT})_h$ subroutine in Appendix 6-1 so that it is capable of computing the corresponding 2-dimensional transforms. Subsequently use it to compute the 2-dimensional $(\text{WHT})_w$ and $(\text{WHT})_h$ of the data matrix

$$[X(m_1, m_2)] = \begin{bmatrix} 1 & \varrho & \varrho^2 & \varrho^3 \\ \varrho & 1 & \varrho & \varrho^2 \\ \varrho^2 & \varrho & 1 & \varrho \\ \varrho^3 & \varrho^2 & \varrho & 1 \end{bmatrix}$$

where $\varrho = 0.9$.

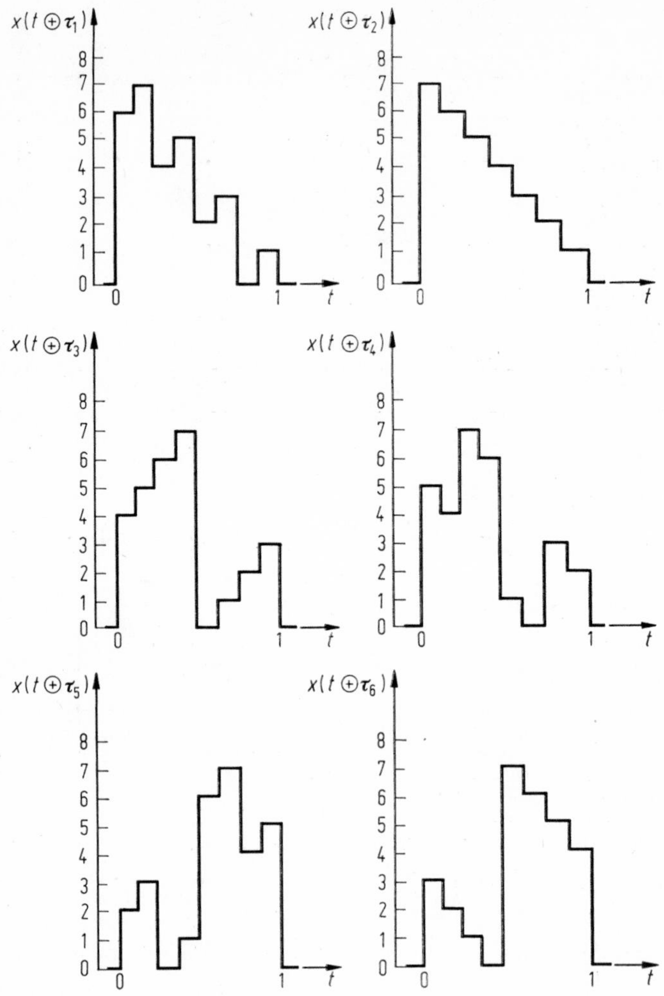

Fig. P6-9-2 Dyadically shifted versions of $x(t)$

6-12 Repeat Problem 6-11 for

$$[X(m_1, m_2)] = \begin{bmatrix} 1 & \varrho & \varrho^2 & \cdots & \varrho^7 \\ \varrho & 1 & \varrho^2 & \cdots & \varrho^6 \\ \varrho^2 & \varrho & 1 & \cdots & \varrho^5 \\ \cdots & \cdots & \cdots & \cdots & \cdots \\ \varrho^7 & \varrho^6 & \varrho^5 & \cdots & 1 \end{bmatrix}$$

where $\varrho = 0.9$.

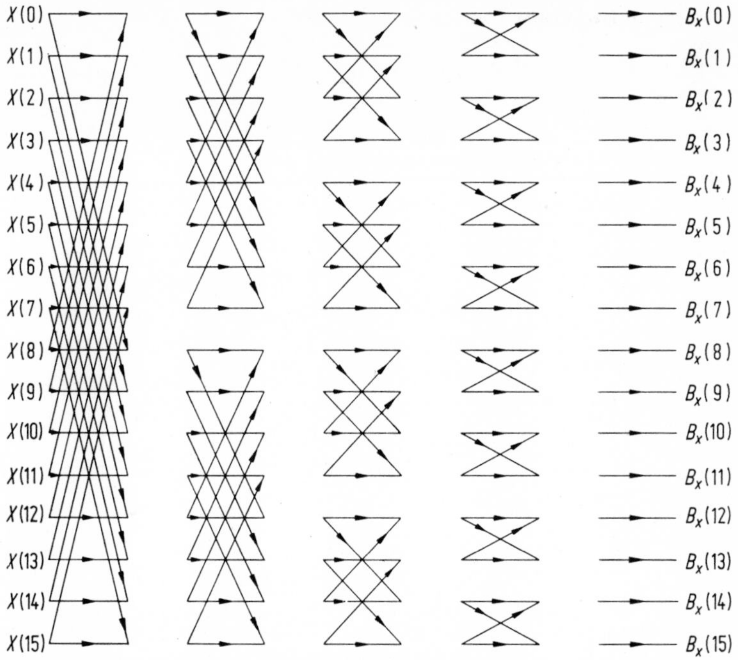

Fig. P6-13-1 $(\text{FWHT})_h$ signal flow graph, $N = 16$

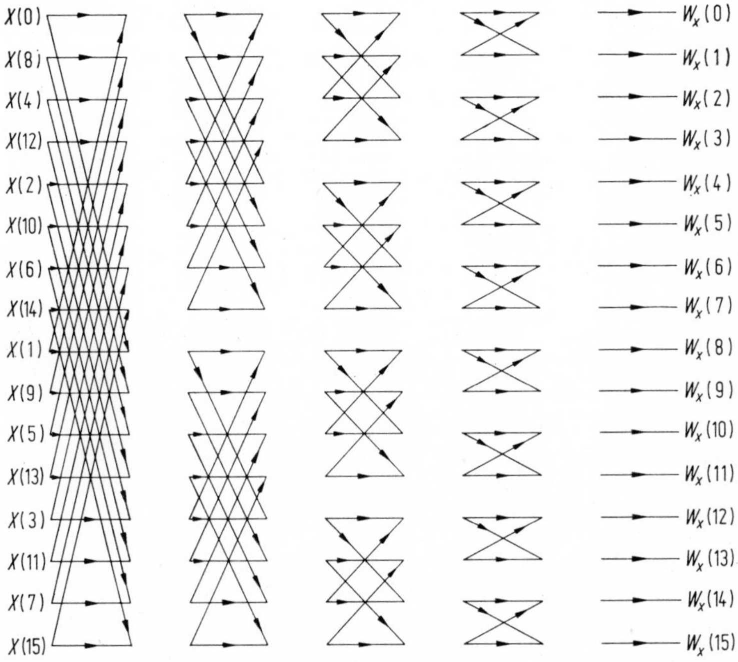

Fig. P6-13-2 $(\text{FWHT})_w$ signal flow graph, $N = 16$

6-13 The transform matrices $\mathbf{H}_h(4)$ and $\tilde{\mathbf{H}}_w(4)$ can be factored as follows:

$$\mathbf{H}_h(4) = \begin{bmatrix} \mathbf{H}_h(1) & & & \\ & \mathbf{H}_h(1) & & \\ & & \ddots & \\ & & & \mathbf{H}_h(1) \end{bmatrix} \begin{bmatrix} \begin{bmatrix} \mathbf{I}_2 & \mathbf{I}_2 \\ \mathbf{I}_2 & -\mathbf{I}_2 \end{bmatrix} & & \\ & \ddots & \\ & & \begin{bmatrix} \mathbf{I}_2 & \mathbf{I}_2 \\ \mathbf{I}_2 & -\mathbf{I}_2 \end{bmatrix} \end{bmatrix} \begin{bmatrix} \mathbf{I}_4 & \mathbf{I}_4 & & \\ \mathbf{I}_4 & -\mathbf{I}_4 & \multicolumn{2}{c}{\mathbf{0}_4} \\ \mathbf{0}_4 & & \mathbf{I}_4 & \mathbf{I}_4 \\ & & \mathbf{I}_4 & -\mathbf{I}_4 \end{bmatrix}$$

$$\times \begin{bmatrix} \mathbf{I}_8 & \mathbf{I}_8 \\ \mathbf{I}_8 & -\mathbf{I}_8 \end{bmatrix},$$

$$\tilde{\mathbf{H}}_w(4) = \operatorname{diag}\left(\begin{bmatrix} 1 & 1 \\ 1 & -1 \end{bmatrix}, \begin{bmatrix} 1 & -1 \\ 1 & 1 \end{bmatrix}, \begin{bmatrix} 1 & 1 \\ 1 & -1 \end{bmatrix}, \begin{bmatrix} 1 & -1 \\ 1 & 1 \end{bmatrix}, \begin{bmatrix} 1 & 1 \\ 1 & -1 \end{bmatrix}, \begin{bmatrix} 1 & -1 \\ 1 & 1 \end{bmatrix}, \begin{bmatrix} 1 & 1 \\ 1 & -1 \end{bmatrix},\right.$$

$$\left.\begin{bmatrix} 1 & -1 \\ 1 & 1 \end{bmatrix}\right) \times \operatorname{diag}\left(\begin{bmatrix} \mathbf{I}_2 & \mathbf{I}_2 \\ \mathbf{I}_2 & -\mathbf{I}_2 \end{bmatrix}, \begin{bmatrix} \mathbf{I}_2 & -\mathbf{I}_2 \\ \mathbf{I}_2 & \mathbf{I}_2 \end{bmatrix}, \begin{bmatrix} \mathbf{I}_2 & \mathbf{I}_2 \\ \mathbf{I}_2 & -\mathbf{I}_2 \end{bmatrix}, \begin{bmatrix} \mathbf{I}_2 & -\mathbf{I}_2 \\ \mathbf{I}_2 & \mathbf{I}_2 \end{bmatrix}\right)$$

$$\times \operatorname{diag}\left(\begin{bmatrix} \mathbf{I}_4 & \mathbf{I}_4 \\ \mathbf{I}_4 & -\mathbf{I}_4 \end{bmatrix}, \begin{bmatrix} \mathbf{I}_4 & -\mathbf{I}_4 \\ \mathbf{I}_4 & \mathbf{I}_4 \end{bmatrix}\right) \begin{bmatrix} \mathbf{I}_8 & \mathbf{I}_8 \\ \mathbf{I}_8 & -\mathbf{I}_8 \end{bmatrix}$$

where $\tilde{\mathbf{H}}_w(4)$ is obtained by rearranging the rows of $\mathbf{H}_w(4)$ in bit-reversed order. Fill in the multipliers in the signal flow graphs shown in Figs. P6-13-1 and P6-13-2 for computing the two Walsh-Hadamard transforms. How many multiplications and additions/subtractions are required for these flow graphs?

6-14 From an inspection of the matrix factors of $\mathbf{H}_h(4)$ and $\tilde{\mathbf{H}}_w(4)$ [see Prob. 6-13], write down the matrix factors of $\mathbf{H}_h(5)$ and $\tilde{\mathbf{H}}_w(5)$. Generalize the matrix factoring of $\mathbf{H}_h(n)$ and $\tilde{\mathbf{H}}_w(n)$ for any, n, $n = \log_2 N$.

6-15 (a) Based on the multipliers shown in the MWHT signal flow graph [see Fig. 6.12], express the transform matrix $\hat{\mathbf{H}}(3)$ [see Eq. (6.10-3)] in terms of its matrix factors.

(b) Repeat part (a) for IMWHT [see Fig. 6.13]. Verify that the product of the matrix factors yields $\hat{\mathbf{H}}(3)'$.

6-16 The transform matrix $\hat{\mathbf{H}}(4)$ can be factored as follows:

$$\hat{\mathbf{H}}(4) = \begin{bmatrix} 1 & 1 & & \mathbf{0} \\ 1 & -1 & & \\ & & \sqrt{2}\,\mathbf{I}_2 & \\ \mathbf{0} & & & \mathbf{I}_{12} \end{bmatrix} \begin{bmatrix} \mathbf{I}_2 & \mathbf{I}_2 & & \mathbf{0} \\ \mathbf{I}_2 & -\mathbf{I}_2 & & \\ & & 2\,\mathbf{I}_4 & \\ \mathbf{0} & & & \mathbf{I}_8 \end{bmatrix} \begin{bmatrix} \mathbf{I}_4 & \mathbf{I}_4 & \mathbf{0}_8 \\ \mathbf{I}_4 & -\mathbf{I}_4 & \\ \mathbf{0}_8 & & 2^{3/2}\,\mathbf{I}_8 \end{bmatrix} \begin{bmatrix} \mathbf{I}_8 & \mathbf{I}_8 \\ \mathbf{I}_8 & -\mathbf{I}_8 \end{bmatrix}$$

Develop the complete flow graph for efficient computation of MWHT, $N = 16$. What are the number of multiplications and additions/subtractions required for this flow graph?

6-17 Repeat Prob. 6-16 for IMWHT, $N = 16$.

6-18 From an inspection of the matrix factors of $\hat{\mathbf{H}}(3)$, and $\hat{\mathbf{H}}(4)$, write down the matrix factors of $\hat{\mathbf{H}}(5)$. Generalize the matrix factoring for any n, $n = \log_2 N$.

6-19 Develop the signal flow graph for $N = 16$, similar to that shown in Fig. 6.10.

Chapter Seven

Miscellaneous Orthogonal Transforms

In addition to the DFT, $(\mathrm{WHT})_w$, $(\mathrm{WHT})_h$, and MWHT, there are several other discrete orthogonal transforms. Of these, we will study the following in this chapter: (1) Generalized transform, (2) Haar transform, (3) slant transform, and (4) the discrete cosine transform.

For a given data sequence $X(m)$, $m = 0, 1, \ldots, N-1$, we show that a class of $\log_2 N$ orthogonal transforms can be defined in terms of the generalized transform $(\mathrm{GT})_r$, $r = 0, 1, \ldots, n-1$, where $n = \log_2 N$. The transform $(\mathrm{GT})_0$ yields the $(\mathrm{WHT})_h$, while $(\mathrm{GT})_{n-1}$ yields the DFT. Consequently the $(\mathrm{GT})_r$ enables a systematic transition from the $(\mathrm{WHT})_h$ to the DFT.

The motivation for studying the Haar, slant, and discrete cosine transforms is that they have been considered for certain applications, some of which will be discussed in Chapter 9. An additional orthogonal transformation called the Karhunen-Loève transform (KLT) will also be considered in Chapter 9. The reason for postponing the discussion of the KLT is that its development requires some material in Chapter 8 as a prerequisite.

7.1 Matrix Factorization

The key to obtaining the definition of the generalized transform is matrix factorization, some aspects of which are discussed in this section.

From the definition of the DFT, it follows that

$$C(n) = \frac{1}{N} \Lambda(n) X(n) \tag{7.1-1}$$

where

$$X(n)' = [X(0)\ X(1) \ldots X(N-1)]$$

is the data vector,

$$C(n)' = [C(0)\ C(1) \ldots C(N-1)]$$

is the DFT vector[1] and

$$\Lambda(n) = \begin{bmatrix} W^0 & W^0 & W^0 & \ldots & W^0 \\ W^0 & W^1 & W^2 & \ldots & W^{(N-1)} \\ W^0 & W^2 & W^4 & \ldots & W^{2(N-1)} \\ \cdot & \cdot & \cdot & \cdot & \cdot \\ W^0 & W^{(N-1)} & W^{2(N-1)} & \ldots & W^{(N-1)(N-1)} \end{bmatrix}$$

with $W = e^{-i\,2\pi/N}$ and $i = \sqrt{-1}$.

[1] For convenience, the subscript x in $C_x(k)$ is omitted.

Let $\tilde{C}(n)$ and $\tilde{\mathbf{\Lambda}}(n)$ denote the vector and matrix obtained by rearranging the *rows* of $C(n)$ and $\mathbf{\Lambda}(n)$ in bit-reversed order. For example, if

$$C(3)' = [C(0)\ C(1)\ C(2)\ C(3)\ C(4)\ C(5)\ C(6)\ C(7)]$$

then

$$\tilde{C}(3)' = [C(0)\ C(4)\ C(2)\ C(6)\ C(1)\ C(5)\ C(3)\ C(7)]$$

Thus corresponding to Eq. (1), we have

$$\tilde{C}(n) = \frac{1}{N}\tilde{\mathbf{\Lambda}}(n)X(n) \tag{7.1-2}$$

From Prob. 4-3 it follows that $\tilde{\mathbf{\Lambda}}(n)$ can be expressed as a product of n sparse matrices, such that

$$\tilde{\mathbf{\Lambda}}(n) = \prod_{j=1}^{n} \mathbf{F}_j(n) \tag{7.1-3}$$

With $N = 16$ for example, Eq. (3) is of the form

$$\tilde{\mathbf{\Lambda}}(4) = \prod_{j=1}^{4} \mathbf{F}_j(4)$$

where

$$\mathbf{F}_1(4) = \operatorname{diag}\left(\begin{bmatrix}1 & 1\\ 1 & -1\end{bmatrix}, \begin{bmatrix}1 & W^4\\ 1 & -W^4\end{bmatrix}, \begin{bmatrix}1 & W^2\\ 1 & -W^2\end{bmatrix}, \begin{bmatrix}1 & W^6\\ 1 & -W^6\end{bmatrix}, \begin{bmatrix}1 & W\\ 1 & -W\end{bmatrix},\right.$$
$$\left.\begin{bmatrix}1 & W^5\\ 1 & -W^5\end{bmatrix}, \begin{bmatrix}1 & W^3\\ 1 & -W^3\end{bmatrix}, \begin{bmatrix}1 & W^7\\ 1 & -W^7\end{bmatrix}\right),$$

$$\mathbf{F}_2(4) = \operatorname{diag}\left(\begin{bmatrix}\mathbf{I}_2 & \mathbf{I}_2\\ \mathbf{I}_2 & -\mathbf{I}_2\end{bmatrix}, \begin{bmatrix}\mathbf{I}_2 & W^4\mathbf{I}_2\\ \mathbf{I}_2 & -W^4\mathbf{I}_2\end{bmatrix}, \begin{bmatrix}\mathbf{I}_2 & W^2\mathbf{I}_2\\ \mathbf{I}_2 & -W^2\mathbf{I}_2\end{bmatrix}, \begin{bmatrix}\mathbf{I}_2 & W^6\mathbf{I}_2\\ \mathbf{I}_2 & -W^6\mathbf{I}_2\end{bmatrix}\right),$$

$$\mathbf{F}_3(4) = \operatorname{diag}\left(\begin{bmatrix}\mathbf{I}_4 & \mathbf{I}_4\\ \mathbf{I}_4 & -\mathbf{I}_4\end{bmatrix}, \begin{bmatrix}\mathbf{I}_4 & W^4\mathbf{I}_4\\ \mathbf{I}_4 & -W^4\mathbf{I}_4\end{bmatrix}\right), \quad \mathbf{F}_4(4) = \begin{bmatrix}\mathbf{I}_8 & \mathbf{I}_8\\ \mathbf{I}_8 & -\mathbf{I}_8\end{bmatrix},$$

and $\mathbf{I}_m$ denotes the $(m \times m)$ identity matrix.

Inspection of the above matrices $\mathbf{F}_j(4)$, $j = 1, 2, 3, 4$ shows that one can define *four* orthogonal transforms as follows:

(i) Set *all* W terms equal to 1. This corresponds to the $(\mathrm{WHT})_h$ which we define as

$$B_0(4) = \frac{1}{16}\tilde{\mathbf{\Lambda}}_0(4)X(4) \tag{7.1-4}$$

where $B_0(4)' = [B_0(0)\ B_0(1)\ \cdots\ B_0(15)]$ denotes the $(\mathrm{WHT})_h$ vector, and $\tilde{\mathbf{\Lambda}}_0(4) = \mathbf{H}_h(4)$. Hence the elements of $\tilde{\mathbf{\Lambda}}_0(4)$ are ± 1.

(ii) Set W^2, W^6, W, W^5, W^3, and W^7 equal to 1. This process results in the complex Walsh-Hadamard or complex BIFORE transform [1–3] which is defined as

$$\boldsymbol{B}_1(4) = \frac{1}{16}\tilde{\boldsymbol{\Lambda}}_1(4)\boldsymbol{X}(4) \tag{7.1-5}$$

where $\boldsymbol{B}_1(4)' = [B_1(0)\ B_1(1) \cdots B_1(15)]$ is the transform vector, and $\tilde{\boldsymbol{\Lambda}}_1(4)$ is the corresponding transform matrix, which consists of the elements $\{\pm 1, \pm W^4\}$.

(iii) Set W, W^5, W^3 and W^7 equal to 1, and denote the resulting transform coefficients by $B_2(k)$, $k = 0, 1, \ldots, 15$. The corresponding transform matrix $\boldsymbol{\Lambda}_2(4)$ consists of the elements $\{\pm 1, \pm W^4, \pm W^2, \pm W^6\}$, and the transform is defined as

$$\boldsymbol{B}_2(4) = \frac{1}{16}\tilde{\boldsymbol{\Lambda}}_2(4)\boldsymbol{X}(4) \tag{7.1-6}$$

(iv) Set none of the W terms equal to 1. The transform that results is given by

$$\boldsymbol{B}_3(4) = \frac{1}{16}\tilde{\boldsymbol{\Lambda}}_3(4)\boldsymbol{X}(4) \tag{7.1-7}$$

where $\boldsymbol{B}_3(4)' = [B_3(0)\ B_3(1) \cdots B_3(15)]$ denotes the transform vector, and $\tilde{\boldsymbol{\Lambda}}_3(4) = \tilde{\boldsymbol{\Lambda}}(4)$. Thus the $B_3(k)$ are related to the DFT coefficients $C(k)$ via a bit-reversal operation. That is

$$B_3(k) = C(\langle k \rangle), \qquad k = 0, 1, \ldots, 15 \tag{7.1-8}$$

where $\langle k \rangle$ is a decimal number obtained from the bit-reversal of a 4-bit binary representation of k.

Table 7.1-1 Multipliers for the signal flow graph shown in Fig. 7.1; $W = e^{-i2\pi/16}$

Multiplier	$B_0(k)$ coefficients	$B_1(k)$ coefficients	$B_2(k)$ coefficients	$B_3(k)$ coefficients
a_1	1	$-i$	W^4	W^4
a_2	-1	i	W^{12}	W^{12}
a_3	1	1	W^2	W^2
a_4	-1	-1	W^{10}	W^{10}
a_5	1	1	W^6	W^6
a_6	-1	-1	W^{14}	W^{14}
a_7	1	1	1	W
a_8	-1	-1	-1	W^9
a_9	1	1	1	W^5
a_{10}	-1	-1	-1	W^{13}
a_{11}	1	1	1	W^3
a_{12}	-1	-1	-1	W^{11}
a_{13}	1	1	1	W^7
a_{14}	-1	-1	-1	W^{15}

The signal flow graph for the above four transforms is shown in Fig. 7.1; the related multipliers are listed in Table 7.1-1. With the exception of the DFT, the flow graph in Fig. 7.1 yields the transform coefficients $B_j(k)$, $j = 0, 1, 2$, in natural order.

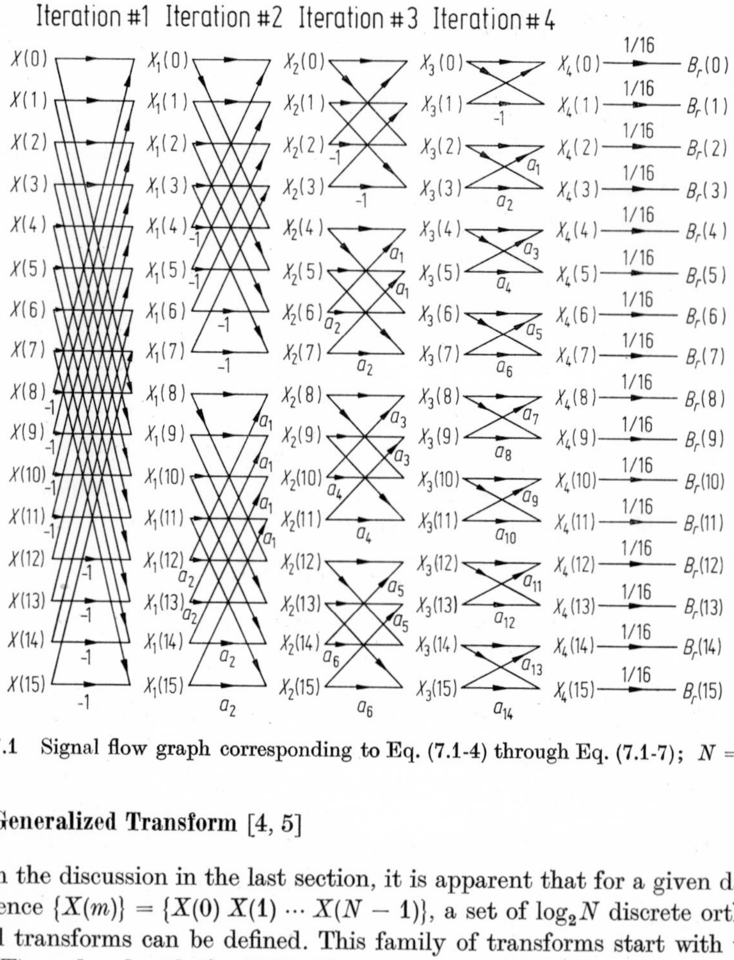

Fig. 7.1 Signal flow graph corresponding to Eq. (7.1-4) through Eq. (7.1-7); $N = 16$

7.2 Generalized Transform [4, 5]

From the discussion in the last section, it is apparent that for a given data sequence $\{X(m)\} = \{X(0)\ X(1) \cdots X(N-1)\}$, a set of $\log_2 N$ discrete orthogonal transforms can be defined. This family of transforms start with the $(\mathrm{WHT})_h$ and end with the DFT. Thus, if $B_r(k)$ denotes the k-th transform coefficient of the r-th transform, $r = 0, 1, \ldots, n-1$, then the generalized transform $(\mathrm{GT})_r$ is defined as

$$\boldsymbol{B}_r(n) = \frac{1}{N}\, \mathbf{G}_r(n)\, \boldsymbol{X}(n), \qquad r = 0, 1, \ldots, n-1 \tag{7.2-1}$$

where

$$\boldsymbol{B}_r(n)' = [B_r(0)\ B_r(1) \cdots B_r(N-1)], \qquad \boldsymbol{X}(n)' = [X(0)\ X(1) \cdots X(N-1)],$$

and $\mathbf{G}_r(n)$ is the transform matrix which can be expressed as a product of n sparse matrices $\mathbf{D}_r^j(n)$, i.e.,

$$\mathbf{G}_r(n) = \prod_{j=1}^{n} \mathbf{D}_r^j(n) \tag{7.2-2}$$

where $\mathbf{D}_r^j(n) = \text{diag}[\mathbf{A}_0^r(j)\ \mathbf{A}_1^r(j) \cdots \mathbf{A}_{2^{n-j}-1}^r(j)]$.

The matrix factor $\mathbf{D}_r^j(n)$ can be generated recursively as follows[1]:

$$\text{diag}\left[\mathbf{A}_0^r(j),\ \mathbf{A}_1^r(j),\ \ldots,\ \mathbf{A}_{2^{n-j}-1}^r(j)\right] \qquad j = 1, 2, \ldots, n$$

where

$$\mathbf{A}_m^r(1) = \begin{cases} \begin{bmatrix} 1 & W^{\langle\langle m \rangle\rangle} \\ 1 & -W^{\langle\langle m \rangle\rangle} \end{bmatrix}, & m = 0, 1, \ldots, 2^r - 1 \\ \begin{bmatrix} 1 & 1 \\ 1 & -1 \end{bmatrix}, & m = 2^r, 2^r + 1, \ldots, 2^{n-j} - 1 \end{cases}$$

$$\mathbf{A}_m^r(j) = \mathbf{A}_m^r(1) \otimes \mathbf{I}_{2^{(j-1)}} \tag{7.2-3}$$

The symbol $\otimes$ denotes Kronecker product[2], and $\langle\langle m \rangle\rangle$ is the decimal number resulting from the bit-reversal of an $(n-1)$-bit binary representation of m. That is, if

$$m = m_{n-2}\, 2^{n-2} + \cdots + m_1 2^1 + m_0 2^0$$

is an $(n-1)$-bit binary representation of m, then

$$\langle\langle m \rangle\rangle = m_0 2^{n-2} + m_1 2^{n-3} + \cdots + m_{n-3} 2^1 + m_{n-2} 2^0$$

The data sequence $\{X(m)\}$ can be recovered using the inverse generalized transform $(\text{IGT})_r$ which is defined as

$$\boldsymbol{X}(n) = \bar{\mathbf{G}}_r(n)' \boldsymbol{B}_r(n), \qquad r = 0, 1, \ldots, n-1 \tag{7.2-4}$$

where $\bar{\mathbf{G}}_r(n)'$ represents the transpose of the complex conjugate of $\mathbf{G}_r(n)$. The $(\text{IGT})_r$ follows from Eq. (1) as a consequence of the property

$$\bar{\mathbf{G}}_r(n)' \mathbf{G}_r(n) = N \mathbf{I}_N$$

From Eq. (2) we obtain

$$\bar{\mathbf{G}}_r(n)' = \prod_{l=0}^{n-1} \bar{\mathbf{D}}_r^{n-l}(n)' \tag{7.2-5}$$

where $\bar{\mathbf{D}}_r^s(n)'$ denotes the transpose of the complex conjugate of $\mathbf{D}_r^s(n)$.

[1] For an illustrative example, see Appendix 7.2.

[2] An introductory discussion of the Kronecker product is available in Appendix 7.1.

The transform matrices $\mathbf{G}_r(n)$ can also be generated recursively [6, 7] i.e.,

$$\mathbf{G}_0(k) = \mathbf{H}_h(k) = \begin{bmatrix} \mathbf{H}_h(k-1) & \mathbf{H}_h(k-1) \\ \mathbf{H}_h(k-1) & -\mathbf{H}_h(k-1) \end{bmatrix}$$

which is Eq. (5.4-10). For $r = 1, 2, \ldots, n-1$

$$\mathbf{G}_r(k) = \begin{bmatrix} \mathbf{G}_r(k-1) & \mathbf{G}_r(k-1) \\ \mathbf{A}_r(k-1) & -\mathbf{A}_r(k-1) \end{bmatrix} \tag{7.2-6}$$

with

$$\mathbf{G}_r(0) = 1, \quad \text{and} \quad \mathbf{G}_r(1) = \begin{bmatrix} 1 & 1 \\ 1 & -1 \end{bmatrix} = \mathbf{H}_h(1)$$

where $\mathbf{A}_r(k-1)$ is described in [6]. For example, the recursion relationship for $r = 2$ is

$$\mathbf{G}_2(k) = \begin{bmatrix} \mathbf{G}_2(k-1) & \mathbf{G}_2(k-1) \\ \mathbf{A}_2(k-1) & -\mathbf{A}_2(k-1) \end{bmatrix} \tag{7.2-7}$$

where

$$\mathbf{A}_2(k-1) = \begin{bmatrix} [1 \quad e^{-i\pi/2}] \otimes \begin{bmatrix} 1 & e^{-i\pi/4} \\ 1 & -e^{-i\pi/4} \end{bmatrix} \\ \\ [1 \quad -e^{-i\pi/2}] \otimes \begin{bmatrix} 1 & e^{-i3\pi/4} \\ 1 & -e^{-i3\pi/4} \end{bmatrix} \end{bmatrix} \otimes \mathbf{H}(k-3)$$

Thus, with respect to the $(\mathrm{GT})_r$, and the $(\mathrm{IGT})_r$, we summarize the following observations.

(i) $r = 0$ yields the $(\mathrm{WHT})_h$.
(ii) $r = 1$ yields the complex BIFORE transform (CBT).
(iii) $r = (n-1)$ yields the DFT in bit-reversed order; i.e.,

$$B_{n-1}(k) = C(\langle k \rangle), \qquad k = 0, 1, \ldots, N-1 \tag{7.2-8}$$

where $\langle k \rangle$ is the decimal number obtained by the bit-reversal of an n-bit binary representation of k.
(iv) As r is varied from 2 through $(n-2)$, an additional $(n-3)$ orthogonal transforms are generated.
(v) The complexity of the $(\mathrm{GT})_r$ increases as r is increased, in the sense that it requires a larger set of powers of W to compute the transform coefficients, as described in Table 7.2-1.

Table 7.2-1 Description of the elements in the $(GT)_r$ family

Transform	Number of different elements	Elements	Elements on the unit circle
$(GT)_0$	2	$e^{-i2\pi}$, $e^{-i\pi}$	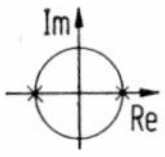
$(GT)_1$	2^2	$e^{-i2\pi}$, $e^{-i\pi}$, $e^{\pm i\pi/2}$	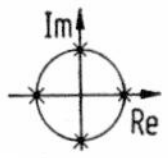
$(GT)_2$	2^3	$e^{-i2\pi}$, $e^{-i\pi}$, $e^{\pm i\pi/2}$, $e^{\pm i\pi/4}$, $e^{\pm i3\pi/4}$	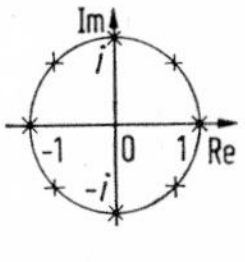
$(GT)_3$	2^4	$e^{-i2\pi}$, $e^{-i\pi}$, $e^{\pm i\pi/2}$, $e^{\pm i\pi/4}$, $e^{\pm i3\pi/4}$, $e^{\pm i\pi/8}$, $e^{\pm i3\pi/8}$, $e^{\pm i5\pi/8}$, $e^{\pm i7\pi/8}$	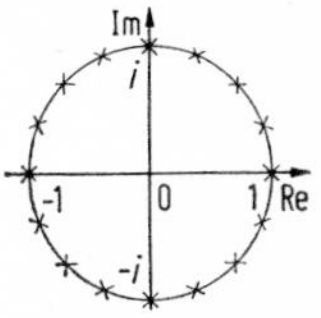
$(GT)_4$	2^5	all the preceding elements plus $e^{\pm ij\pi/16}$, $j = 1, 3, 5, 7, 9, 11, 13, 15$	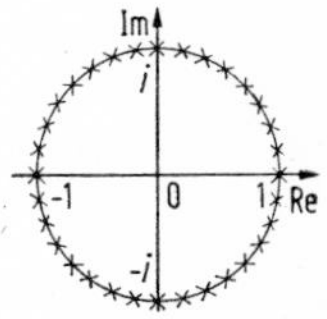
⋮	⋮	⋮	⋮
$(GT)_{n-1}$	$2^n = N$	$e^{\pm(i2\pi j/N)}$, $j = 0, 1, 2, \ldots, N/2$	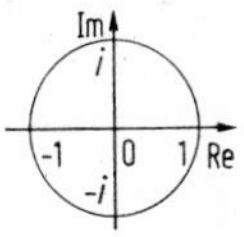

7.3 Haar Transform

The Haar transform (HT) coefficients $Y_x(k)$, $k = 0, 1, \ldots, N-1$ corresponding to a data sequence $\{X(m)\} = \{X(0)\ X(1) \cdots X(N-1)\}$, are obtained by computing the transformation

$$\boldsymbol{Y}_x(n) = \frac{1}{N} \mathbf{H}^*(n) \boldsymbol{X}(n) \tag{7.3-1}$$

where $\mathbf{H}^*(n)$ is the $(N \times N)$ Haar matrix [8]. $\mathbf{H}^*(n)$ is obtained by sampling the set of Haar functions $\{\mathrm{har}(r, m, t)\}$ defined in Eq. (5.3-2). For example, the (8×8) Haar matrix is given by [see Fig. 5.4b]

$$\mathbf{H}^*(3) = \begin{bmatrix} 1 & 1 & 1 & 1 & 1 & 1 & 1 & 1 \\ 1 & 1 & 1 & 1 & -1 & -1 & -1 & -1 \\ \sqrt{2} & \sqrt{2} & -\sqrt{2} & -\sqrt{2} & 0 & 0 & 0 & 0 \\ 0 & 0 & 0 & 0 & \sqrt{2} & \sqrt{2} & -\sqrt{2} & -\sqrt{2} \\ 2 & -2 & 0 & 0 & 0 & 0 & 0 & 0 \\ 0 & 0 & 2 & -2 & 0 & 0 & 0 & 0 \\ 0 & 0 & 0 & 0 & 2 & -2 & 0 & 0 \\ 0 & 0 & 0 & 0 & 0 & 0 & 2 & -2 \end{bmatrix} \begin{matrix} \} N/N \\ \} N/N \\ \left.\begin{matrix} \\ \end{matrix}\right\} N/4 \\ \left.\begin{matrix} \\ \\ \\ \\ \end{matrix}\right\} N/2 \end{matrix} \tag{7.3-2}$$

Examining $\mathbf{H}^*(3)$ we observe that $N/2$ coefficients in the Haar domain measure the adjacent correlation of coordinates in the data space taken two at a time, $N/4$ measure coordinates taken 4 at a time, etc., up to N/N coefficients measuring all the N coordinates of the data space. This implies that the HT provides a domain that is both *locally* sensitive as well as *globally* sensitive. In the case of the discrete Fourier and Walsh-Hadamard transforms, each transform coefficient is a function of *all* coordinates in the original data space (global), whereas this is true only for the first two Haar coefficients.

7.4 Algorithms to Compute the HT [31]

The HT is implementable in $2(N-1)$ additions/subtractions and N multiplications as illustrated in Fig. 7.2a for the case $N = 8$. This algorithm to compute the HT was developed by Andrews [9]. The corresponding algorithm to compute the inverse HT is shown in Fig. 7.2b.

From Fig. 7.2 it is apparent that Andrews' algorithm is not of the Cooley-Tukey variety [10]. However, we will now show that the HT can also be computed using a Cooley-Tukey type algorithm. The motivation for seeking such an algorithm is that a single Cooley-Tukey type FFT processor can be used to compute the $(\mathrm{WHT})_h$, $(\mathrm{WHT})_w$, $(\mathrm{GT})_r$ and the HT, in addition to the DFT.

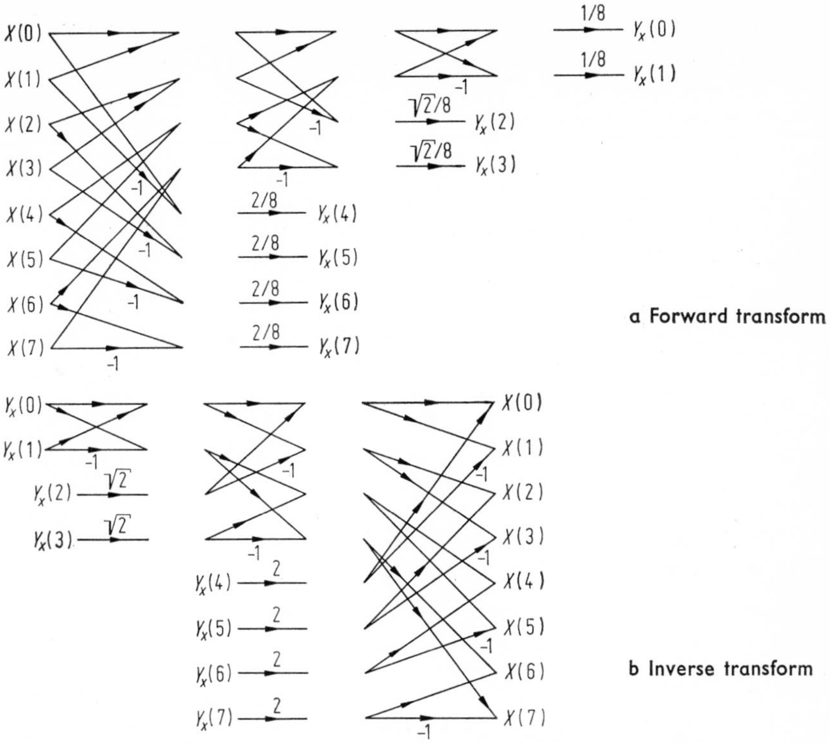

Fig. 7.2 Signal flow graphs for Andrews' algorithm to compute the HT and its inverse; $N = 8$. (a) Forward transform; (b) Inverse transform

Cooley-Tukey type algorithm [8, 31]. The development is best illustrated for the case $N = 8$. Then the Haar matrix given by Eq. (7.3-2) is

$$
\begin{array}{r}
\text{column } \# \\
\\
\\
\mathbf{H}^*(3) =
\end{array}
\begin{array}{cccccccc}
0 & 1 & 2 & 3 & 4 & 5 & 6 & 7 \\
\downarrow & \downarrow & \downarrow & \downarrow & \downarrow & \downarrow & \downarrow & \downarrow
\end{array}
$$

$$
\mathbf{H}^*(3) = \begin{bmatrix}
1 & 1 & 1 & 1 & 1 & 1 & 1 & 1 \\
1 & 1 & 1 & 1 & -1 & -1 & -1 & -1 \\
\sqrt{2} & \sqrt{2} & -\sqrt{2} & -\sqrt{2} & 0 & 0 & 0 & 0 \\
0 & 0 & 0 & 0 & \sqrt{2} & \sqrt{2} & -\sqrt{2} & -\sqrt{2} \\
2 & -2 & 0 & 0 & 0 & 0 & 0 & 0 \\
0 & 0 & 2 & -2 & 0 & 0 & 0 & 0 \\
0 & 0 & 0 & 0 & 2 & -2 & 0 & 0 \\
0 & 0 & 0 & 0 & 0 & 0 & 2 & -2
\end{bmatrix}
$$

We rearrange the columns of $\mathbf{H}^*(3)$ using successive bit-reversals for $N = 8$, $N = 4$, and $N = 2$ as illustrated below.

Step 1: Rearrange the columns of $\mathbf{H}^*(3)$ in bit-reversed order for $N = 8$; i.e., $\{0, 1, 2, 3, 4, 5, 6, 7\} \to \{0, 4, 2, 6, 1, 5, 3, 7\}$ and hence we obtain

$$\mathbf{H}_1^*(3) = \begin{bmatrix} 1 & 1 & 1 & 1 & 1 & 1 & 1 & 1 \\ 1 & -1 & 1 & -1 & 1 & -1 & 1 & -1 \\ \sqrt{2} & 0 & -\sqrt{2} & 0 & \sqrt{2} & 0 & -\sqrt{2} & 0 \\ 0 & \sqrt{2} & 0 & -\sqrt{2} & 0 & \sqrt{2} & 0 & -\sqrt{2} \\ 2 & 0 & 0 & 0 & -2 & 0 & 0 & 0 \\ 0 & 0 & 2 & 0 & 0 & 0 & -2 & 0 \\ 0 & 2 & 0 & 0 & 0 & -2 & 0 & 0 \\ 0 & 0 & 0 & 2 & 0 & 0 & 0 & -2 \end{bmatrix} \tag{7.4-1}$$

$$\begin{matrix} & \uparrow & \uparrow & \uparrow & \uparrow & \uparrow & \uparrow & \uparrow & \uparrow \\ \text{column \#} & 0 & 1 & 2 & 3 & 0 & 1 & 2 & 3 \end{matrix}$$

Step 2: Rearrange the columns of the (4×4) matrices enclosed by squares in $N = 4$ bit-reversed order; i.e., $\{0, 1, 2, 3\} \to \{0, 2, 1, 3\}$. This yields

$$\mathbf{H}_2^*(3) = \begin{bmatrix} 1 & 1 & 1 & 1 & 1 & 1 & 1 & 1 \\ 1 & -1 & 1 & -1 & 1 & -1 & 1 & -1 \\ \sqrt{2} & 0 & -\sqrt{2} & 0 & \sqrt{2} & 0 & -\sqrt{2} & 0 \\ 0 & \sqrt{2} & 0 & -\sqrt{2} & 0 & \sqrt{2} & 0 & -\sqrt{2} \\ 2 & 0 & 0 & 0 & -2 & 0 & 0 & 0 \\ 0 & 2 & 0 & 0 & 0 & -2 & 0 & 0 \\ 0 & 0 & 2 & 0 & 0 & 0 & -2 & 0 \\ 0 & 0 & 0 & 2 & 0 & 0 & 0 & -2 \end{bmatrix} \tag{7.4-2}$$

Step 3: Rearrange the columns of the (2×2) matrices enclosed by squares in $N = 2$ bit-reversed order; i.e., $\{0, 1\} \to \{0, 1\}$, and hence the resulting matrix is identical to $\mathbf{H}_2^*(3)$. Thus

$$\mathbf{H}_3^*(3) = \begin{bmatrix} 1 & 1 & 1 & 1 & 1 & 1 & 1 & 1 \\ 1 & -1 & 1 & -1 & 1 & -1 & 1 & -1 \\ \sqrt{2} & 0 & -\sqrt{2} & 0 & \sqrt{2} & 0 & -\sqrt{2} & 0 \\ 0 & \sqrt{2} & 0 & -\sqrt{2} & 0 & \sqrt{2} & 0 & -\sqrt{2} \\ 2 & 0 & 0 & 0 & -2 & 0 & 0 & 0 \\ 0 & 2 & 0 & 0 & 0 & -2 & 0 & 0 \\ 0 & 0 & 2 & 0 & 0 & 0 & -2 & 0 \\ 0 & 0 & 0 & 2 & 0 & 0 & 0 & -2 \end{bmatrix} \tag{7.4-3}$$

We now observe that $\mathbf{H}_3^*(3)$ in Eq. (3) and the MWHT matrix $\hat{\mathbf{H}}(3)$ in Eq. (6.10-3) are identical. Thus it follows that for $N = 8$, the HT can be computed using the MWHT signal flow graph with a simple modification as shown in Fig. 7.3. We observe that this flow graph in effect is a simplified version of the $(\text{FWHT})_w$ flow graph in Fig. 6.6. The development of the corresponding signal flow graph for the Cooley-Tukey type algorithm to compute the inverse HT is left as an exercise [see Prob. 7-1].

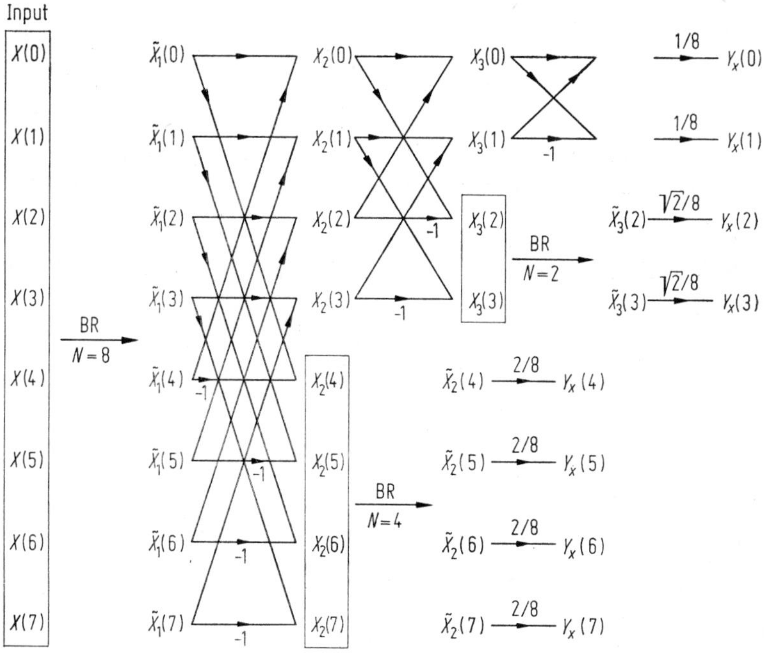

Fig. 7.3 Cooley-Tukey type HT signal flow graph, $N = 8$

From the discussion it follows that in general, the Cooley-Tukey type algorithm to compute the HT requires $\log_2 N$ bit-reversals, and is implementable in $2(N - 1)$ additions/subtractions and N multiplications.

7.5 Slant Matrices

The notion of an orthogonal transformation containing "slant" basis vectors was introduced by Enomoto and Shibata [11]. The slant vector is a discrete sawtooth waveform decreasing in uniform steps over its length, as illustrated in Fig. 7.4.

It has been shown that slant vectors are suitable for efficiently representing gradual brightness changes in an image line. The work of Enomoto and Shibata was restricted to slant vector lengths of 4 and 8. A generalization by

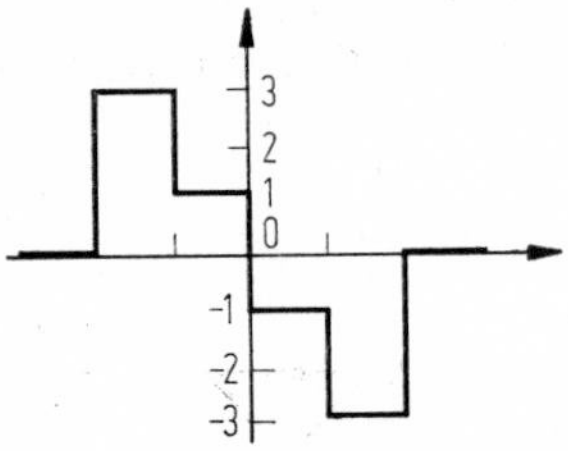

Fig. 7.4 A slant vector for $N = 4$ and step size of 2 units

Pratt, Welch, and Chen [12, 13] leads to the definition of the slant transform. The slant transform has been successfully used in image coding [12–15].

Slant matrix construction. If $\mathbf{S}(n)$ denotes the $(N \times N)$ slant matrix $(N = 2^n)$, then

$$\mathbf{S}(1) = \frac{1}{\sqrt{2}}\begin{bmatrix} 1 & 1 \\ 1 & -1 \end{bmatrix} \tag{7.5-1}$$

The slant matrix for $N = 4$ can be written as

$$\mathbf{S}(2) = \frac{1}{\sqrt{4}}\begin{bmatrix} 1 & 1 & 1 & 1 \\ a+b & a-b & -a+b & -a-b \\ 1 & -1 & -1 & 1 \\ a-b & -a-b & a+b & -a+b \end{bmatrix} \tag{7.5-2}$$

where a and b are real constants to be determined subject to the following conditions:

(i) step size must be uniform, and
(ii) $\mathbf{S}(2)$ must be orthogonal.

The step size between the first two elements of the slant vector [see second row of $\mathbf{S}(2)$] is

$$(a + b) - (a - b) = 2b, \tag{7.5-3}$$

and the step size between the second and third elements is

$$(a - b) - (-a + b) = 2a - 2b \tag{7.5-4}$$

which leads to

$$a = 2b$$

Hence

$$\mathbf{S}(2) = \frac{1}{\sqrt{4}}\begin{bmatrix} 1 & 1 & 1 & 1 \\ 3b & b & -b & -3b \\ 1 & -1 & -1 & 1 \\ b & -3b & 3b & -b \end{bmatrix} \qquad (7.5\text{-}5)$$

Using the orthogonal condition

$$\frac{1}{\sqrt{4}}[3b \; b \; -b \; -3b]\frac{1}{\sqrt{4}}[3b \; b \; -b \; -3b]' = 1$$

we obtain

$$b = \frac{1}{\sqrt{5}}, \qquad a = \frac{2}{\sqrt{5}}$$

Thus the slant matrix in Eq. (2) becomes

$$\mathbf{S}(2) = \frac{1}{\sqrt{4}}\begin{bmatrix} 1 & 1 & 1 & 1 \\ \frac{3}{\sqrt{5}} & \frac{1}{\sqrt{5}} & \frac{-1}{\sqrt{5}} & \frac{-3}{\sqrt{5}} \\ 1 & -1 & -1 & 1 \\ \frac{1}{\sqrt{5}} & \frac{-3}{\sqrt{5}} & \frac{3}{\sqrt{5}} & \frac{-1}{\sqrt{5}} \end{bmatrix} \qquad (7.5\text{-}6)$$

We observe that $\mathbf{S}(2)$ possesses the sequency property. It is easily seen that the sequencies of the rows of $\mathbf{S}(2)$ are 0, 1, 1 and 2, which equal the sequencies of the corresponding rows of the Walsh-Hadamard matrix

$$\mathbf{H}_w(2) = \begin{bmatrix} 1 & 1 & 1 & 1 \\ 1 & 1 & -1 & -1 \\ 1 & -1 & -1 & 1 \\ 1 & -1 & 1 & -1 \end{bmatrix}$$

Now, $\mathbf{S}(2)$ can be expressed in terms of $\mathbf{S}(1)$ such that

$$\mathbf{S}(2) = \frac{1}{\sqrt{2}}\begin{bmatrix} 1 & 0 & 1 & 0 \\ a_4 & b_4 & -a_4 & b_4 \\ 0 & 1 & 0 & -1 \\ -b_4 & a_4 & b_4 & a_4 \end{bmatrix}\begin{bmatrix} \mathbf{S}(1) & \mathbf{0}_2 \\ \mathbf{0}_2 & \mathbf{S}(1) \end{bmatrix}$$

where $a_4 = 2/\sqrt{5}$, and $b_4 = 1/\sqrt{5}$.

Similarly the relation between $\mathbf{S}(2)$ and $\mathbf{S}(3)$ is given by

$$\mathbf{S}(3) = \frac{1}{\sqrt{8}} \left[\begin{array}{cccc:cccc} 1 & 0 & 0 & 0 & 1 & 0 & 0 & 0 \\ a_8 & b_8 & 0 & 0 & -a_8 & b_8 & 0 & 0 \\ 0 & 0 & 1 & 0 & 0 & 0 & 1 & 0 \\ 0 & 0 & 0 & 1 & 0 & 0 & 0 & 1 \\ \hdashline 0 & 1 & 0 & 0 & 0 & -1 & 0 & 0 \\ -b_8 & a_8 & 0 & 0 & b_8 & a_8 & 0 & 0 \\ 0 & 0 & 1 & 0 & 0 & 0 & -1 & 0 \\ 0 & 0 & 0 & 1 & 0 & 0 & 0 & -1 \end{array}\right] \left[\begin{array}{cccc:cccc} 1 & 1 & 1 & 1 & & & & \\ \frac{3}{\sqrt{5}} & \frac{1}{\sqrt{5}} & \frac{-1}{\sqrt{5}} & \frac{-3}{\sqrt{5}} & & \mathbf{0}_4 & & \\ 1 & -1 & -1 & 1 & & & & \\ \frac{1}{\sqrt{5}} & \frac{-3}{\sqrt{5}} & \frac{3}{\sqrt{5}} & \frac{-1}{\sqrt{5}} & & & & \\ \hdashline & & & & 1 & 1 & 1 & 1 \\ & \mathbf{0}_4 & & & \frac{3}{\sqrt{5}} & \frac{1}{\sqrt{5}} & \frac{-1}{\sqrt{5}} & \frac{-3}{\sqrt{5}} \\ & & & & 1 & -1 & -1 & 1 \\ & & & & \frac{1}{\sqrt{5}} & \frac{-3}{\sqrt{5}} & \frac{3}{\sqrt{5}} & \frac{-1}{\sqrt{5}} \end{array}\right] \tag{7.5-7}$$

where a_8 and b_8 are constants. In $\mathbf{S}(3)$ the slant vector is obtained by a simple scaling operation of $\mathbf{S}(2)$, while the remaining terms serve to obtain the sequency and orthogonal properties.

Equation (7) can be generalized to give the slant matrix of order N in terms of the slant matrix of order $N/2$ by the construction

$$\mathbf{S}(n) = \frac{1}{\sqrt{2}} \left[\begin{array}{cccc|cccc} \begin{bmatrix} 1 & 0 \\ a_N & b_N \end{bmatrix} & & & \mathbf{0} & \begin{bmatrix} 1 & 0 \\ -a_N & b_N \end{bmatrix} & & & \mathbf{0} \\ & \mathbf{I}_2 & & & & \mathbf{I}_2 & & \\ & & \ddots & & & & \ddots & \\ \mathbf{0} & & & \mathbf{I}_2 & \mathbf{0} & & & \mathbf{I}_2 \\ \hline \begin{bmatrix} 0 & 1 \\ -b_N & a_N \end{bmatrix} & & & \mathbf{0} & \begin{bmatrix} 0 & -1 \\ b_N & a_N \end{bmatrix} & & & \mathbf{0} \\ & \mathbf{I}_2 & & & & -\mathbf{I}_2 & & \\ & & \ddots & & & & \ddots & \\ \mathbf{0} & & & \mathbf{I}_2 & \mathbf{0} & & & -\mathbf{I}_2 \end{array}\right] \left[\begin{array}{c|c} \mathbf{S}(n-1) & \mathbf{0} \\ \hline \mathbf{0} & \mathbf{S}(n-1) \end{array}\right] \tag{7.5-8}$$

where $\mathbf{I}_2$ denotes the (2×2) identity matrix.

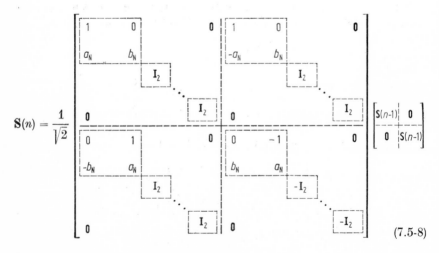

The coefficients (a_N, b_N) can be computed using the following relations [12]:

$$a_2 = 1$$

$$b_N = \frac{1}{(1 + 4a_{N/2}^2)^{1/2}}$$

and

$$a_N = 2b_N a_{N/2}, \qquad N = 4, 8, 16, \ldots \tag{7.5-9}$$

7.6 Definition of the Slant Transform (ST)

The above slant matrices are used to define the ST as

$$\boldsymbol{D}_x(n) = \mathbf{S}(n)\boldsymbol{X}(n) \tag{7.6-1}$$

where

$$\boldsymbol{D}_x(n)' = [D_x(0)\ D_x(1) \cdots D_x(N-1)] \text{ is the ST vector,}$$

$$\boldsymbol{X}(n)' = [X(0)\ X(1) \cdots X(N-1)] \text{ is the data vector,}$$

and $\mathbf{S}(n)$ is the $(N \times N)$ slant matrix.

Clearly, the basis vectors of the ST are the rows of the slant matrix $\mathbf{S}(n)$. For the purposes of illustration, the basis vectors of the ST for $N = 16$ are shown in Fig. 7.5 along with the $(\mathrm{WHT})_w$ basis vectors. It is worthwhile noting that many of the mid-sequency vectors are identical for the two transforms.

Computational considerations. The ST can be computed using a fast algorithm whose development is best illustrated for the case $N = 4$. Then Eq. (1) yields

$$\boldsymbol{D}_x(2) = \mathbf{S}(2)\boldsymbol{X}(2) \tag{7.6-2}$$

where $\mathbf{S}(2)$ is given by Eq. (7.5-6).

The fast computational property is apparent from the matrix decomposition

$$\mathbf{S}(2) = \frac{1}{\sqrt{4}} \begin{bmatrix} 1 & 0 & 0 & 0 \\ 0 & \frac{3}{\sqrt{5}} & 0 & 0 \\ 0 & 0 & 1 & 0 \\ 0 & 0 & 0 & \frac{3}{\sqrt{5}} \end{bmatrix} \begin{bmatrix} 1 & 1 & 0 & 0 \\ 0 & 0 & 1 & \frac{1}{3} \\ 1 & -1 & 0 & 0 \\ 0 & 0 & \frac{1}{3} & -1 \end{bmatrix} \begin{bmatrix} 1 & 0 & 0 & 1 \\ 0 & 1 & 1 & 0 \\ 1 & 0 & 0 & -1 \\ 0 & 1 & -1 & 0 \end{bmatrix} \tag{7.6-3}$$

Using these matrix factors the ST can be computed as shown in Fig. 7.6. Examining Fig. 7.6 it is found that the algorithm requires 8 additions/subtractions and 6 multiplications to compute the ST coefficients $D_x(k)$, $k = 0, 1, 2, 3$.

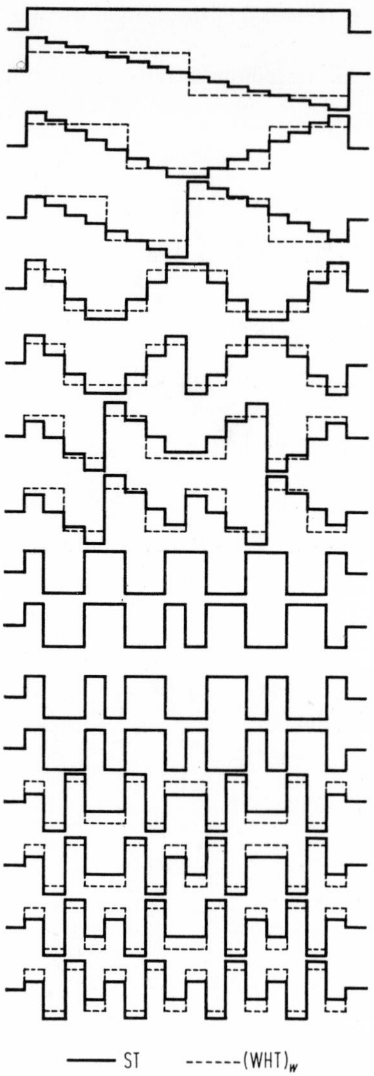

Fig. 7.5 Comparison of $(\mathrm{WHT})_w$ and ST basis vectors for $N = 16$

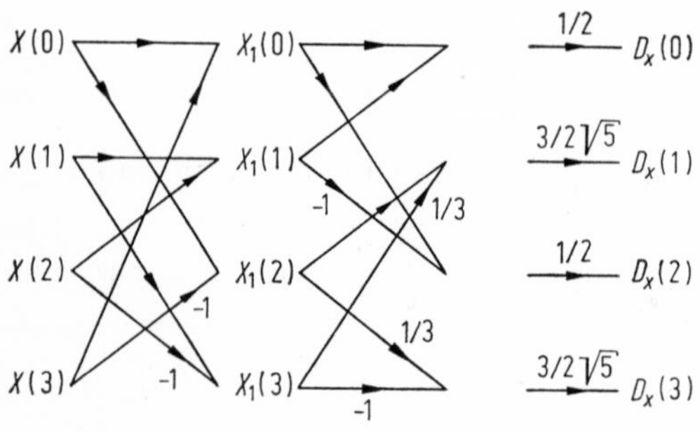

Fig. 7.6 Signal flow graph for the ST, $N = 4$

7.7 Discrete Cosine Transform (DCT)

The DCT of a data sequence $X(m)$, $m = 0, 1, \ldots, N-1$ is defined as [16]

$$L_x(0) = \frac{1}{\sqrt{N}} \sum_{m=0}^{N-1} X(m)$$

and

$$L_x(k) = \sqrt{\frac{2}{N}} \sum_{m=0}^{N-1} X(m) \cos \frac{(2m+1)k\pi}{2N}, \qquad k = 1, 2, \ldots, N-1 \qquad (7.7\text{-}1)$$

where $L_x(k)$ is the k-th DCT coefficient.

It is worthwhile noting that the set of basis vector elements

$$\left\{\frac{1}{\sqrt{N}}, \sqrt{\frac{2}{N}} \cos \frac{(2m+1)k\pi}{2N}\right\}$$

is actually a class of discrete Chebyshev polynomials. This is easily seen by examining the following definition of Chebyshev polynomials [17][1].

$$T_0(p) = \frac{1}{\sqrt{N}}$$

and

$$T_k(Z_m) = \sqrt{\frac{2}{N}} \cos[k \cos^{-1}(Z_m)], \qquad k, m = 1, 2, \ldots, N-1 \qquad (7.7\text{-}2)$$

where $T_k(Z_m)$ is the k-th Chebyshev polynomial.

Now, the zeros of the N-th polynomial $T_N(Z_m)$ are given by [17]

$$Z_m = \cos \frac{(2m+1)\pi}{2N}, \qquad m = 0, 1, \ldots, N-1 \qquad (7.7\text{-}3)$$

Substituting Eq. (3) in Eq. (2), we evaluate $\{T_l(Z_m)\}$, $l = 0, 1, \ldots, N-1$ at the zeros of $T_N(Z_m)$. This results in the set of Chebyshev polynomials

$$T_0(m) = \frac{1}{\sqrt{N}}$$

and

$$T_k(m) = \sqrt{\frac{2}{N}} \cos \frac{(2m+1)k\pi}{2N}, \qquad m = 0, 1, \ldots, N-1 \qquad (7.7\text{-}4)$$

which are equivalent to the basis set of the DCT.

Again, the inverse discrete cosine transform (IDCT) is defined as

$$X(m) = \frac{1}{\sqrt{N}} L_x(0) + \sqrt{\frac{2}{N}} \sum_{k=1}^{N-1} L_x(k) \cos \frac{(2m+1)k\pi}{2N}, \qquad m = 0, 1, \ldots, N-1$$
$$(7.7\text{-}5)$$

[1] In particular see pp. 95–98 and 125–130.

It can be shown that application of the orthogonal property [17]

$$\sum_{m=0}^{N-1} T_p(m)\, T_q(m) = \begin{cases} 1, & p = q \\ 0, & p \neq q \end{cases} \tag{7.7-6}$$

to Eq. (5) results in the definition of the DCT [see Prob. 7-3]. If Eq. (1) is written in matrix form and $\mathbf{\Gamma}(n)$ denotes the $(N \times N)$ DCT matrix, then this orthogonal property can be expressed as

$$\mathbf{\Gamma}(n)'\,\mathbf{\Gamma}(n) = \mathbf{I}_N \tag{7.7-7}$$

For the purposes of illustration, we evaluate $\mathbf{\Gamma}(3)$ to obtain

$$\mathbf{\Gamma}(3) = \left[\begin{array}{cccc:cccc} 0.354 & 0.354 & 0.354 & 0.354 & 0.354 & 0.354 & 0.354 & 0.354 \\ 0.490 & 0.416 & 0.278 & 0.098 & -0.098 & -0.278 & -0.416 & -0.490 \\ 0.462 & 0.191 & -0.191 & -0.462 & -0.462 & -0.191 & 0.191 & 0.462 \\ 0.416 & -0.098 & -0.490 & -0.278 & 0.278 & 0.490 & 0.098 & -0.416 \\ \hdashline 0.354 & -0.354 & -0.354 & 0.354 & 0.354 & -0.354 & -0.354 & 0.354 \\ 0.278 & -0.490 & 0.098 & 0.416 & -0.416 & -0.098 & 0.490 & -0.278 \\ 0.191 & -0.462 & 0.462 & -0.191 & -0.191 & 0.462 & -0.462 & 0.191 \\ 0.098 & -0.278 & 0.416 & -0.490 & 0.490 & -0.416 & 0.278 & -0.098 \end{array}\right] \tag{7.7-8}$$

It can be easily verified that $\mathbf{\Gamma}(3)'\,\mathbf{\Gamma}(3) = \mathbf{I}_8$.

Computational considerations. It can be shown that [see Prob. 7-4] the DCT can be equivalently expressed as

$$L_x(0) = \frac{1}{\sqrt{N}} \sum_{m=0}^{N-1} X(m)$$

and

$$L_x(k) = \sqrt{\frac{2}{N}}\, \mathrm{Re}\left\{ e^{-\frac{ik\pi}{2N}} \sum_{m=0}^{2N-1} X(m)\, W^{km} \right\}, \qquad k = 1, 2, \ldots, N-1 \tag{7.7-9}$$

where $W = e^{-i\,2\pi/2N}$, $i = \sqrt{-1}$, $X(m) = 0$, $m = N, N+1, \ldots, 2N-1$, and $\mathrm{Re}\{\cdot\}$ denotes the real part of the term enclosed.

From Eq. (9) it follows that all the N coefficients of the DCT can be computed using a $2N$-point FFT. Similarly it can be shown that a $2N$-point IFFT yields all the IDCT coefficients [see Prob. 7-5].

A basic property. The DCT has the important property that its basis vectors closely approximate the eigenvectors of a class of matrices called

Toeplitz matrices [18, 19]. This class of matrices is defined as

$$\psi = \begin{bmatrix} 1 & \varrho & \varrho^2 & \cdots \varrho^{N-1} \\ \varrho & 1 & \varrho & \cdots \varrho^{N-2} \\ \cdots & \cdots & \cdots & \cdots \\ \varrho^{N-1} & \varrho^{N-2} & \varrho^{N-3} & \cdots 1 \end{bmatrix}, \quad 0 < \varrho < 1 \qquad (7.7\text{-}10)$$

For the purposes of illustration Fig. 7.7 shows the eigenvectors of ψ for $N = 8$ and $\varrho = 0.9$, plotted against the corresponding DCT basis set given by the rows of $\Gamma(3)$ in Eq. (8). The close resemblence (aside from the 180° phase shift) between the eigenvectors and the DCT basis set is apparent. In Chapter 9 we will show that as a consequence of this property, the DCT can be used effectively in the area of image processing.

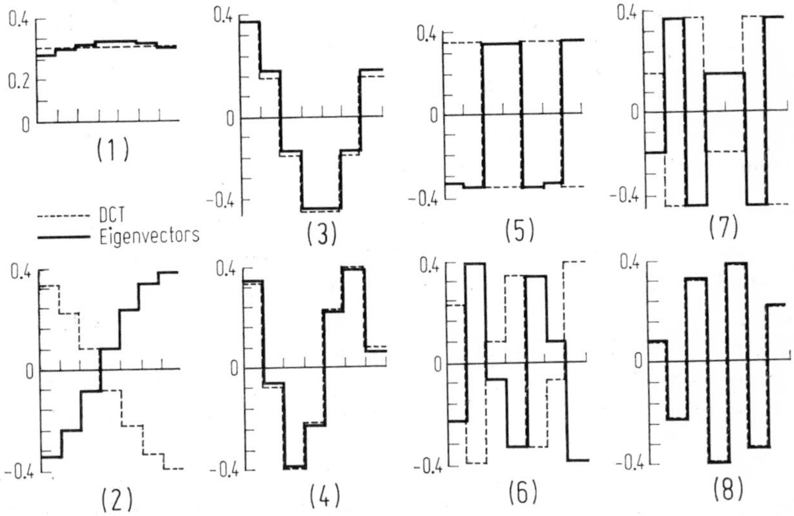

Fig. 7.7 Eigenvectors of (8×8) Toeplitz matrix ($\varrho = 0.9$), and the DCT basis vectors

7.8 2-dimensional Transform Considerations

The transforms discussed thus far can easily be extended to the 2-dimensional case. The matrix form definitions for the 2-dimensional Haar, slant, discrete cosine transforms and the corresponding inverse transforms are summarized in Table 7.8-1. In this Table, $[X(m_1, m_2)]$ denotes the $(N_1 \times N_2)$ data matrix while $[Y_{xx}(u_1, u_2)]$, $[D_{xx}(u_1, u_2)]$, and $[L_{xx}(u_1, u_2)]$ denote the Haar, slant, and discrete cosine transform matrices.

Table 7.8-1 2-dimensional transform definitions

Name of the transform	Definition
Haar	$[Y_{xx}(u_1, u_2)] = \frac{1}{N_1 N_2} \mathbf{H}^*(n_1)[X(m_1, m_2)]\mathbf{H}^*(n_2)'$ $[X(m_1, m_2)] = \mathbf{H}^*(n_1)'[Y_{xx}(u_1, u_2)]\mathbf{H}^*(n_2)$
slant	$[D_{xx}(u_1, u_2)] = \mathbf{S}(n_1)[Xm_1, m_2)]\mathbf{S}(n_2)'$ $[X(m_1, m_2)] = \mathbf{S}(n_1)'[D_{xx}(u_1, u_2)]\mathbf{S}(n_2)$
discrete cosine	$[L_{xx}(u_1, u_2)] = \mathbf{\Gamma}(n_1)[X(m_1, m_2)]\mathbf{\Gamma}(n_2)'$ $[X(m_1, m_2)] = \mathbf{\Gamma}(n_1)'[L_{xx}(u_1, u_2)]\mathbf{\Gamma}(n_2)$

As was the case with the discrete Fourier and Walsh-Hadamard transforms, the above 2-dimensional transforms and their inverses can be computed by means of N_1N_2 applications of the algorithms used to compute the corresponding 1-dimensional transforms.

7.9 Summary

The Haar, slant, discrete cosine, and generalized transforms were introduced in this chapter. It was shown that for a given data sequence $X(m)$, $m = 0, 1, \ldots, N-1$, the $(\text{GT})_r$ defines a class of $\log_2 N$ orthogonal transforms, and hence provides a systematic transition from the $(\text{WHT})_h$ to the DFT. It can be shown that the transforms $(\text{GT})_r$, $r = 1, 2, \ldots, n-2$ also possess shift-invariant power spectra [4–7]. Efficient implementation of these transforms is accomplished either by Kronecker product of a set of matrices [9, 20, 21] or by matrix factorization [22–25].

Fast algorithms to compute the Haar, slant, and discrete cosine transforms were derived. It was shown that Cooley-Tukey algorithm can be used to compute the HT in addition to the DFT, $(\text{WHT})_w$ and $(\text{WHT})_h$.

Appendix 7.1 Kronecker Products

Let **A** and **B** be two matrices such that

$$\mathbf{A} = \begin{bmatrix} a_{11} & a_{12} & \cdots & a_{1n} \\ a_{21} & a_{22} & \cdots & a_{2n} \\ \cdots & \cdots & \cdots & \cdots \\ a_{m1} & a_{m2} & \cdots & a_{mn} \end{bmatrix}, \quad \text{and} \quad \mathbf{B} = \begin{bmatrix} b_{11} & b_{12} & \cdots & b_{1l} \\ b_{21} & b_{22} & \cdots & b_{2l} \\ \cdots & \cdots & \cdots & \cdots \\ b_{k1} & b_{k2} & \cdots & b_{kl} \end{bmatrix}$$

Then their *Kronecker product* is defined as

$$\mathbf{A} \otimes \mathbf{B} = \begin{bmatrix} a_{11}\mathbf{B} & a_{12}\mathbf{B} & \cdots & a_{1n}\mathbf{B} \\ a_{21}\mathbf{B} & a_{22}\mathbf{B} & \cdots & a_{2n}\mathbf{B} \\ \cdots & \cdots & \cdots & \cdots \\ a_{m1}\mathbf{B} & a_{m2}\mathbf{B} & \cdots & a_{mn}\mathbf{B} \end{bmatrix}, \quad \mathbf{B} \otimes \mathbf{A} = \begin{bmatrix} \mathbf{A}b_{11} & \mathbf{A}b_{12} & \cdots & \mathbf{A}b_{1l} \\ \mathbf{A}b_{21} & \mathbf{A}b_{22} & \cdots & \mathbf{A}b_{2l} \\ \cdots & \cdots & \cdots & \cdots \\ \mathbf{A}b_{k1} & \mathbf{A}b_{k2} & \cdots & \mathbf{A}b_{kl} \end{bmatrix},$$

where $\otimes$ denotes the Kronecker product operator.

From the above definition it is apparent that $\mathbf{A} \otimes \mathbf{B}$ is an $(mk \times nl)$ matrix. As an illustrative example, we consider the recursive definition of Hadamard matrices $\mathbf{H}_h(k)$ which is [see Eq. (6.1-6)]

$$\mathbf{H}_h(k) = \begin{bmatrix} \mathbf{H}_h(k-1) & \mathbf{H}_h(k-1) \\ \mathbf{H}_h(k-1) & -\mathbf{H}_h(k-1) \end{bmatrix}, \qquad k = 1, 2, \ldots, n$$

It is easily verified that this recursion can be equivalently expressed as

$$\mathbf{H}_h(k) = \begin{bmatrix} 1 & 1 \\ 1 & -1 \end{bmatrix} \otimes \mathbf{H}_h(k-1),$$

or

$$\mathbf{H}_h(k) = \mathbf{H}_h(1) \otimes \mathbf{H}_h(k-1), \qquad k = 1, 2, \ldots, n$$

The following identities [18] can be shown to be valid.

$$\mathbf{A} \otimes \mathbf{B} \otimes \mathbf{C} = (\mathbf{A} \otimes \mathbf{B}) \otimes \mathbf{C} = \mathbf{A} \otimes (\mathbf{B} \otimes \mathbf{C})$$

$$(\mathbf{A} + \mathbf{B}) \otimes (\mathbf{C} + \mathbf{D}) = \mathbf{A} \otimes \mathbf{C} + \mathbf{A} \otimes \mathbf{D} + \mathbf{B} \otimes \mathbf{C} + \mathbf{B} \otimes \mathbf{D}$$

and

$$(\mathbf{A} \otimes \mathbf{B})(\mathbf{C} \otimes \mathbf{D}) = (\mathbf{AC}) \otimes (\mathbf{BD}),$$

where $\mathbf{AC}$ and $\mathbf{BD}$ denote the conventional matrix product of $\mathbf{A}$, $\mathbf{C}$ and $\mathbf{B}$, $\mathbf{D}$ respectively.

Appendix 7.2 Matrix Factorization

Factoring of a transform matrix is illustrated for the case $N = 16$ and $r = 2$. Then from Eqs. (7.2-1) and (7.2-2) we have

$$\boldsymbol{B}_2(4) = \frac{1}{16}\left(\prod_{j=1}^{4} \mathbf{D}_2^j(4)\right)\mathbf{X}(4) \qquad \text{(A7.2-1)}$$

The matrices $\mathbf{D}_2^j(4)$ are generated recursively [see Eq. (7.2-3)] as shown below.

Case 1: $j = 1$

$$\mathbf{D}_2^1(4) = \text{diag}[\mathbf{A}_0^2(1)\,\mathbf{A}_1^2(1)\,\ldots\,\mathbf{A}_7^2(1)] \tag{A7.2-2}$$

where

$$\mathbf{A}_0^2(1) = \begin{bmatrix} 1 & W^{\langle\langle 0\rangle\rangle} \\ 1 & -W^{\langle\langle 0\rangle\rangle} \end{bmatrix} = \begin{bmatrix} 1 & 1 \\ 1 & -1 \end{bmatrix}$$

$$\mathbf{A}_1^2(1) = \begin{bmatrix} 1 & W^{\langle\langle 1\rangle\rangle} \\ 1 & -W^{\langle\langle 1\rangle\rangle} \end{bmatrix} = \begin{bmatrix} 1 & W^4 \\ 1 & -W^4 \end{bmatrix}$$

$$\mathbf{A}_2^2(1) = \begin{bmatrix} 1 & W^{\langle\langle 2\rangle\rangle} \\ 1 & -W^{\langle\langle 2\rangle\rangle} \end{bmatrix} = \begin{bmatrix} 1 & W^2 \\ 1 & -W^2 \end{bmatrix}$$

$$\mathbf{A}_3^2(1) = \begin{bmatrix} 1 & W^{\langle\langle 3\rangle\rangle} \\ 1 & -W^{\langle\langle 3\rangle\rangle} \end{bmatrix} = \begin{bmatrix} 1 & W^6 \\ 1 & -W^6 \end{bmatrix}$$

$$\mathbf{A}_m^2(1) = \begin{bmatrix} 1 & 1 \\ 1 & -1 \end{bmatrix}, \qquad m = 4, 5, 6, 7$$

where $W = e^{-i\,2\pi/16}$.

Case 2: $j = 2$

$$\mathbf{D}_2^2(4) = \text{diag}\,[\mathbf{A}_0^2(2)\,\mathbf{A}_1^2(2)\,\mathbf{A}_2^2(2)\,\mathbf{A}_3^2(2)]$$

where

$$\mathbf{A}_0^2(2) = \begin{bmatrix} 1 & 1 \\ 1 & -1 \end{bmatrix} \otimes \mathbf{I}_2 = \begin{bmatrix} \mathbf{I}_2 & \mathbf{I}_2 \\ \mathbf{I}_2 & -\mathbf{I}_2 \end{bmatrix}$$

$$\mathbf{A}_1^2(2) = \begin{bmatrix} 1 & W^4 \\ 1 & -W^4 \end{bmatrix} \otimes \mathbf{I}_2 = \begin{bmatrix} \mathbf{I}_2 & W^4\mathbf{I}_2 \\ \mathbf{I}_2 & -W^4\mathbf{I}_2 \end{bmatrix}$$

$$\mathbf{A}_2^2(2) = \begin{bmatrix} 1 & W^2 \\ 1 & -W^2 \end{bmatrix} \otimes \mathbf{I}_2 = \begin{bmatrix} \mathbf{I}_2 & W^2\mathbf{I}_2 \\ \mathbf{I}_2 & -W^2\mathbf{I}_2 \end{bmatrix}$$

$$\mathbf{A}_3^2(2) = \begin{bmatrix} 1 & W^6 \\ 1 & -W^6 \end{bmatrix} \otimes \mathbf{I}_2 = \begin{bmatrix} \mathbf{I}_2 & W^6\mathbf{I}_2 \\ \mathbf{I}_2 & -W^6\mathbf{I}_2 \end{bmatrix}$$

Case 3: $j = 3$

$$\mathbf{D}_2^3(4) = \text{diag}[\mathbf{A}_0^2(3)\,\mathbf{A}_1^2(3)]$$

As before,

$$\mathbf{A}_0^2(3) = \begin{bmatrix} 1 & 1 \\ 1 & -1 \end{bmatrix} \otimes \mathbf{I}_4 = \begin{bmatrix} \mathbf{I}_4 & \mathbf{I}_4 \\ \mathbf{I}_4 & -\mathbf{I}_4 \end{bmatrix}$$

$$\mathbf{A}_1^2(3) = \begin{bmatrix} 1 & W^4 \\ 1 & -W^4 \end{bmatrix} \otimes \mathbf{I}_4 = \begin{bmatrix} \mathbf{I}_4 & W^4\mathbf{I}_4 \\ \mathbf{I}_4 & -W^4\mathbf{I}_4 \end{bmatrix}$$

Hence

$$\mathbf{D}_2^3(4) = \left[\begin{array}{cc:cc} \mathbf{I}_4 & \mathbf{I}_4 & \multicolumn{2}{c}{\mathbf{0}_8} \\ \mathbf{I}_4 & -\mathbf{I}_4 & & \\ \hdashline & & \mathbf{I}_4 & W^4\mathbf{I}_4 \\ \multicolumn{2}{c:}{\mathbf{0}_8} & \mathbf{I}_4 & -W^4\mathbf{I}_4 \end{array}\right]$$

Similarly, it can be shown that

$$\mathbf{D}_2^4(4) = \begin{bmatrix} \mathbf{I}_8 & \mathbf{I}_8 \\ \mathbf{I}_8 & -\mathbf{I}_8 \end{bmatrix}$$

References

1. Ohnsorg, F. R.: Application of Walsh Functions to Complex Signals. *Proc. 1970 Symp. Applications of Walsh Functions*, 123–127.
2. Ohnsorg, F. R.: Properties of Complex Walsh Functions. *Proc. IEEE Fall Electronics Conf.*, Chicago, Oct. 18–20, 1971, 383–385.
3. Rao, K. R., and Ahmed, N.: Complex BIFORE Transform. *Int. J. Systems Sci.* 2 (1971) 149–162.
4. Ahmed, N., and Rao, K. R.: Generalized Transform. *Proc. 1971 Symp. Applications of Walsh Functions*, 60–67.
5. Ahmed, N., and Rao, K. R.: A Generalized Discrete Transform. *Proc. IEEE* 59 (1971) 1360–1362.
6. Rao, K. R., Ahmed, N., and Schultz, R. B.: A Class of Discrete Orthogonal Transforms. Intl. Symp. Circuit Theory, April 9–11, 1973, Toronto, Canada. Published in the Symp. Digest, 189–192.
7. Rao, K. R., Mrig, L. C., and Ahmed, N.: A Modified Generalized Discrete Transform. *Proc. IEEE* 61 (1973) 668–669.
8. Ahmed, N., Natarajan, T., and Rao, K. R.: Some Considerations of the Modified Walsh-Hadamard and Haar Transforms. *Proc. 1973 Symp. Applications of Walsh Functions*, 91–95.
9. Andrews, H. C., and Caspari, K. L.: A Generalized Technique for Spectral Analysis. *IEEE Trans. Computers* C-19 (1970) 16–25.
10. Cooley, J. W., and Tukey, J. W.: An Algorithm for the Machine Calculation of Complex Fourier Series. *Math. Computation* 19 (1965) 297–301.
11. Enomoto, H., and Shibata, K.: Orthogonal Transform Coding System for Television Signals. *Proc. 1971 Symp. Applications of Walsh Functions*, 11–17.
12. Pratt, W. K., Welch, L. R., and Chen, W. H.: Slant Transforms for Image Coding. *Proc. 1972 Symp. Applications of Walsh Functions*, 229–234.
13. Chen, W. H., and Pratt, W. K.: Color Image Coding with the Slant Transform. *Proc. 1973 Symp. Applications of Walsh Functions*, 155–161.
14. Shibata, K.: Waveform Analysis of Image Signals by Orthogonal Transformation. *Proc. 1972 Symp. Applications of Walsh Functions*, 210–215.
15. Shibata, K.: Block Waveform Coding of Image Signals by Orthogonal Transformation. *Proc. 1973 Symp. Applications of Walsh Functions*, 137–143.
16. Ahmed, N., Natarajan, T., and Rao, K. R.: Discrete Cosine Transform. *IEEE Trans. Computers* C-23 (1974) 90–93.

17. Fike, C. T.: *Computer Evaluation of Mathematical Functions.* Englewood Cliffs, N. J.: Prentice Hall, 1968.
18. Bellman, R.: *Introduction to Matrix Analysis.* New York: McGraw-Hill, 1960.
19. Grenander, V., and Szego, G.: *Toeplitz Forms and Their Applications.* Berkeley and Los Angeles: University of California Press, 1958.
20. Andrews, H. C., and Kane, J.: Kronecker Matrices, Computer Implementation, and Generalized Spectra. *J. Assoc. Comput. Mach.* 17 (1970) 260–268.
21. Whelchel, J. E., and Guinn, D. F.: The Fast Fourier-Hadamard Transform and Its Use in Signal Representation and Classification. EASCON *'68 Record,* 1968, 561–573.
22. Brigham, E. O., and Morrow, R. E.: The Fast Fourier Transform. *IEEE Spectrum 4,* Dec. 1967, 63–70.
23. Glassman, J. A.: A Generalization of the Fast Fourier Transform. *IEEE Trans. Computers* C-19 (1970) 105–116.
24. Theilheimer, F.: A Matrix Version of the Fast Fourier Transform. *IEEE Trans. Audio and Electroacoustics* AU-17 (1969) 158–161.
25. Gentleman, W. M.: Matrix Multiplication and the Fast Fourier Transform. *Bell System Tech. J.* 47 (1968) 1099–1103.
26. Rao, K. R., and Ahmed, N.: Modified Complex BIFORE Transform. *Proc. IEEE* 60 (1972) 1010–1012.
27. Rao, K. R., Mrig, L. C., and Ahmed, N.: A Modified Generalized Discrete Transform. Proc. Sixth Asilomar Conf. on Circuits and Systems, 1972, 189–195.
28. Fino, B. J.: Relations Between Haar and Walsh-Hadamard Transforms. *Proc. IEEE* 60 (1972) 647–648.
29. Gibbs, J. E.: Discrete Complex Walsh Functions. *Proc. 1970 Symp. Applications of Walsh Functions,* 106–122.
30. Elliott, D. F.: A Class of Generalized Continuous Orthogonal Transforms. *IEEE Trans. Acoustics, Speech and Signal Processing* ASSP-23 (1974) 245–254.
31. Ahmed, N., Natarajan, T., and Rao, K. R.: Cooley-Tukey type Algorithm for the Haar Transform. *Electronics Letters* 9 (1973) 276–278.
32. Revuluri, K., et al.: Complex Haar Transform. Proc. Seventh Asilomar Conference on Circuits, Systems, and Computers. Pacific Grove, California, Nov. 27–29, 1973, 729–733.
33. Elliott, D. F.: A Transform Class Governed by Signed-Bit Dyadic Time Shift. 1974 International Symposium on Circuits and Systems, San Francisco, Calif. April 22–24, 1974.
34. Pratt, W. K., et al.: Slant Transform Image Coding. *IEEE Trans. Communications* COM-22 (1974) 1075–1093.

Problems

7-1 Develop the Cooley-Tukey type algorithm to compute the IHT, for the case $N = 8$.

7-2 Obtain the signal flow graph to compute the ST for $N = 8$. Use the same to find the ST of the data sequence

$$\{X(m)\} = \{1\ 2\ 1\ 1\ 3\ 2\ 1\ 2\}$$

7-3 Use Eqs. (7.7-5) and (7.7-6) to obtain Eq. (7.7-1).

7-4 Starting with Eq. (7.7-1) derive Eq. (7.7-9).

7-5 Show that the IDCT given by Eq. (7.7-5) can be equivalently expressed as follows:

$$X(m) = \frac{1}{\sqrt{N}}\,\hat{L}_x(0) + \\ + \sqrt{\frac{2}{N}}\,\mathrm{Re}\left\{\sum_{k=0}^{2N-1} \hat{L}_x(k)\,\overline{W}^{km}\right\}, \qquad m = 0, 2, \ldots, N-1 \qquad \text{(P7-5-1)}$$

where

$$\hat{L}_x(k) = \begin{cases} L_x(k)\,e^{ik\pi/2N}, & k = 0, 1, \ldots, N-1 \\ 0, & k = N, N+1, \ldots, 2N-1, \end{cases}$$

and $W = e^{-i\,2\pi/2N}$ while $\overline{W}$ is its complex conjugate.

Comment. From Eq. (P7-5-1) it is apparent that the IDCT can be computed using a $2N$-point IFFT.

7-6 Consider the data sequence

$$\{X(m)\} = \{1\ 2\ 1\ 1\ 2\ 3\ 4\ 2\ 1\ 2\ 3\ 1\ 2\ 4\ 5\ 3\}$$

(a) Using Andrews' algorithm, show that the HT of $\{X(m)\}$ is given by

$$\{Y_x(m)\} = \frac{1}{16}\begin{Bmatrix} 37 & -5 & -6\sqrt{2} & -7\sqrt{2} & 2 & -2 & -2 & -4 \\ -2\sqrt{2} & 0 & -2\sqrt{2} & 4\sqrt{2} & -2\sqrt{2} & 4\sqrt{2} & -4\sqrt{2} & 4\sqrt{2} \end{Bmatrix}$$

(b) Verify that the same $\{Y_x(m)\}$ is obtained using the Cooley-Tukey type algorithm.

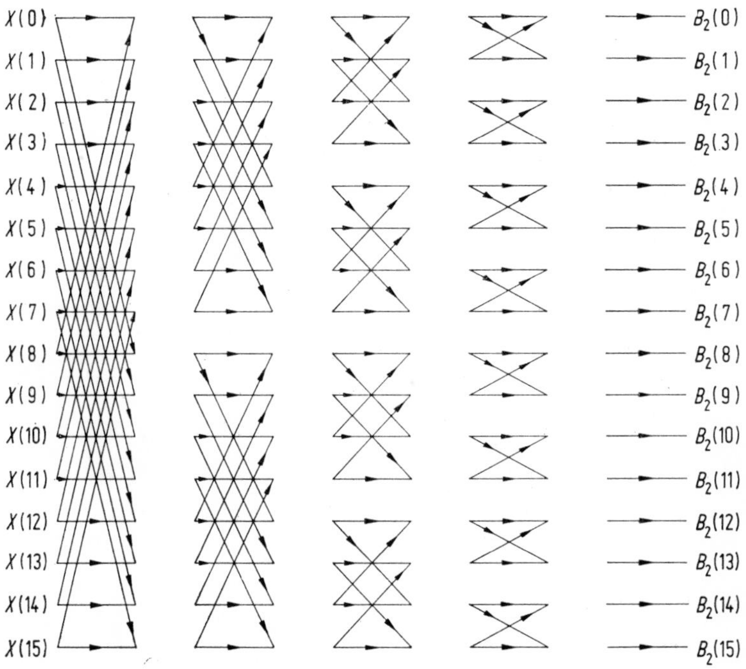

Fig. P7-8-1 Signal flow graph for $(\mathrm{GT})_r$, $r = 2$, $N = 16$, $(n = 4)$

7-7 (a) Develop the signal flow graph to compute the *inverse* ST for $N = 4$ using the flow graph in Fig. 7.6.
(b) Use the information in (a) and Prob. 7-2 to develop the inverse ST signal flow graph for $N = 8$.

7-8 The $(\mathrm{GT})_r$ transform matrix $\mathbf{G}_r(n)$ is developed in terms of its matrix factors for $r = 2$ and $N = 16$, $(n = 4)$, i.e., $\mathbf{G}_2(4) = \prod_{j=1}^{4} \mathbf{D}_2^j(4)$. (See Appendix 7.2.) Based on these matrix factors identify the multipliers on the flow graph shown in Fig. P7-8-1. Verify this flow graph with that shown in Fig. 7.1.

7-9 The transform matrix $\mathbf{G}_r(n)$ can also be generated recursively as described in [6]. Develop $\mathbf{G}_2(4)$ using this reference and verify that this is equal to the product of the matrix factors described in Prob. 7-8.

7-10 (a) A modification of the $(\mathrm{GT})_r$ called modified generalized discrete transform $(\mathrm{MGT})_r$ has also been developed [6, 7, 27]. The transform matrix $\mathbf{M}_r(n)$ for this case has the recursion relation

$$\mathbf{M}_r(k) = \begin{bmatrix} \mathbf{M}_r(k-1) & \mathbf{M}_r(k-1) \\ \mathbf{C}_r(k-1) & -\mathbf{C}_r(k-1) \end{bmatrix} \tag{P7-10-1}$$

where $\mathbf{M}_r(0) = 1$, and $\mathbf{M}_r(1) = \begin{bmatrix} 1 & 1 \\ 1 & 1 \end{bmatrix} = \mathbf{H}_h(1)$.

The submatrix $\mathbf{C}_r(k-1)$ is described in [6]. Obtain the transform matrix of $(\mathrm{MGT})_r$ for $r = 2$ and $N = 16$, $(n = 4)$, i.e., obtain $\mathbf{M}_2(4)$.

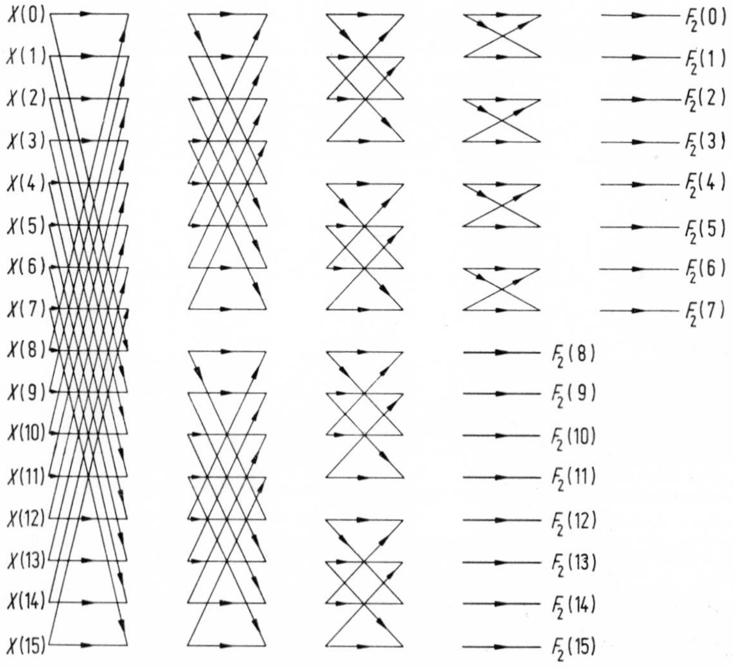

Fig. P7-10-1 Signal flow graph for $(\mathrm{MGT})_r$, $r = 2$, $N = 16$, $(n = 4)$

(b) The transform matrix of the $(\text{MGT})_r$ can also be expressed in terms of its matrix factors based on slight modification of Eq. (7.2-3), (see [7]). Using this, develop the matrix factors of $\mathbf{M}_2(4)$. Based on these matrix factors identify the multipliers on the flow graph shown in Fig. P7-10-1.

(c) Show that $\mathbf{M}_2(4)$ as obtained in parts (a) and (b) above are one and the same.

7-11 Repeat Problem 7-8 for $N = 16$ and $r = 0, 1,$ and 3.

7-12 Repeat Problem 7-9 for $N = 16$ and $r = 0, 1,$ and 3.

7-13 Repeat Problem 7-10 for $N = 16$ and $r = 0, 1,$ and 3.

7-14 For $N = 16$ and $r = 2$, develop the shift-invariant power spectra of $(\text{GT})_r$ and $(\text{MGT})_r$ and show that they are one and the same; see [4–6, 7].

7-15 Relations between HT and $(\text{WHT})_w$ have been developed by Fino [28]. Verify the relations between the submatrices of $\mathbf{H}_w(3)$ (see Fig. 5.5b) and the submatrices of $\mathbf{H}^*(3)$ [see Eq. (7.3-2)] using [28].

7-16 Obtain $\mathbf{H}^*(4)$ and $\mathbf{H}_w(4)$ from $\mathbf{H}^*(3)$ and $\mathbf{H}_w(3)$ respectively using the recursive relations developed by Fino [28].

7-17 Develop the flow graphs similar to Fig. 7.2 based on the Andrews' algorithm to compute HT and its inverse for $N = 16$.

7-18 Rearrange the columns of $\mathbf{H}^*(4)$ similar to the steps applied to $\mathbf{H}^*(3)$ [see Eqs. (7.4-1), (7.4-2) and (7.4-3)] and obtain the Cooley-Tukey type signal flow graph to compute HT for $N = 16$ (see Fig. 7.3).

7-19 Develop the Cooley-Tukey type signal flow graph to compute the IHT for $N = 16$.

7-20 (a) Develop $\mathbf{S}(4)$ using $\mathbf{S}(3)$ [see Eq. (7.5-7)] and the recursion relations described in Eqs. (7.5-8) and (7.5-9).
(b) Obtain the matrix factors of $\mathbf{S}(4)$ and then develop the signal flow graph to compute the ST. Using this signal flow graph obtain $\{D_x(k)\}$, the ST of $\{X(m)\}$ described in Prob. 7-6.

7-21 Develop the signal flow graph to compute the IST for $N = 16$. Recover $\{X(m)\}$ from $\{D_x(k)\}$ obtained in Prob. 7-20(b).

Chapter Eight

Generalized Wiener Filtering

In Chapter 1 it was stated that three applications of orthogonal transform would be studied. The application studied in this chapter concerns the classical signal processing technique known as Wiener filtering [1]. We will show that orthogonal transforms can be used to extend Wiener filtering to the processing of discrete signals with emphasis on reduction of computational requirements.

8.1 Some Basic Matrix Operations

It is instructive to consider some matrix operations which will be required for the generalized Wiener filtering development to be discussed in the next section. We define a vector $\boldsymbol{X}$ and a matrix $\mathbf{A}$ as follows:

$$\boldsymbol{X}' = [x_1\ x_2\ \cdots\ x_d] \tag{8.1-1}$$

and

$$\mathbf{A} = \begin{bmatrix} a_{11} & a_{12} & \cdots & a_{1d} \\ a_{21} & a_{22} & \cdots & a_{2d} \\ \vdots & \vdots & \ddots & \vdots \\ a_{d1} & a_{d2} & \cdots & a_{dd} \end{bmatrix} \tag{8.1-2}$$

We now demonstrate that

$$\nabla_{\mathbf{A}}\{\boldsymbol{X}'\mathbf{A}'\mathbf{A}\boldsymbol{X}\} = 2\mathbf{A}(\boldsymbol{X}\boldsymbol{X}') \tag{8.1-3}$$

where $\nabla_{\mathbf{A}}\{\boldsymbol{X}'\mathbf{A}'\mathbf{A}\boldsymbol{X}\}$ is the gradient (i.e., generalization of the derivative operation) of $\boldsymbol{X}'\mathbf{A}'\mathbf{A}\boldsymbol{X}$ with respect to the matrix $\mathbf{A}$. For the purposes of illustration, we consider the case when $d = 2$. Then from Eqs. (1) and (2) it follows that

$$\begin{aligned} P &= [x_1\ x_2] \begin{bmatrix} a_{11} & a_{21} \\ a_{12} & a_{22} \end{bmatrix} \begin{bmatrix} a_{11} & a_{12} \\ a_{21} & a_{22} \end{bmatrix} \begin{bmatrix} x_1 \\ x_2 \end{bmatrix} \\ &= (a_{11}x_1 + a_{12}x_2)^2 + (a_{21}x_1 + a_{22}x_2)^2 \end{aligned}$$

where $P = \boldsymbol{X}'\mathbf{A}'\mathbf{A}\boldsymbol{X}$.

Now, the gradient of P with respect to $\mathbf{A}$ is defined as

$$\nabla_{\mathbf{A}} P = \begin{bmatrix} \dfrac{\partial P}{\partial a_{11}} & \dfrac{\partial P}{\partial a_{12}} \\ \dfrac{\partial P}{\partial a_{21}} & \dfrac{\partial P}{\partial a_{22}} \end{bmatrix} \tag{8.1-4}$$

Evaluation of the elements of $\nabla_{\mathbf{A}} P$ leads to

$$\begin{aligned} \frac{\partial P}{\partial a_{11}} &= 2(a_{11}x_1 + a_{12}x_2)\, x_1 \\ \frac{\partial P}{\partial a_{12}} &= 2(a_{11}x_1 + a_{12}x_2)\, x_2 \\ \frac{\partial P}{\partial a_{21}} &= 2(a_{21}x_1 + a_{22}x_2)\, x_1 \\ \frac{\partial P}{\partial a_{22}} &= 2(a_{21}x_1 + a_{22}x_2)\, x_2 \end{aligned} \tag{8.1-5}$$

Substitution of Eq. (5) in Eq. (4) leads to

$$\begin{aligned} \nabla_{\mathbf{A}} P &= 2 \begin{bmatrix} a_{11} & a_{12} \\ a_{21} & a_{22} \end{bmatrix} \begin{bmatrix} x_1^2 & x_1 x_2 \\ x_1 x_2 & x_2^2 \end{bmatrix} \\ &= 2 \begin{bmatrix} a_{11} & a_{12} \\ a_{21} & a_{22} \end{bmatrix} \begin{bmatrix} x_1 \\ x_2 \end{bmatrix} [x_1 \quad x_2] \end{aligned} \tag{8.1-6}$$

Equation (6) yields the desired result

$$\nabla_{\mathbf{A}} P = \nabla_{\mathbf{A}} \{\boldsymbol{X}'\mathbf{A}'\mathbf{A}\boldsymbol{X}\} = 2\mathbf{A}(\boldsymbol{X}\boldsymbol{X}')$$

Similarly, if $\boldsymbol{V}' = [v_1\ v_2 \cdots v_d]$, then it can be shown that [see Prob. 8-1]

$$\nabla_{\mathbf{A}} \{\boldsymbol{X}'\mathbf{A}'\boldsymbol{V}\} = \boldsymbol{V}\boldsymbol{X}' \tag{8.1-7}$$

and

$$\nabla_{\mathbf{A}} \{\boldsymbol{V}'\boldsymbol{V}\} = 0 \tag{8.1-8}$$

8.2 Mathematical Model [2]

Figure 8.1 is the block diagram of a 1-dimensional generalized Wiener filtering system. $\boldsymbol{Z}$ is an input N-vector which is the sum of a data vector $\boldsymbol{X}$ and a noise vector $\boldsymbol{W}$. The Wiener filter $\mathbf{A}$ is in the form of an $(N \times N)$ matrix. $\mathbf{T}$ and $\mathbf{T}^{-1}$ are $(N \times N)$ matrices which represent an orthogonal transform and its inverse respectively. $\hat{\boldsymbol{X}}$ is the estimate of $\boldsymbol{X}$. The goal is to design the filter $\mathbf{A}$ in such a way that the expected value of the mean-square error between $\boldsymbol{X}$ and $\hat{\boldsymbol{X}}$ is minimized.

We observe that for the special case when $\mathbf{T}$ is the identity matrix $\mathbf{I}$, the model in Fig. 8.1 reduces to that originally proposed by Wiener. In this sense, the model in Fig. 8.1 is more general. Hence the name generalized Wiener filtering. The case $\mathbf{T} = \mathbf{I}$ will be referred to as the *identity transform* (IT) in order to emphasize that it is the simplest orthogonal transform.

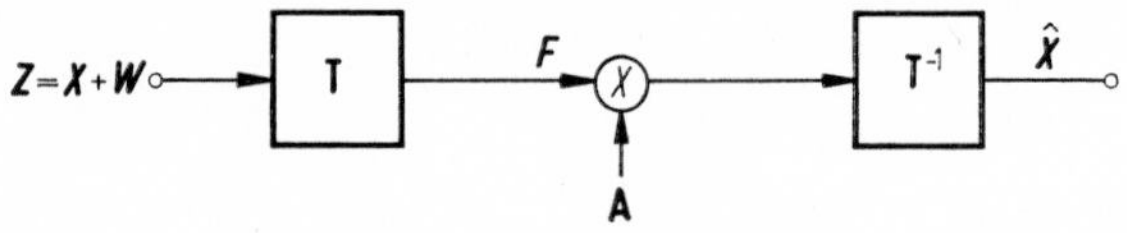

Fig. 8.1 Generalized Wiener filtering model

Statistical representation. For convenience we will assume that the elements of $\mathbf{T}$ are real numbers. The definitions which follow can easily be extended to the case when complex-valued data is present [see Prob. 8-3].

For many classes of input data, the vector $\boldsymbol{Z}$ can be considered as a sample of a random process (stationary or nonstationary). If the mean vector of the process is denoted by $\bar{\boldsymbol{Z}}$, then the *data domain covariance matrix*[1] is defined as

$$\boldsymbol{\Sigma}_z = \mathrm{E}\{(\boldsymbol{Z} - \bar{\boldsymbol{Z}})(\boldsymbol{Z} - \bar{\boldsymbol{Z}})'\} \tag{8.2-1}$$

where E represents the expected value operator. Using Eq. (1) it can be shown that [see Appendix 8.1, Eq. (A8.1-4)]

$$\boldsymbol{\Sigma}_z = \mathrm{E}\{\boldsymbol{Z}\boldsymbol{Z}'\} - \bar{\boldsymbol{Z}}\bar{\boldsymbol{Z}}' \tag{8.2-2}$$

Similarly, if the *transform domain covariance matrix* is denoted by $\tilde{\boldsymbol{\Sigma}}_z$, then

$$\tilde{\boldsymbol{\Sigma}}_z = \mathrm{E}\{(\boldsymbol{F} - \bar{\boldsymbol{F}})(\boldsymbol{F} - \bar{\boldsymbol{F}})'\} = \mathrm{E}\{\boldsymbol{F}\boldsymbol{F}'\} - \bar{\boldsymbol{F}}\bar{\boldsymbol{F}}' \tag{8.2-3}$$

where $\boldsymbol{F} = \mathbf{T}\boldsymbol{Z}$.

If the orthogonal matrix $\mathbf{T}$ is such that

$$\mathbf{T}'\mathbf{T} = \mathbf{I}$$

where $\mathbf{I}$ is the identity matrix, then $\tilde{\boldsymbol{\Sigma}}_z$ in Eq. (3) reduces to [see Prob. 8-2]

$$\tilde{\boldsymbol{\Sigma}}_z = \mathbf{T}\boldsymbol{\Sigma}_z\mathbf{T}' = \mathbf{T}\boldsymbol{\Sigma}_z\mathbf{T}^{-1} \tag{8.2-4}$$

From Eq. (4) it follows that the data and transform domain covariance matrices are related by a similarity transform. Further, we observe that $\tilde{\boldsymbol{\Sigma}}_z$ can be regarded as a 2-dimensional transform of $\boldsymbol{\Sigma}_z$ for computational purposes.

[1] A brief discussion of correlation and covariance matrices is included in Appendix 8.1, for the reader who is not familiar with related terminology and definitions.

Example 8.2-1

Consider the set of data vectors

$$\boldsymbol{Z}_1 = \begin{bmatrix} 0 \\ 2 \end{bmatrix}, \quad \boldsymbol{Z}_2 = \begin{bmatrix} 1 \\ 0 \end{bmatrix}, \quad \boldsymbol{Z}_3 = \begin{bmatrix} 1 \\ 1 \end{bmatrix}, \quad \boldsymbol{Z}_4 = \begin{bmatrix} 1 \\ -1 \end{bmatrix}, \quad \boldsymbol{Z}_5 = \begin{bmatrix} 2 \\ 3 \end{bmatrix}$$

(a) Evaluate $\boldsymbol{\Sigma}_z$.

(b) Use the result in (a) to find $\tilde{\boldsymbol{\Sigma}}_z$ with $\mathbf{T} = \frac{1}{\sqrt{2}} \begin{bmatrix} 1 & 1 \\ 1 & -1 \end{bmatrix}$

Solution. (a) We compute $\bar{\boldsymbol{Z}}\bar{\boldsymbol{Z}}'$, and $\mathrm{E}(\boldsymbol{Z}\boldsymbol{Z}')$ as follows:

$$\bar{\boldsymbol{Z}} = \frac{1}{5} \sum_{j=1}^{5} \boldsymbol{Z}_j = \begin{bmatrix} 1 \\ 1 \end{bmatrix}$$

$$\bar{\boldsymbol{Z}}\bar{\boldsymbol{Z}}' = \begin{bmatrix} 1 \\ 1 \end{bmatrix} [1 \;\; 1] = \begin{bmatrix} 1 & 1 \\ 1 & 1 \end{bmatrix}$$

$$\mathrm{E}(\boldsymbol{Z}\boldsymbol{Z}') = \frac{1}{5}\left(\begin{bmatrix} 0 \\ 2 \end{bmatrix} [0 \;\; 2] + \begin{bmatrix} 1 \\ 0 \end{bmatrix} [1 \;\; 0] + \begin{bmatrix} 1 \\ 1 \end{bmatrix} [1 \;\; 1] + \begin{bmatrix} 1 \\ -1 \end{bmatrix} [1 \;\; -1] + \begin{bmatrix} 2 \\ 3 \end{bmatrix} [2 \;\; 3] \right)$$

$$= \frac{1}{5} \begin{bmatrix} 7 & 6 \\ 6 & 15 \end{bmatrix}$$

Substitution for $\bar{\boldsymbol{Z}}\bar{\boldsymbol{Z}}'$ and $\mathrm{E}\{\boldsymbol{Z}\boldsymbol{Z}'\}$ in Eq. (2) leads to

$$\boldsymbol{\Sigma}_z = \begin{bmatrix} 0.4 & 0.2 \\ 0.2 & 2 \end{bmatrix}$$

(b) Equation (4) yields

$$\tilde{\boldsymbol{\Sigma}}_z = \frac{1}{2} \begin{bmatrix} 1 & 1 \\ 1 & -1 \end{bmatrix} \begin{bmatrix} 0.4 & 0.2 \\ 0.2 & 2 \end{bmatrix} \begin{bmatrix} 1 & 1 \\ 1 & -1 \end{bmatrix} = \begin{bmatrix} 1.4 & -0.8 \\ -0.8 & 2 \end{bmatrix}$$

8.3 Filter Design

Without loss of generality, we make the following assumptions:

1. The data and noise processes have zero means; i.e., $\bar{\boldsymbol{X}} = \bar{\boldsymbol{W}} = 0$ which implies that

$$\bar{\boldsymbol{Z}} = 0$$

$$\boldsymbol{\Sigma}_x = \mathrm{E}(\boldsymbol{X}\boldsymbol{X}')$$

$$\boldsymbol{\Sigma}_w = \mathrm{E}(\boldsymbol{W}\boldsymbol{W}')$$

and

$$\boldsymbol{\Sigma}_z = \mathrm{E}(\boldsymbol{Z}\boldsymbol{Z}') \qquad (8.3\text{-}1)$$

2. The data and noise processes are uncorrelated; i.e.,

$$\mathrm{E}(\boldsymbol{X}\boldsymbol{W}') = \mathrm{E}(\boldsymbol{W}\boldsymbol{X}') = 0 \tag{8.3-2}$$

3. $\mathbf{T}'\mathbf{T} = \mathbf{I}$ which implies that [see Eq. (8.2-4)]

$$\tilde{\mathbf{\Sigma}}_x = \mathbf{T}\mathbf{\Sigma}_x\mathbf{T}'$$

$$\tilde{\mathbf{\Sigma}}_w = \mathbf{T}\mathbf{\Sigma}_w\mathbf{T}'$$

and

$$\tilde{\mathbf{\Sigma}}_z = \mathbf{T}\mathbf{\Sigma}_z\mathbf{T}' \tag{8.3-3}$$

Derivation of the filter matrix. Since $\hat{\mathbf{X}}$ is the estimate of $\boldsymbol{X}$, the corresponding error vector is

$$\boldsymbol{\varepsilon} = \hat{\boldsymbol{X}} - \boldsymbol{X} \tag{8.3-4}$$

Thus the total mean-square error in estimating $\boldsymbol{X}$ is given by

$$\varepsilon = \mathrm{E}\{\|\hat{\boldsymbol{X}} - \boldsymbol{X}\|^2\} = \mathrm{E}\{(\hat{\boldsymbol{X}} - \boldsymbol{X})'(\hat{\boldsymbol{X}} - \boldsymbol{X})\} \tag{8.3-5}$$

From the model in Fig. 8.1, it follows that

$$\hat{\boldsymbol{X}} = \mathbf{T}^{-1}\mathbf{A}\mathbf{T}\boldsymbol{Z} = \mathbf{T}'\mathbf{A}\mathbf{T}\boldsymbol{Z} \tag{8.3-6}$$

since $\mathbf{T}^{-1} = \mathbf{T}'$.

Substitution of Eq. (6) in Eq. (5) leads to

$$\varepsilon = \mathrm{E}\{\boldsymbol{Z}'\mathbf{T}'\mathbf{A}'\mathbf{T}\mathbf{T}'\mathbf{A}\mathbf{T}\boldsymbol{Z}\} - 2\mathrm{E}\{\boldsymbol{Z}'\mathbf{T}'\mathbf{A}'\mathbf{T}\boldsymbol{X}\} + \mathrm{E}\{\|\boldsymbol{X}\|^2\} \tag{8.3-7}$$

Since $\mathbf{T}\mathbf{T}' = \mathbf{I}$, and $\boldsymbol{F} = \mathbf{T}\boldsymbol{Z}$, Eq. (7) can be expressed as

$$\varepsilon = \mathrm{E}\{\boldsymbol{F}'\mathbf{A}'\mathbf{A}\boldsymbol{F}\} - 2\mathrm{E}\{\boldsymbol{F}'\mathbf{A}'\mathbf{T}\boldsymbol{X}\} + \mathrm{E}\{\|\boldsymbol{X}\|^2\} \tag{8.3-8}$$

Now, $\mathbf{A}$ must be chosen in such a way that ε is *minimum*. Hence it must satisfy the necessary condition

$$\nabla_{\mathbf{A}}\,\varepsilon = 0$$

which yields

$$\mathrm{E}\{\nabla_{\mathbf{A}}(\boldsymbol{F}'\mathbf{A}'\mathbf{A}\boldsymbol{F})\} - 2\mathrm{E}\{\nabla_{\mathbf{A}}(\boldsymbol{F}'\mathbf{A}'\mathbf{T}\boldsymbol{X})\} + \mathrm{E}\{\nabla_{\mathbf{A}}(\|\boldsymbol{X}\|^2)\} = 0 \tag{8.3-9}$$

From Eqs. (8.1-3), (8.1-7), and (8.1-8) it follows that

$$\nabla_{\mathbf{A}}(\boldsymbol{F}'\mathbf{A}'\mathbf{A}\boldsymbol{F}) = 2\mathbf{A}\boldsymbol{F}\boldsymbol{F}'$$

$$\nabla_{\mathbf{A}}(\boldsymbol{F}'\mathbf{A}'\mathbf{T}\boldsymbol{X}) = \mathbf{T}\boldsymbol{X}\boldsymbol{F}'$$

and

$$\nabla_{\mathbf{A}}(\|\boldsymbol{X}\|^2) = 0$$

Thus Eq. (9) becomes

$$\mathbf{A}\mathrm{E}\{\boldsymbol{F}\boldsymbol{F}'\} = \mathbf{T}\mathrm{E}\{\boldsymbol{X}\boldsymbol{F}'\}$$

which with $\boldsymbol{F} = \mathbf{T}\boldsymbol{Z}$ yields

$$\mathbf{A}\mathbf{T}\mathrm{E}\{\boldsymbol{Z}\boldsymbol{Z}'\}\mathbf{T}' = \mathbf{T}\mathrm{E}\{\boldsymbol{X}\boldsymbol{Z}'\}\mathbf{T}' \tag{8.3-10}$$

Recalling that $\boldsymbol{Z} = \boldsymbol{X} + \boldsymbol{W}$, and using Eq. (1), the following relations are readily obtained.

$$\mathrm{E}(\boldsymbol{X}\boldsymbol{Z}') = \mathrm{E}(\boldsymbol{X}\boldsymbol{X}') = \boldsymbol{\Sigma}_x$$

and

$$\mathrm{E}(\boldsymbol{Z}\boldsymbol{Z}') = \boldsymbol{\Sigma}_x + \boldsymbol{\Sigma}_w \tag{8.3-11}$$

Substitution of Eq. (11) in Eq. (10) leads to

$$\mathbf{AT}(\boldsymbol{\Sigma}_x + \boldsymbol{\Sigma}_w)\mathbf{T}' = \mathbf{T}\boldsymbol{\Sigma}_x\mathbf{T}' \tag{8.3-12}$$

Equation (12) yields the desired optimum filter matrix in the following two forms.

(i) In terms of data domain covariance matrices $\boldsymbol{\Sigma}_x$ and $\boldsymbol{\Sigma}_w$:

$$\mathbf{A}_0 = \mathbf{T}\boldsymbol{\Sigma}_x\mathbf{T}'[\mathbf{T}(\boldsymbol{\Sigma}_x + \boldsymbol{\Sigma}_w)\mathbf{T}']^{-1} \tag{8.3-13}$$

which simplifies to yield

$$\mathbf{A}_0 = \mathbf{T}\mathbf{A}_r\mathbf{T}' \tag{8.3-14}$$

where

$$\mathbf{A}_r = \boldsymbol{\Sigma}_x(\boldsymbol{\Sigma}_x + \boldsymbol{\Sigma}_w)^{-1}$$

The matrix $\mathbf{A}_r$ is called the *response matrix* [2] in analogy with the impulse response of a linear system.

(ii) In terms of transform covariance matrices $\tilde{\boldsymbol{\Sigma}}_x$ and $\tilde{\boldsymbol{\Sigma}}_w$:

$$\mathbf{A}_0 = \tilde{\boldsymbol{\Sigma}}_x[\tilde{\boldsymbol{\Sigma}}_x + \tilde{\boldsymbol{\Sigma}}_w]^{-1} \tag{8.3-15}$$

where

$$\tilde{\boldsymbol{\Sigma}}_x = \mathbf{T}\boldsymbol{\Sigma}_x\mathbf{T}'$$

and

$$\tilde{\boldsymbol{\Sigma}}_w = \mathbf{T}\boldsymbol{\Sigma}_w\mathbf{T}'$$

Minimum mean-square error. To calculate the mean-square error associated with the above optimum filter, it is convenient to express ε in Eq. (5) as

$$\varepsilon = \mathrm{tr}\{\mathrm{E}[(\hat{\boldsymbol{X}} - \boldsymbol{X})(\hat{\boldsymbol{X}} - \boldsymbol{X})']\} \tag{8.3-16}$$

where "tr" denotes the trace of a matrix; i.e., the sum of the elements along the main diagonal. It can be shown that [see Prob. 8-4]

$$\mathrm{tr}\{\mathrm{E}[(\hat{\boldsymbol{X}} - \boldsymbol{X})(\hat{\boldsymbol{X}} - \boldsymbol{X})']\} = \mathrm{E}\{\|\hat{\boldsymbol{X}} - \boldsymbol{X}\|^2\} \tag{8.3-17}$$

From Eq. (16) we have

$$\varepsilon = \mathrm{tr}\left\{\mathrm{E}\{\hat{\boldsymbol{X}}\hat{\boldsymbol{X}}'\} - \mathrm{E}\{\hat{\boldsymbol{X}}\boldsymbol{X}'\} - \mathrm{E}\{\boldsymbol{X}\hat{\boldsymbol{X}}'\} + \mathrm{E}\{\boldsymbol{X}\boldsymbol{X}'\}\right\} \tag{8.3-18}$$

Recalling that $\hat{X} = \mathbf{T}'\mathbf{A}\mathbf{T}Z$, it can be shown that [see Prob. 8-5]

$$E\{\hat{X}\hat{X}'\} = E\{X\hat{X}'\}$$

and

$$E\{\hat{X}X'\} = \mathbf{T}'\mathbf{A}\mathbf{T}\boldsymbol{\Sigma}_x \tag{8.3-19}$$

Substituting Eq. (19) in Eq. (18) and using the relation $\boldsymbol{\Sigma}_x = E\{XX'\}$, we obtain

$$\varepsilon = \mathrm{tr}[\boldsymbol{\Sigma}_x - \mathbf{T}'\mathbf{A}\mathbf{T}\boldsymbol{\Sigma}_x] \tag{8.3-20}$$

Now, the expression for the optimum filter is given by Eq. (14) to be

$$\mathbf{A}_0 = \mathbf{T}\boldsymbol{\Sigma}_x(\boldsymbol{\Sigma}_x + \boldsymbol{\Sigma}_w)^{-1}\mathbf{T}'$$

Substitution of $\mathbf{A}_0$ for $\mathbf{A}$ in Eq. (20) leads to the following expression for the *minimum* mean-square error.

$$\varepsilon_{min} = \mathrm{tr}[\boldsymbol{\Sigma}_x - \boldsymbol{\Sigma}_x(\boldsymbol{\Sigma}_x + \boldsymbol{\Sigma}_w)^{-1}\boldsymbol{\Sigma}_x] \tag{8.3-21}$$

ε_{min} in Eq. (21) is now exclusively in terms of the data domain covariance matrices $\boldsymbol{\Sigma}_x$ and $\boldsymbol{\Sigma}_w$ and hence can be computed. Alternately, ε_{min} can be expressed in terms of the transform domain covariance matrices $\tilde{\boldsymbol{\Sigma}}_x$ and $\tilde{\boldsymbol{\Sigma}}_w$ to obtain [see Prob. 8-7]

$$\varepsilon_{min} = \mathrm{tr}[\tilde{\boldsymbol{\Sigma}}_x - \tilde{\boldsymbol{\Sigma}}_x(\tilde{\boldsymbol{\Sigma}}_x + \tilde{\boldsymbol{\Sigma}}_w)^{-1}\tilde{\boldsymbol{\Sigma}}_x] \tag{8.3-22}$$

The important result associated with the above analysis is conveyed by Eqs. (21) and (22). These equations state that the minimum mean-square error ε_{min} is *independent* of the orthogonal transform employed.

8.4 Suboptimal Wiener Filtering

The conclusion that follows from ε_{min} being independent of transform $\mathbf{T}$ is that one is free to choose the transform that minimizes the computational processes entailed.

Examination of Fig. 8.1 shows that the multiplications required to perform the Wiener filtering operations of Eq. (8.3-6) is as summarized in Table 8.4-1. Clearly, the main hindrance to real-processing is that the number of multiplications required is proportional to N^2. Thus, as a compromise, we consider the possibility of developing filter matrices that contain a relatively large number of zero entries. With such matrices, the filter operation $\boldsymbol{AF}$ [see Fig. 8.1] can be performed with reduced number of multiplications. The goal in the filter design, of course, is to maintain the mean-square error performance as close to the optimum filter level as possible.

Table 8.4-1 Optimal Wiener Filtering Multiplication Requirements

Transform	Approximate number of multiplications
IT	N^2
DFT[1]	$N^2 + 2N \log_2 N$
$(\mathrm{WHT})_w$, $(\mathrm{WHT})_h$, MWHT, or HT	N^2
DCT[1]	$N^2 + 4N \log_2 N$

[1] complex number multiplications

The suboptimal Wiener filtering design problem can be formulated as an optimization problem whereby the filter matrix **A** is chosen to minimize [see Eq. (8.3-20)]

$$\varepsilon = \mathrm{tr}[\mathbf{\Sigma}_x - \mathbf{T}'\mathbf{A}\mathbf{T}\mathbf{\Sigma}_x]$$

under the constraint that certain selected elements of **A** are zero. We consider the following possibilities.

1. Constrain **A** to be a *diagonal* matrix. This class of filters will be considered in Section 8.6.

2. Determine the filter design when the filter matrix contains two elements per row, and continue with additional matrix elements until the desired mean-square error is achieved. However, this approach rapidly becomes quite complex.

3. Obtain the suboptimal filter matrix from the optimum filter matrix by retaining only those elements of relatively large magnitude. Set all remaining elements to zero.

In what follows we study some aspects of 3. by means of an example.

Suboptimal filtering example. Consider the mean-square estimation of a random signal in the presence of noise where the signal and noise covariance matrices are given by

$$\mathbf{\Sigma}_x = \begin{bmatrix} 1 & \varrho & \varrho^2 & \cdots & \varrho^{N-1} \\ \varrho & 1 & \varrho & \cdots & \varrho^{N-2} \\ \varrho^2 & \varrho & 1 & \cdots & \varrho^{N-3} \\ \cdots & \cdots & \cdots & \cdots & \cdots \\ \varrho^{N-1} & \varrho^{N-2} & \varrho^{N-3} & \cdots & 1 \end{bmatrix}, \quad 0 < \varrho < 1 \tag{8.4-1}$$

and

$$\mathbf{\Sigma}_w = k_0 \cdot \mathbf{I}_N \tag{8.4-2}$$

Note that $\boldsymbol{\Sigma}_x$ is a Toeplitz matrix [see Eq. (7.7-10)]. The above $\boldsymbol{\Sigma}_x$ and $\boldsymbol{\Sigma}_w$ are said to be the covariance matrices of a first-order Markov process and white noise process respectively.

To determine the effect of orthogonal transforms on the structure of the corresponding filter matrix, we compute the filter matrices for the identity and discrete cosine transforms using Eq. (8.3-14) with $N = 16$, $\varrho = 0.9$, and $k_0 = 0.1$. Since $k_0 = 0.1$, the signal-to-noise ratio is 10. The filter matrices obtained are conveniently displayed as shown in Fig. 8.2. The hatched region represents those elements in a filter matrix whose magnitude is greater than or equal to 1% of the element with the largest magnitude in the filter matrix. From Fig. 8.2 it is apparent that the effect of the DCT is to cause the corresponding filter matrix to contain appreciably less elements with relatively large magnitudes.

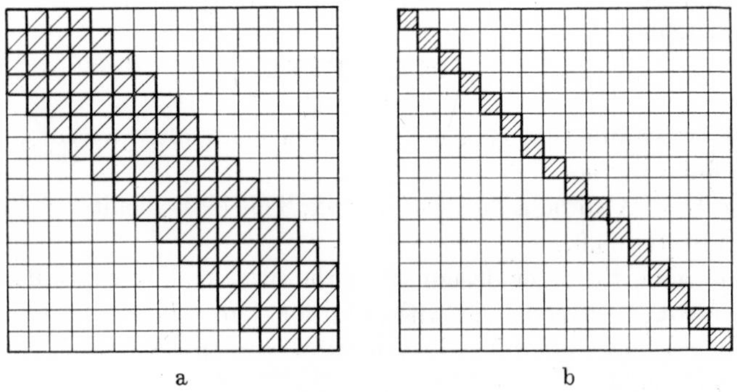

Fig. 8.2 Display of (a) IT, and (b) DCT Wiener filter matrices

Next, we consider an example which concerns the mean-square error performance of suboptimal Wiener filters. For the purposes of illustration, we consider a $(\mathrm{WHT})_w$ or $(\mathrm{WHT})_h$ filter with $N = 16$ and $\varrho = 0.9$. The signal-to-noise ratio is chosen to be unity; i.e., $k_0 = 1$. Figure 8.3 shows the mean-square error performance of the identity and Walsh-Hadamard filters as a function of the number of largest magnitude filter elements not set to zero in the optimum filter matrix [2]. It is apparent that the IT filter performs quite poorly compared to the $(\mathrm{WHT})_h$ filter since it requires a substantially larger number of non-zero elements for a specified mean-square error. Furthermore, the performance of the $(\mathrm{WHT})_h$ filter for as few as 10 nonzero terms is quite close to its optimum performance when all 256 terms are employed.

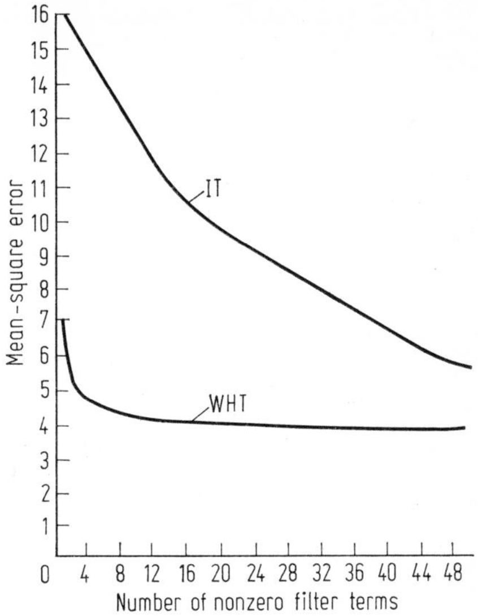

Fig. 8.3
Mean-square error performance of IT and $(\mathrm{WHT})_h$ Wiener filters

8.5 Optimal Diagonal Filters

A logical question which follows from the above discussion is: "What is the orthogonal transform $\mathbf{T}$, which results in an *optimal diagonal filter* $\mathbf{A}_{od}$?" The answer to this question is readily obtained from the following theorem, a proof of which is available elsewhere[1]. The key to the proof is that the eigenvalues of a real symmetric matrix are real and distinct. We note that the response matrix $\mathbf{A}_r = \mathbf{\Sigma}_x(\mathbf{\Sigma}_x + \mathbf{\Sigma}_w)^{-1}$ in Eq. (8.3-14) is a real symmetric matrix.

Theorem. If λ_i and $\boldsymbol{\phi}_i$, $i = 1, 2, \ldots, N$ are the eigenvalues and N-eigenvectors respectively of a real symmetric matrix $\mathbf{\Gamma}$, then

$$\mathbf{\Phi\Gamma\Phi}' = \mathbf{\Lambda} \tag{8.5-1}$$

where

$$\mathbf{\Phi} = [\boldsymbol{\phi}_1 \ \boldsymbol{\phi}_2 \ \cdots \ \boldsymbol{\phi}_N]$$

is the $(N \times N)$ eigenvector matrix such that $\mathbf{\Phi}'\mathbf{\Phi} = \mathbf{I}$, and

$$\mathbf{\Lambda} = \operatorname{diag}(\lambda_1, \lambda_2, \ldots, \lambda_N)$$

is the eigenvalue matrix.

From Eq. (1) it is apparent that if $\mathbf{T}$ is chosen to be the eigenvector matrix of the response matrix $\mathbf{A}_r$, then $\mathbf{A}_{od} = \mathbf{T}\mathbf{A}_r\mathbf{T}'$ is the desired optimal diagonal filter.

[1] For example, see Fukunaga, K.: *Introduction to Statistical Pattern Recognition.* New York, London: Academic Press 1973, 29–31.

It is instructive to illustrate the above theorem by means of a simple example. Let the data domain covariance matrices be as follows:

$$\mathbf{\Sigma}_x = \begin{bmatrix} 1 & \varrho \\ \varrho & 1 \end{bmatrix}, \quad 0 < \varrho < 1$$

and

$$\mathbf{\Sigma}_w = \begin{bmatrix} 1 & 0 \\ 0 & 1 \end{bmatrix}$$

Now, the response matrix $\mathbf{A}_r$ is given by Eq. (8.3-14) as

$$\mathbf{A}_r = \mathbf{\Sigma}_x (\mathbf{\Sigma}_x + \mathbf{\Sigma}_w)^{-1}$$

Evaluating $\mathbf{A}_r$, one obtains

$$\mathbf{A}_r = \begin{bmatrix} \alpha & \beta \\ \beta & \alpha \end{bmatrix}$$

where

$$\alpha = \frac{2 - \varrho^2}{4 - \varrho^2}$$

and

$$\beta = \frac{\varrho}{4 - \varrho^2}$$

To obtain the eigenvalues, we evaluate

$$|\mathbf{A}_r - \lambda \mathbf{I}| = 0$$

which yields

$$\lambda^2 - 2\alpha\lambda + (\alpha^2 - \beta^2) = 0 \tag{8.5-2}$$

Solving Eq. (2) one obtains the eigenvalues

$$\lambda_1 = \alpha + \beta, \quad \text{and} \quad \lambda_2 = \alpha - \beta$$

It is left as an exercise to show that [see Prob. 8-8] the normalized eigenvectors corresponding to λ_1 and λ_2 respectively are given by

$$\boldsymbol{\phi}_1 = \frac{1}{\sqrt{2}} \begin{bmatrix} 1 \\ 1 \end{bmatrix} \quad \text{and} \quad \boldsymbol{\phi}_2 = \frac{1}{\sqrt{2}} \begin{bmatrix} 1 \\ -1 \end{bmatrix}$$

Thus the desired orthogonal transform is

$$\mathbf{T} = [\boldsymbol{\phi}_1 \boldsymbol{\phi}_2] = \frac{1}{\sqrt{2}} \begin{bmatrix} 1 & 1 \\ 1 & -1 \end{bmatrix}$$

It is easily verified that the corresponding optimum Wiener filter matrix is given by [see Eq. (8.3-14)]

$$\mathbf{A}_{od} = \mathbf{T}\mathbf{A}_r\mathbf{T}' = \begin{bmatrix} \alpha + \beta & 0 \\ 0 & \alpha - \beta \end{bmatrix}$$

which is a diagonal matrix.

The above orthogonal transform whose basis vectors are eigenvectors of specified covariance matrices is called the *Karhunen-Loève transform*

(KLT)[1]. We make the following observations regarding the KLT and the KLT filter.

(i) The KLT filter is an optimal filter because its mean-square error is given by ε_{min} in Eq. (8.3-21) or Eq. (8.3-22).
(ii) Since the KLT depends on specified covariance matrices, no general fast algorithm exists to compute it or its inverse. This implies that $\boldsymbol{F} = \mathbf{T}\boldsymbol{Z}$, and $\hat{\boldsymbol{X}} = \mathbf{T}^{-1}(\mathbf{A}\boldsymbol{F})$ in Fig. 8.1 will require approximately $2N^2$ multiplications. Thus the total number of multiplications required to obtain $\hat{\boldsymbol{X}}$ is $(2N^2 + N) \simeq 2N^2$.
(iii) As N increases, the task of computing eigenvectors rapidly becomes a formidable one.

From observations (ii) and (iii) it is apparent that when N is large, it is not feasible to use the KLT. However, since it is optimum [see observation (i)], it serves as a useful tool to assess the performance of other suboptimal filters.

8.6 Suboptimal Diagonal Filters [2–4]

Since the KLT can not be computed using fast algorithms, it is natural to consider the possibility of developing suboptimal diagonal filter matrices for transforms that do possess fast computational algorithms. In this section we develop a class of such filters which consists of the discrete Fourier, Walsh-Hadamard, Haar, and discrete cosine filters.

Filters with diagonal filter matrices are sometimes referred to as *scalar* filters, while the term *vector* filters refers to a more general class that consists of filter matrices which can have non-zero off diagonal elements also.

The desired scalar filter is obtained by constraining $\mathbf{A}$ in Eq. (8.3-12) to be a diagonal matrix. Thus, if

$$\mathbf{A}_d = \operatorname{diag}(a_{11}, a_{22}, \ldots, a_{NN}) \tag{8.6-1}$$

is the scalar filter matrix, then from Eq. (8.3-12) we obtain

$$a_{ii} = \frac{\tilde{\Sigma}_x(i, i)}{\tilde{\Sigma}_x(i, i) + \tilde{\Sigma}_w(i, i)}, \qquad i = 1, 2, \ldots, N \tag{8.6-2}$$

where $\tilde{\mathbf{\Sigma}}_x = \mathbf{T}\mathbf{\Sigma}_x\mathbf{T}'$, $\tilde{\mathbf{\Sigma}}_w = \mathbf{T}\mathbf{\Sigma}_w\mathbf{T}'$, and $\tilde{\Sigma}_x(i, i)$, $\tilde{\Sigma}_w(i, i)$ denote the i-th diagonal elements of $\tilde{\mathbf{\Sigma}}_x$ and $\tilde{\mathbf{\Sigma}}_w$ respectively.

The mean-square error associated with $\mathbf{A}_d$ is obtained from Eq. (8.3-20) as

$$\varepsilon_d = \operatorname{tr}[\mathbf{\Sigma}_x - \mathbf{T}'\mathbf{A}_d\mathbf{T}\mathbf{\Sigma}_x] = \operatorname{tr}(\mathbf{\Sigma}_x) - \operatorname{tr}(\mathbf{T}'\mathbf{A}_d\mathbf{T}\mathbf{\Sigma}_x) \tag{8.6-3}$$

[1] This terminology is a consequence of the KL expansion which will be discussed in Chapter 9.

Since $\tilde{\mathbf{\Sigma}}_x = \mathbf{T}\mathbf{\Sigma}_x\mathbf{T}'$ and $\mathbf{T}'\mathbf{T} = \mathbf{I}$, it follows that

$$\mathbf{\Sigma}_x = \mathbf{T}'\tilde{\mathbf{\Sigma}}_x\mathbf{T} \tag{8.6-4}$$

Substitution of Eq. (4) in Eq. (3) leads to

$$\varepsilon_d = \mathrm{tr}\,(\mathbf{\Sigma}_x) - \mathrm{tr}\,(\mathbf{T}'\mathbf{A}_d\tilde{\mathbf{\Sigma}}_x\mathbf{T}) \tag{8.6-5}$$

Now, since $\mathbf{T}$ is an orthonormal transform it is known that [see Prob. 8-6, Eq. (P8-6-1)]

$$\mathrm{tr}\,(\mathbf{\Sigma}_x) = \mathrm{tr}\,(\mathbf{T}\mathbf{\Sigma}_x\mathbf{T}') = \mathrm{tr}\,(\tilde{\mathbf{\Sigma}}_x)$$

and

$$\mathrm{tr}\,(\mathbf{T}'\mathbf{A}_d\tilde{\mathbf{\Sigma}}_x\mathbf{T}) = \mathrm{tr}\,(\mathbf{A}_d\tilde{\mathbf{\Sigma}}_x) \tag{8.6-6}$$

Combining Eqs. (5) and (6) we obtain

$$\varepsilon_d = \mathrm{tr}\,(\tilde{\mathbf{\Sigma}}_x) - \mathrm{tr}\,(\mathbf{A}_d\tilde{\mathbf{\Sigma}}_x)$$

However, since $\mathbf{A}_d$ is a diagonal matrix, ε_d simplifies to yield

$$\varepsilon_d = \mathrm{tr}\,(\tilde{\mathbf{\Sigma}}_x) - \sum_{i=1}^{N} a_{ii}\tilde{\mathbf{\Sigma}}_x(i, i) \tag{8.6-7}$$

Substitution for a_{ii} using Eq. (2) results in the following mean-square error for the scalar filter $\mathbf{A}_d$ and orthogonal transform $\mathbf{T}$:

$$\varepsilon_d = \mathrm{tr}\,(\tilde{\mathbf{\Sigma}}_x) - \sum_{i=1}^{N} \frac{\tilde{\mathbf{\Sigma}}_x^2(i, i)}{\tilde{\mathbf{\Sigma}}_x(i, i) + \tilde{\mathbf{\Sigma}}_w(i, i)} \tag{8.6-8}$$

where $\tilde{\mathbf{\Sigma}}_x = \mathbf{T}\mathbf{\Sigma}_x\mathbf{T}'$, and $\tilde{\mathbf{\Sigma}}_w = \mathbf{T}\mathbf{\Sigma}_w\mathbf{T}'$.

In particular for the KLT, Eq. (8) can alternately be expressed as [see Eq. (8.5-1)]

$$\varepsilon_{KLT} = \sum_{i=1}^{N} \xi_i - \sum_{i=1}^{N} (\xi_i^2/\hat{\xi}_i) \tag{8.6-9}$$

where ξ_i and $\hat{\xi}_i$ are the eigenvalues of $\mathbf{\Sigma}_x$ and $(\mathbf{\Sigma}_x + \mathbf{\Sigma}_w)$ respectively.

In order to compare the performances of various scalar filters, we again consider the problem of estimating a Markov process signal in the presence of white noise. The corresponding signal and noise covariance matrices are given by Eqs. (8.4-1) and (8.4-2) respectively. Table 8.6-1 lists the values of ε_d for different values of N for the case $\varrho = 0.9$ and a signal-to-noise ratio of unity (i.e., $k_0 = 1$) [5]. From Table 8.6-1 it is evident that the DCT compares very closely to the KLT. This information is also presented in terms of a set of performance curves in Fig. 8.4. We observe that the DFT performance curve tends to approach the KLT curve asymptotically. This result is a special case of a theorem by Toeplitz which states that [3, 6]

$$\lim_{N\to\infty} [\mathbf{T}_{DFT}\mathbf{\Sigma}\mathbf{T}_{DFT}^{-1}] = \mathbf{\Lambda} \tag{8.6-10}$$

where

$$\Sigma = \begin{bmatrix} \varrho_0 & \varrho_1 & \varrho_2 & \cdots & \varrho_{N-1} \\ \varrho_1 & \varrho_0 & \varrho_1 & \cdots & \varrho_{N-2} \\ \varrho_2 & \varrho_1 & \varrho_0 & \cdots & \varrho_{N-3} \\ \cdots & \cdots & \cdots & \cdots & \cdots \\ \varrho_{N-1} & \varrho_{N-2} & \varrho_{N-3} & \cdots & \varrho_0 \end{bmatrix}$$

is the covariance matrix of a wide-sense stationary process [see Appendix 8.1, Eq. (A8.1-7)], and

$$\Lambda = \text{diag}\,(\lambda_1, \lambda_2, \ldots, \lambda_N)$$

is the eigenvalue matrix of Σ.

Table 8.6-1 Mean-square error performance of various scalar filters; $\varrho = 0.9$, $k_0 = 1$

transform ↓ $N \rightarrow$	2	4	8	16	32	64
KLT	5.9680	4.6640	4.0528	3.7696	3.6288	3.5584
DCT	5.9680	4.6720	4.0736	3.7894	3.6512	3.5712
DFT	5.9680	4.7424	4.3926	4.1472	3.9056	3.7120
$(\text{WHT})_h$	5.9680	4.7072	4.2384	4.1312	4.1312	4.0944
HT	5.9680	4.7072	4.2400	4.1424	4.1312	4.1296

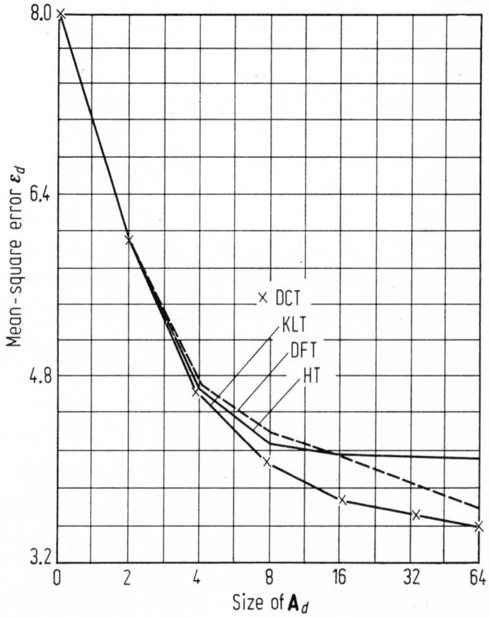

Fig. 8.4 Mean-square error performance of scalar filters

8.7 2-dimensional Wiener Filtering Considerations

Although our attention was restricted to 1-dimensional filtering, the techniques we have developed so far can be used for 2-dimensional Wiener filtering purposes. In what follows, we list the steps involved to achieve the 2-dimensional filtering by sequential application of 1-dimensional filtering techniques.

Let $[Z(x, y)]$ and $[F(u, v)]$ represent $(N \times N)$ data and transform matrices respectively. If $\mathbf{T}$ denotes an $(N \times N)$ transform matrix such that $\mathbf{T}'\mathbf{T} = \mathbf{I}$, then

$$[F(u, v)] = \mathbf{T}[Z(x, y)]\mathbf{T}' \tag{8.7-1}$$

and

$$[Z(x, y)] = \mathbf{T}'[F(u, v)]\mathbf{T} \tag{8.7-2}$$

The input data matrix $[Z(x, y)]$ is the sum of a signal matrix $[X(x, y)]$, and a noise matrix $[W(x, y)]$; that is

$$[Z(x, y)] = [X(x, y)] + [W(x, y)] \tag{8.7-3}$$

The transform matrix $[F(u, v)]$ is then modified to obtain the filtered data transform matrix $[G(u, v)]$ as follows:

$$[G(u, v)] = \mathbf{A}_y[F(u, v)]\mathbf{A}_x' \tag{8.7-4}$$

where $\mathbf{A}_y$ and $\mathbf{A}_x$ are Wiener filter matrices for the *rows* and *columns* respectively, of the input data matrix $[Z(x, y)]$. We emphasize that $\mathbf{A}_x$ and $\mathbf{A}_y$ are obtained using the 1-dimensional design methods developed earlier.

Finally, the estimate $[\hat{X}(x, y)]$ of $[X(x, y)]$ is obtained by taking the inverse transform of $[G(u, v)]$ in Eq. (4). Thus

$$[\hat{X}(x, y)] = \mathbf{T}'[G(u, v)]\mathbf{T} \tag{8.7-5}$$

From the above discussion it is apparent that 2-dimensional Wiener filtering can be treated as a sequential application of 1-dimensional filtering to the rows and columns of the data matrix $[Z(x, y)]$. Two-dimensional generalized Wiener filtering has been used successfully for image enhancement purposes — i.e., enhancing the quality of noisy images [2].

8.8 Summary

We have shown that classical Wiener filtering operations can be performed through the use of orthogonal transforms. The amount of computation required depends upon the transform employed. It has been demonstrated that the computational load can be reduced significantly with only a small degradation in mean-square error performance.

A brief discussion of 2-dimensional Wiener filtering was included merely to illustrate that it can be carried out by sequential application of 1-dimensional filtering techniques. Orthogonal transforms can also be used for other 2-dimensional filtering applications in the area of image processing [7–10]. Such applications have been reviewed by Pratt [11] in terms of the role of Walsh-Hadamard transforms in image processing and 2-dimensional filtering.

Appendix 8.1 Some Terminology and Definitions

If z_i, $i = 1, 2, \ldots, N$ are a set of N random variables, then $\boldsymbol{Z}' = [z_1 \; z_2 \cdots z_N]$ is a random vector. The *covariance matrix* of $\boldsymbol{Z}$ is defined as

$$\boldsymbol{\Sigma}_z = \mathrm{E}\{(\boldsymbol{Z} - \bar{\boldsymbol{Z}})(\boldsymbol{Z} - \bar{\boldsymbol{Z}})'\} \tag{A8.1-1}$$

where E represents the expected value operator, and $\bar{\boldsymbol{Z}}$ denotes the mean vector $\mathrm{E}\{\boldsymbol{Z}\}$. Using matrix notation, $\boldsymbol{\Sigma}_z$ can be expressed as

$$\begin{aligned}
\boldsymbol{\Sigma}_z &= \mathrm{E}\left\{\begin{bmatrix}(z_1 - \bar{z}_1)\\ \vdots \\ (z_N - \bar{z}_N)\end{bmatrix}[(z_1 - \bar{z}_1) \cdots (z_N - \bar{z}_N)]\right\}\\
&= \mathrm{E}\left\{\begin{bmatrix}(z_1 - \bar{z}_1)(z_1 - \bar{z}_1) & \cdots & (z_1 - \bar{z}_1)(z_N - \bar{z}_N)\\ \vdots & & \vdots \\ (z_N - \bar{z}_N)(z_1 - \bar{z}_1) & \cdots & (z_N - \bar{z}_N)(z_N - \bar{z}_N)\end{bmatrix}\right\}\\
&= \begin{bmatrix}\mathrm{E}\{(z_1 - \bar{z}_1)(z_1 - \bar{z}_1)\} & \cdots & \mathrm{E}\{(z_1 - \bar{z}_1)(z_N - \bar{z}_N)\}\\ \vdots & & \vdots \\ \mathrm{E}\{(z_N - \bar{z}_N)(z_1 - \bar{z}_1)\} & \cdots & \mathrm{E}\{(z_N - \bar{z}_N)(z_N - \bar{z}_N)\}\end{bmatrix}
\end{aligned} \tag{A8.1-2}$$

That is

$$\boldsymbol{\Sigma}_z = \begin{bmatrix}\sigma_{11}^2 & \sigma_{12}^2 & \cdots & \sigma_{1N}^2\\ \sigma_{21}^2 & \sigma_{22}^2 & \cdots & \sigma_{2N}^2\\ \cdots & \cdots & \cdots & \cdots \\ \sigma_{N1}^2 & \sigma_{N2}^2 & \cdots & \sigma_{NN}^2\end{bmatrix} \tag{A8.1-3}$$

where

$$\sigma_{ij}^2 = \mathrm{E}\{(z_i - \bar{z}_i)(z_j - \bar{z}_j)\}, \qquad i, j = 1, 2, \ldots, N$$

From Eq. (A8.1-3) it follows that the diagonal terms of the covariance matrix are the *variances* of individual random variables while each off-diagonal term is the *covariance* of two random variables z_i and z_j. Again, we observe that covariance matrices are symmetric. This property enables one to employ the results pertaining to the theory of symmetric matrices to the analysis of covariance matrices.

Equation (A8.1-1) is often expressed in the following alternative form:

$$\boldsymbol{\Sigma}_z = \mathrm{E}\{\boldsymbol{Z}\boldsymbol{Z}' - \boldsymbol{Z}\bar{\boldsymbol{Z}}' - \bar{\boldsymbol{Z}}\boldsymbol{Z}' + \bar{\boldsymbol{Z}}\bar{\boldsymbol{Z}}'\} = \mathrm{E}\{\boldsymbol{Z}\boldsymbol{Z}'\} - \mathrm{E}\{\boldsymbol{Z}\}\bar{\boldsymbol{Z}}' - \bar{\boldsymbol{Z}}\,\mathrm{E}\{\boldsymbol{Z}'\} + \mathrm{E}\{\bar{\boldsymbol{Z}}\bar{\boldsymbol{Z}}'\}$$

which yields

$$\boldsymbol{\Sigma}_z = \mathbf{S} - \bar{\boldsymbol{Z}}\bar{\boldsymbol{Z}}' \tag{A8.1-4}$$

where

$$\mathbf{S} = \mathrm{E}\{\boldsymbol{Z}\boldsymbol{Z}'\} = \begin{bmatrix} \mathrm{E}\{z_1^2\} & \cdots & \mathrm{E}\{z_1 z_N\} \\ \vdots & & \vdots \\ \mathrm{E}\{z_N z_1\} & \cdots & \mathrm{E}\{z_N^2\} \end{bmatrix}.$$

S is called the *autocorrelation matrix*, or occasionally the *scatter matrix* of **Z**.

It is sometimes convenient to express $\boldsymbol{\Sigma}_z$ in terms of *correlation coefficients* which are given by

$$r_{ij} = \frac{\sigma_{ij}^2}{\sigma_{ii}\sigma_{jj}} \tag{A8.1-5}$$

Then, substitution of Eq. (A8.1-5) in Eq. (A8.1-3) leads to

$$\boldsymbol{\Sigma}_z = \boldsymbol{\Gamma}\mathbf{R}\boldsymbol{\Gamma} \tag{A8.1-6}$$

where

$$\boldsymbol{\Gamma} = \begin{bmatrix} \sigma_{11} & 0 & \cdots & 0 \\ 0 & \sigma_{22} & \cdots & 0 \\ \cdot & \cdot & \cdot & \cdot \\ 0 & 0 & \cdots & \sigma_{NN} \end{bmatrix}$$

and

$$\mathbf{R} = \begin{bmatrix} 1 & r_{12} & \cdots & r_{1N} \\ r_{21} & 1 & \cdots & r_{2N} \\ \cdot & \cdot & \cdot & \cdot \\ r_{N1} & r_{N2} & \cdots & 1 \end{bmatrix}, \quad |r_{ij}| \leq 1$$

The matrix **R** is called the *correlation matrix*.

Special case. If $\boldsymbol{\Sigma}_z$ represents a wide-sense stationary random process, then it is given by

$$\boldsymbol{\Sigma}_z = \begin{bmatrix} \varrho_0 & & \varrho_1 & & \varrho_2 & & \cdots & \varrho_{N-1} \\ & \searrow & & \searrow & & \searrow & & \\ \varrho_1 & & \varrho_0 & & \varrho_1 & & \cdots & \varrho_{N-2} \\ & \searrow & & \searrow & & & & \\ \varrho_2 & & \varrho_1 & & \varrho_0 & & \cdots & \varrho_{N-3} \\ & \searrow & & & & & & \\ \cdot & \cdot & \cdot & \cdot & \cdot & \cdot & \cdot & \cdot \\ \varrho_{N-1} & & \varrho_{N-2} & & \varrho_{N-3} & & \cdots & \varrho_0 \end{bmatrix} \tag{A8.1-7}$$

It is observed that $\mathbf{\Sigma}_z$ in Eq. (A8.1-7) is completely specified by any one of its rows or columns. This is because each element α_{ij} of $\mathbf{\Sigma}_z$ is such that

$$\alpha_{ij} = \varrho_{|i-j|}, \qquad i, j = 1, 2, \ldots, N \tag{A8.1-8}$$

References

1. Davenport, W. B., Jr.: *Random Signals and Noise.* New York: McGraw-Hill, 1968.
2. Pratt, W. K.: Generalized Wiener Filtering Computation Techniques. *IEEE Trans. Computers* C-21 (1972) 636–641.
3. Pearl, J.: Basis Restricted Transformations and Performance Measures for Spectral Representation. *IEEE Trans. Information Theory* IT-17 (1971) 751–752.
4. —, Walsh Processing of Random Signals. *IEEE Trans. Electromagnetic Compatability* (special issue) EMC-13 (1971) 137–141.
5. Ahmed, N., Natarajan, T., and Rao, K. R.: Discrete Cosine Transform. *IEEE Trans. Computers* C-23 (1974) 90–93.
6. Grenander, V., and Szego, G.: *Toeplitz Forms and Their Applications.* Berkeley, Los Angeles: Univ. of California Press, 1958.
7. Andrews, H. C.: *Computer Techniques in Image Processing.* New York, London: Academic Press, 1970.
8. Pratt, W. K.: Linear and Nonlinear Filtering in the Walsh Domain. *IEEE Trans. Electromagnetic Compatability* (special issue) EMC-13 (1971) 38–42.
9. Several articles in the Proceedings of the 1970–74 Symposia on Applications of Walsh Functions.
10. Andrews, H. C., et al.: Image Processing by Digital Computer. *IEEE Spectrum* 9, July 1972, 20–32.
11. Pratt, W. K.: Walsh Functions in Image Processing and Two-Dimensional Filtering. *Proc. 1972 Symp. Applications of Walsh Functions,* 14–22.
12. Fukunaga, K.: *Introduction to Statistical Pattern Recognition.* New York, London: Academic Press, 1972.
13. Hildebrand, F. B.: *Methods of Applied Mathematics.* Englewood Cliffs, N. J.: Prentice-Hall, 1952.
14. Bellman, R.: *Introduction to Matrix Analysis.* New York, London: McGraw-Hill, 1960.

Problems

8-1 If $\boldsymbol{X}' = [x_1\, x_2 \cdots x_d]$, $\boldsymbol{V}' = [v_1\, v_2 \cdots v_d]$, and

$$\mathbf{A} = \begin{bmatrix} a_{11} & a_{12} & \cdots & a_{1d} \\ a_{21} & a_{22} & \cdots & a_{2d} \\ \cdots & \cdots & \cdots & \cdots \\ a_{d1} & a_{d2} & \cdots & a_{dd} \end{bmatrix}$$

then, show that:

(a) $$\nabla_{\mathbf{A}} \{\boldsymbol{V}'\,\boldsymbol{V}\} = 0$$

(b) $$\nabla_{\mathbf{A}} \{\boldsymbol{X}'\mathbf{A}'\boldsymbol{V}\} = \boldsymbol{V}\boldsymbol{X}'$$

8-2 Starting with Eq. (8.2-3), derive Eq. (8.2-4) using the relation $\mathbf{T}'\mathbf{T} = \mathbf{I}$.

8-3 Let $\mathbf{T}$ be a transformation matrix whose elements are complex numbers, such that $(\mathbf{T}^*)'\mathbf{T} = \mathbf{I}$. Then the definition in Eq. (8.2-3) is modified as follows:

$$\begin{aligned}\tilde{\mathbf{\Sigma}}_Z &= \mathrm{E}\{(\boldsymbol{F} - \bar{\boldsymbol{F}})[(\boldsymbol{F} - \bar{\boldsymbol{F}})^*]'\} \\ &= \mathrm{E}\{(\boldsymbol{F} - \bar{\boldsymbol{F}})[(\boldsymbol{F}^*)' - (\bar{\boldsymbol{F}}^*)']\}\end{aligned}$$

where $\boldsymbol{F}^*$ denotes the complex conjugate of $\boldsymbol{F}$.

Using the above information, show that the relation corresponding to that in Eq. (8.2-4) is

$$\tilde{\mathbf{\Sigma}}_z = \mathbf{T}\mathbf{\Sigma}_z(\mathbf{T}^*)'$$

where $\mathbf{\Sigma}_z$ is as defined in Eq. (8.2-1).

8-4 Let $\boldsymbol{\varepsilon} = \hat{\boldsymbol{X}} - \boldsymbol{X}$ be a random 2-vector such that

$$\boldsymbol{\varepsilon}_1 = \begin{bmatrix}\varepsilon_{11}\\ \varepsilon_{12}\end{bmatrix}, \quad \boldsymbol{\varepsilon}_2 = \begin{bmatrix}\varepsilon_{21}\\ \varepsilon_{22}\end{bmatrix}, \ldots, \quad \boldsymbol{\varepsilon}_N = \begin{bmatrix}\varepsilon_{N1}\\ \varepsilon_{N2}\end{bmatrix}$$

Show that

$$\mathrm{tr}\{\mathrm{E}[\hat{\boldsymbol{X}} - \boldsymbol{X})(\hat{\boldsymbol{X}} - \boldsymbol{X})']\} = \mathrm{E}\{\|\hat{\boldsymbol{X}} - \boldsymbol{X}\|^2\} = \frac{1}{N}\sum_{k=1}^{N}(\varepsilon_{k1}^2 + \varepsilon_{k2}^2)$$

8-5 Use $\hat{\boldsymbol{X}} = \mathbf{T}'\mathbf{A}\mathbf{T}\boldsymbol{Z}$ and Eqs. (8.3-11) and (8.3-12) to derive Eq. (8.3-19). That is, show that

$$\mathrm{E}\{\hat{\boldsymbol{X}}\hat{\boldsymbol{X}}'\} = \mathrm{E}\{\boldsymbol{X}\hat{\boldsymbol{X}}'\} = \mathbf{\Sigma}_x\mathbf{T}'\mathbf{A}'\mathbf{T}$$

and

$$\mathrm{E}\{\hat{\boldsymbol{X}}\boldsymbol{X}'\} = \mathbf{T}'\mathbf{A}\mathbf{T}\mathbf{\Sigma}_x$$

8-6 Consider the transformation matrix $\mathbf{T} = \dfrac{1}{\sqrt{5}}\begin{bmatrix}1 & -2\\ 2 & 1\end{bmatrix}$.

(a) Verify that $\mathbf{T}$ is orthonormal; i.e., $\mathbf{T}'\mathbf{T} = \mathbf{I}$.

(b) If $\mathbf{Q} = \begin{bmatrix}q_{11} & q_{12}\\ q_{12} & q_{22}\end{bmatrix}$, then show that

$$\mathrm{tr}(\mathbf{Q}) = \mathrm{tr}(\mathbf{T}\mathbf{Q}\mathbf{T}') = q_{11} + q_{22} \tag{P8-6-1}$$

It can be shown that the general form of Eq. (P8-6-1) can be stated in the form of a theorem as follows:

If Q is a real symmetric matrix and $\mathbf{T}$ is such that $\mathbf{T}'\mathbf{T} = \mathbf{I}$, then

$$\mathrm{tr}(\mathbf{Q}) = \mathrm{tr}(\mathbf{T}\mathbf{Q}\mathbf{T}') \tag{P8-6-2}$$

That is, the trace of $\mathbf{Q}$ is invariant under any orthonormal transformation.

8-7 From Eq. (8.3-3) it is known that $\tilde{\mathbf{\Sigma}}_z = \mathbf{T}\mathbf{\Sigma}_z\mathbf{T}'$.

(a) Show that $\mathbf{\Sigma}_z = \mathbf{T}'\tilde{\mathbf{\Sigma}}_z\mathbf{T}$. (P8-7-1)

(b) Starting with Eq. (8.3-20), derive Eq. (8.3-22).

Hint: Use Eqs. (P8-7-1), (P8-6-2), and (8.3-15).

8-8 If the eigenvalues of the matrix

$$\mathbf{A}_r = \begin{bmatrix}\alpha & \beta\\ \beta & \alpha\end{bmatrix}$$

are $\lambda_1 = (\alpha + \beta)$ and $\lambda_2 = (\alpha - \beta)$, show that the corresponding normalized eigenvectors are given by

$$\boldsymbol{\phi}_1 = \frac{1}{\sqrt{2}}\begin{bmatrix}1\\ 1\end{bmatrix} \quad \text{and} \quad \boldsymbol{\phi}_2 = \frac{1}{\sqrt{2}}\begin{bmatrix}1\\ -1\end{bmatrix}$$

8-9 Given the symmetric matrix

$$\mathbf{Q} = \begin{bmatrix} 1 & -1 & 0 & 0 \\ -1 & 2 & -1 & 0 \\ 0 & -1 & 2 & -1 \\ 0 & 0 & -1 & 1 \end{bmatrix}$$

Show that $\mathrm{tr}(\mathbf{Q}) = 6 = \sum_{i=1}^{4} \lambda_i$, where λ_i are the eigenvalues of $\mathbf{Q}$.

Answer: $\lambda_1 = 0$, $\lambda_2 = 2$, $\lambda_3 = 2 + \sqrt{2}$, $\lambda_4 = 2 - \sqrt{2}$.

Remark: In general it can be shown that if $\mathbf{Q}$ is an $(N \times N)$ real symmetric matrix with eigenvalues λ_i, $i = 1, 2, \ldots, N$, then

$$\mathrm{tr}(\mathbf{Q}) = \sum_{i=1}^{N} \lambda_i \tag{P8-9-1}$$

Chapter Nine

Data Compression

An important application of the orthogonal transforms is *data compression.* The key to securing data compression is *signal representation,* which concerns the representation of a given class (or classes) of signals in an efficient manner. If a discrete signal is comprised of N sampled values, then it can be thought of as being a point in an N-dimensional space. Each sampled value is then a component of the data N-vector $\boldsymbol{X}$ which represents the signal in this space. For more efficient representation, one obtains an orthogonal transform of $\boldsymbol{X}$ which results in $\boldsymbol{Y} = \mathbf{T}\boldsymbol{X}$, where $\boldsymbol{Y}$ and $\mathbf{T}$ denote the transform vector and transform matrix respectively. The objective is to select a subset of M components of $\boldsymbol{Y}$, where M is substantially less than N. The remaining $(N - M)$ components can then be discarded without introducing objectionable error, when the signal is reconstructed using the retained M components of $\boldsymbol{Y}$. The orthogonal transforms must therefore be compared with respect to some error criterion. One such often used criterion is the mean-square error criterion.

A natural sequel to the above considerations is data compression, in the sense that the signal representation can be used to reduce the amount of redundant information. Hence we will first discuss signal representation via orthogonal transforms. This will be followed by data compression which will be illustrated by examples pertaining to the processing of electrocardiographic and image data.

9.1 Search for the Optimum Transform

We seek the orthogonal transform that enables signal representation and yet is optimum with respect to the mean-square error criterion.

Let $\mathbf{T}$ be an orthogonal transform given by

$$\mathbf{T}' = [\boldsymbol{\phi}_1 \boldsymbol{\phi}_2 \ldots \boldsymbol{\phi}_N] \tag{9.1-1}$$

where the $\boldsymbol{\phi}_i$ are N-vectors. For convenience, the basis vectors $\{\boldsymbol{\phi}_m\}$ are assumed to be real-valued and orthonormal – i.e.,

$$\boldsymbol{\phi}_i' \boldsymbol{\phi}_j = \begin{cases} 1, & i = j \\ 0, & i \neq j \end{cases} \tag{9.1-2}$$

For each vector $\boldsymbol{X}$ belonging to a given class of data vectors, we obtain

$$\boldsymbol{Y} = \mathbf{T}\boldsymbol{X} \tag{9.1-3}$$

where $\boldsymbol{X}' = [x_1\, x_2 \cdots x_N]$ and $\boldsymbol{Y}' = [y_1\, y_2 \cdots y_N]$.

From Eqs. (1) and (2) it follows that $\mathbf{T}'\mathbf{T} = \mathbf{I}$, and hence

$$\boldsymbol{X} = \mathbf{T}'\boldsymbol{Y} = [\boldsymbol{\phi}_1 \boldsymbol{\phi}_2 \cdots \boldsymbol{\phi}_N]\boldsymbol{Y}$$

which yields

$$\boldsymbol{X} = y_1\boldsymbol{\phi}_1 + y_2\boldsymbol{\phi}_2 + \cdots + y_N\boldsymbol{\phi}_N = \sum_{i=1}^{N} y_i\boldsymbol{\phi}_i \tag{9.1-4}$$

We wish to retain a subset $\{y_1, y_2, \ldots, y_M\}$ of the components of $\boldsymbol{Y}$ and yet estimate $\boldsymbol{X}$. This can be done by replacing the remaining $N - M$ components of $\boldsymbol{Y}$ by preselected constants b_i to obtain

$$\tilde{\boldsymbol{X}}(M) = \sum_{i=1}^{M} y_i\boldsymbol{\phi}_i + \sum_{i=M+1}^{N} b_i\boldsymbol{\phi}_i \tag{9.1-5}$$

where $\tilde{\boldsymbol{X}}(M)$ denotes the estimate of $\boldsymbol{X}$.

The error introduced by neglecting the $N - M$ terms can be represented as

$$\Delta\boldsymbol{X} = \boldsymbol{X} - \tilde{\boldsymbol{X}}(M)$$

where $\Delta\boldsymbol{X}$ is the error vector. That is

$$\Delta\boldsymbol{X} = \boldsymbol{X} - \sum_{i=1}^{M} y_i\boldsymbol{\phi}_i - \sum_{i=M+1}^{N} b_i\boldsymbol{\phi}_i \tag{9.1-6}$$

From Eqs. (4) and (6) it follows that

$$\Delta\boldsymbol{X} = \sum_{i=M+1}^{N} (y_i - b_i)\,\boldsymbol{\phi}_i \tag{9.1-7}$$

Thus the mean-square error $\varepsilon(M)$ is given by

$$\varepsilon(M) = \mathrm{E}\{\|\Delta\boldsymbol{X}\|^2\} = \mathrm{E}\{(\Delta\boldsymbol{X})'\,(\Delta\boldsymbol{X})\} \tag{9.1-8}$$

Substitution of Eq. (7) in Eq. (8) leads to

$$\varepsilon(M) = \mathrm{E}\left\{\sum_{i=M+1}^{N}\ \sum_{j=M+1}^{N} (y_i - b_i)\,(y_j - b_j)\,\boldsymbol{\phi}_i'\boldsymbol{\phi}_j\right\}$$

which simplifies to yield [see Eq. (2)]

$$\varepsilon(M) = \sum_{i=M+1}^{N} \mathrm{E}\{(y_i - b_i)^2\} \tag{9.1-9}$$

From Eq. (9) it follows that for every choice of $\boldsymbol{\phi}_i$ and b_i, we obtain a value for $\varepsilon(M)$. We seek the choice which minimizes $\varepsilon(M)$. The process of choosing the optimum b_i and $\boldsymbol{\phi}_i$ respectively is carried out in two steps.

Step 1. The optimum b_i are obtained as follows:

$$\frac{\partial}{\partial b_i} \mathrm{E}\{(y_i - b_i)^2\} = -2[\mathrm{E}\{y_i\} - b_i] = 0$$

which yields

$$b_i = \mathrm{E}\{y_i\} \tag{9.1-10}$$

Now, from Eqs. (2) and (4) one obtains

$$y_i = \boldsymbol{\phi}_i' \boldsymbol{X} \tag{9.1-11}$$

Thus we have

$$b_i = \boldsymbol{\phi}_i' \mathrm{E}\{\boldsymbol{X}\} = \boldsymbol{\phi}_i' \bar{\boldsymbol{X}}$$

where $\bar{\boldsymbol{X}} = \mathrm{E}\{\boldsymbol{X}\}$.

Again, since the quantity $(y_i - b_i)$ in Eq. (9) is a scalar, we can write $\varepsilon(M)$ as

$$\varepsilon(M) = \sum_{i=M+1}^{N} \mathrm{E}\{(y_i - b_i)(y_i - b_i)'\} \tag{9.1-12}$$

Substitution of $y_i = \boldsymbol{\phi}_i' \boldsymbol{X}$ and $b_i = \boldsymbol{\phi}_i' \bar{\boldsymbol{X}}$ in Eq. (12) results in

$$\varepsilon(M) = \sum_{i=M+1}^{N} \boldsymbol{\phi}_i' \mathrm{E}\{(\boldsymbol{X} - \bar{\boldsymbol{X}})(\boldsymbol{X} - \bar{\boldsymbol{X}})'\} \boldsymbol{\phi}_i$$

Since $\boldsymbol{\Sigma}_x = \mathrm{E}\{(\boldsymbol{X} - \bar{\boldsymbol{X}})(\boldsymbol{X} - \bar{\boldsymbol{X}})'\}$ is the covariance matrix of $\boldsymbol{X}$, we have

$$\varepsilon(M) = \sum_{i=M+1}^{N} \boldsymbol{\phi}_i' \Sigma_x \boldsymbol{\phi}_i \tag{9.1-13}$$

Step 2. To obtain the optimum $\boldsymbol{\phi}_i$ we must not only minimize $\varepsilon(M)$ with respect to $\boldsymbol{\phi}_i$, but also satisfy the constraint $\boldsymbol{\phi}_i' \boldsymbol{\phi}_i = 1$. Thus we use the method of Lagrange multipliers[1] and minimize

$$\hat{\varepsilon}(M) = \varepsilon(M) - \sum_{i=M+1}^{N} \beta_i [\boldsymbol{\phi}_i' \boldsymbol{\phi}_i - 1] = \sum_{i=M+1}^{N} \{\boldsymbol{\phi}_i' \Sigma_x \boldsymbol{\phi}_i - \beta_i [\boldsymbol{\phi}_i' \boldsymbol{\phi}_i - 1]\} \tag{9.1-14}$$

with respect to $\boldsymbol{\phi}_i$, where β_i denotes the Lagrange multiplier.

Now, it can be shown that [see Prob. 9-1]

$$\nabla_{\phi_i}[\boldsymbol{\phi}_i' \Sigma_x \boldsymbol{\phi}_i] = 2\Sigma_x \boldsymbol{\phi}_i$$

and

$$\nabla_{\phi_i}[\boldsymbol{\phi}_i' \boldsymbol{\phi}_i] = 2\boldsymbol{\phi}_i$$

Thus from Eq. (14) we have

$$\nabla_{\phi_i}[\hat{\varepsilon}(M)] = 2\Sigma_x \boldsymbol{\phi}_i - 2\beta_i \boldsymbol{\phi}_i = 0$$

which yields

$$\boldsymbol{\Sigma}_x \boldsymbol{\phi}_i = \beta_i \boldsymbol{\phi}_i \tag{9.1-15}$$

[1] A brief discussion of the method of Lagrange multipliers is included in Appendix 9.1.

By definition, Eq. (15) implies that $\boldsymbol{\phi}_i$ is the *eigenvector* of the covariance matrix $\boldsymbol{\Sigma}_x$ and β_i is the i-th corresponding *eigenvalue*. Denoting β_i by λ_i, and substituting Eq. (15) in Eq. (13), the *minimum* mean-square error is given by

$$\varepsilon_{min}(M) = \sum_{i=M+1}^{N} \lambda_i \tag{9.1-16}$$

The expansion defined in Eq. (4) is thus in terms of the eigenvectors of the covariance matrix. This expansion is called the *Karhunen-Loève expansion*. The $\boldsymbol{\phi}_i$ which comprise $\mathbf{T}$ in Eq. (1) are eigenvectors of $\boldsymbol{\Sigma}_x$. Hence the transformation $\boldsymbol{Y} = \mathbf{T}\boldsymbol{X}$ is called the Karhunen-Loève transform (KLT). We recall that the KLT was initially referred to in Section 8.5.

In the literature on statistics, the problem of minimizing $\varepsilon(M)$ is generally called *factor analysis* or *principal component analysis*. Two important conclusions pertaining to the above analysis are as follows:

(i) The KLT is the *optimum* transform for signal representation with respect to the mean-square error criterion.

(ii) Since $\boldsymbol{Y} = \mathbf{T}\boldsymbol{X}$, the transform domain covariance matrix $\boldsymbol{\Sigma}_y$ is given by [see Eq. (8.2-4)]

$$\boldsymbol{\Sigma}_y = \mathbf{T}\boldsymbol{\Sigma}_x\mathbf{T}^{-1} = \mathbf{T}\boldsymbol{\Sigma}_x\mathbf{T}'$$

However, since $\mathbf{T}$ is comprised of the eigenvectors of $\boldsymbol{\Sigma}_x$, it follows that [see Eq. (8.5-1)]

$$\boldsymbol{\Sigma}_y = \text{diag}\,(\lambda_1, \lambda_2, \ldots, \lambda_N) \tag{9.1-17}$$

where λ_i, $i = 1, 2, \ldots, N$ are the eigenvalues of $\boldsymbol{\Sigma}_x$.

Since $\boldsymbol{\Sigma}_y$ is a *diagonal matrix*, we conclude that the transform vector components y_i in Eq. (11) are *uncorrelated*.

9.2 Variance Criterion and Variance Distribution [1]

Clearly, Eq. (9.1-16) states that the effectiveness of a transform component y_i for representing the data vector $\boldsymbol{X}$, is determined by its corresponding eigenvalue. If a component y_k is deleted, then the mean-square error increases by λ_k, the corresponding eigenvalue. Thus the set of y_i with the M largest eigenvalues should be selected, and the remaining y_i discarded since they can be replaced by the constants b_i, $i = M+1, \ldots, N$.[1]

Now, since the eigenvalues are the main diagonal terms of $\boldsymbol{\Sigma}_y$, they correspond to the *variances* of the transform components y_i, $i = 1, 2, \ldots, N$. For all other transforms, however, $\boldsymbol{\Sigma}_y$ has non-zero off diagonal terms. Thus a logical criterion for selecting transform components is to retain the set

[1] Since $b_i = \boldsymbol{\phi}_i'\bar{\boldsymbol{X}}$, we observe that the remaining y_i, $i = M+1, \ldots, N$ can be set to zero if the given data is pre-processed such that $\bar{\boldsymbol{X}} = \boldsymbol{O}$.

i.e. X is forward transformed where its mean = O

of components with the M largest variances—the remaining $(N - M)$ components can be discarded. We shall refer to this process of component selection as the *variance criterion*.

A graphical representation of the variance criterion is a plot of variance versus transform components, where the variances are arranged in decreasing order of magnitude, normalized to the *trace*[1] of Σ_x or Σ_y. The motivation for normalization is that the ratio of the variance to the sum of variances (i.e., trace) provides a measure of the percentage of the mean-square error introduced by eliminating the corresponding $\boldsymbol{\phi}_i$ in Eq. (9.1-4). This plot will be referred to as the *variance distribution*. Figure 9.1 shows an example of the variance distributions associated with four transforms. The area under each curve for a given number of transform components is an indication of the amount of energy contained in those components. The total area under each curve is unity as a consequence of the normalization to the trace. For example, with 20 components, Fig. 9.1 shows that the ranking of the most effective to the least effective is as follows:

$$\text{transform } 4 > \text{transform } 2 > \text{transform } 3 > \text{transform } 1$$

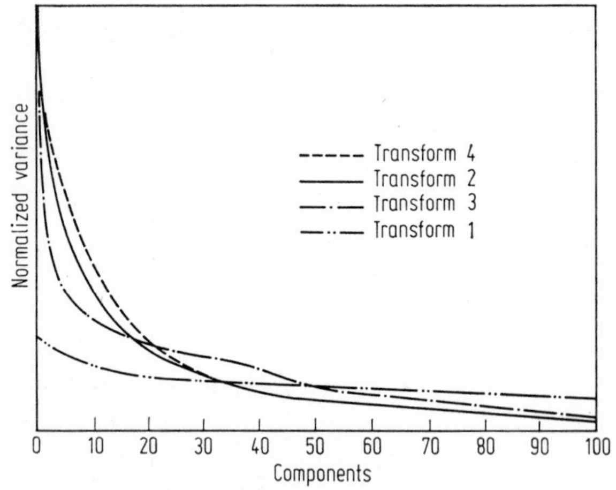

Fig. 9.1 A variance distribution

Throughout the above discussion we have assumed that X belongs to one class of signals. The developments are easily extended to the case when several classes are present. The only difference is that the basis vectors for

[1] Recall that the trace is invariant under any orthonormal transformation (see Prob. 8-6).

the corresponding KLT are the eigenvectors of the overall matrix $\boldsymbol{\Sigma}_x^*$ rather than $\boldsymbol{\Sigma}_x$. The matrix $\boldsymbol{\Sigma}_x^*$ is given by

$$\boldsymbol{\Sigma}_x^* = P_1 \boldsymbol{\Sigma}_{x_1} + P_2 \boldsymbol{\Sigma}_{x_2} + \cdots + P_K \boldsymbol{\Sigma}_{x_K} \tag{9.2-1}$$

where "K" is the number of classes of signals, and $\boldsymbol{\Sigma}_{x_i}$ denotes the covariance matrix of the i-th class with a priori probability P_i.

In what follows, we illustrate how the variance distribution can be used to achieve data compression by considering an example pertaining to electrocardiograph data processing.

9.3 Electrocardiographic Data Compression [2]

Before defining the problem of electrocardiographic data compression, it is instructive to consider some elementary aspects of the electrocardiographic signal.

The electrocardiogram (ECG) is a measure of the electrical activity of the heart. This electrical signal is generated during the cardiac cycle by the depolarization and repolarization of the heart muscle cells during their process of contraction and relaxation. The actual potential measured across the cell membrane is due to an ionic gradient. The changes in this ionic gradient occur with the muscle action of the heart and these changes in the ionic potential are recorded as the ECG [see Fig. 9.2].

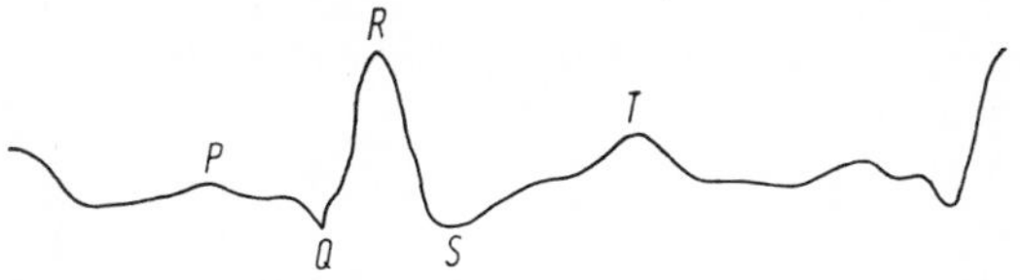

Fig. 9.2 A normal ECG

The heart is a cyclic pump. The cardiac cycle includes pumping blood from the ventricles to the body and to the lungs as well as the return of blood from the body and the lungs to the heart. The right and left ventricles are the pumping chambers of the heart with the right sending oxygen-deficient blood to the lungs and the left sending oxygen-laden blood to the body.

The atria are the receiving chambers of the heart, with the left receiving oxygen-laden blood from the lungs and the right receiving oxygen-deficient blood from the body.

The ECG under most circumstances may be assumed to be periodic since it is generated during the cardiac cycle. The sino-atrial node initiates the stimulus for the contraction of the heart muscle. The stimulus travels across the atria causing them to contract and then after a short delay in passing through the atrio-venticular node, it passes down the septum and on through the ventricles which then contract. The depolarizations of the atria and of the ventricles are evidenced by the P wave and the QRS complex respectively as shown in Fig. 9.2. After these muscle cells contract they return to their initial state through the process of repolarization. The repolarization of the atria is masked by the QRS complex while the repolarization of the ventricles produces the T wave of the ECG.

Statement of the problem. Suppose the ECG in Fig. 9.2 is sampled to obtain the data vector $\boldsymbol{X}' = [x_1\, x_2 \cdots x_N]$. If each x_i is stored as a word in memory, then N words per ECG are required for storage purposes. An $m:1$ data compression implies that the number of words required to store ECG data in the transform domain is N/m words per ECG.

In what follows, ECG data compression is demonstrated using canine data. The canine was chosen because of the similarity of its heart and ECG to those of the human. This similarity would then allow the extension of the techniques developed to the human ECG.

Data acquisition. ECG waveforms are dependent upon the placement of the recording electrodes on the body. This arrangement is called a lead system. There are several different accepted lead systems used to record the ECG. The one used for this study is the so-called standard lead system which consists of three leads or pairs of connections. Its configuration is as shown in Fig. 9.3.

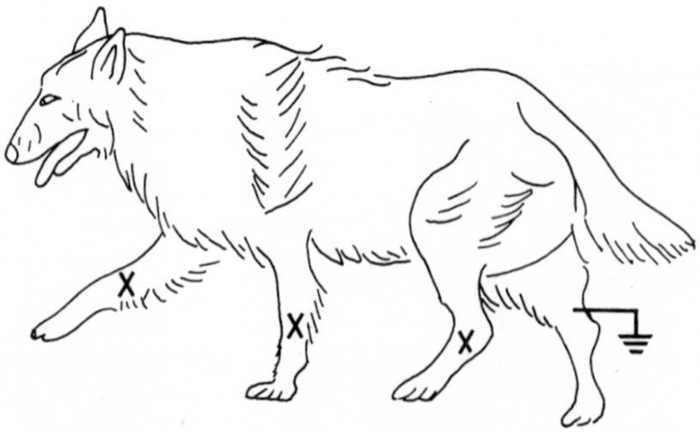

Fig. 9.3 Standard 3-lead system configuration

The data was separated into two classes, normal and abnormal.[1] The abnormals showed ventricular defects that were induced by both chemical and mechanical means. These ECG signals were then digitized[2] using a sampling rate of 400 samples per second, which implies that the bandwidth of the ECG signal was considered to be 200 Hz.

Each digital ECG was represented by 128 samples. These samples were chosen such that the QRS complex and the T wave would always appear. Thus if 128 samples would not span the entire ECG, those in the P wave portion were neglected. The resulting digital data was displayed using a CALCOMP plotter as illustrated in Fig. 9.4. Individual ECG's were then chosen for the data compression study. Three hundred ECG's were chosen, of which 150 were normal and the rest were abnormal.

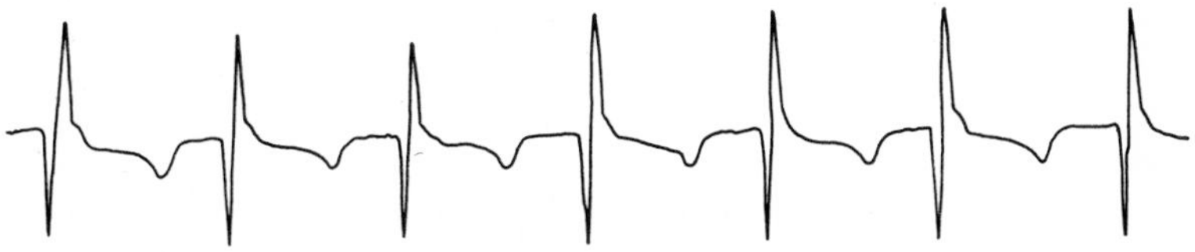

Fig. 9.4 CALCOMP plot of some lead 1 normal ECG's

Experimental results. For the purposes of illustration, we consider the DCT and lead 1 ECG's. The overall data domain covariance matrix is computed as [see Eq. (9.2-1)][3]

$$\mathbf{\Sigma}_x^* = \frac{1}{2}[\mathbf{\Sigma}_{x_1} + \mathbf{\Sigma}_{x_2}] \tag{9.3-1}$$

where $\mathbf{\Sigma}_{x_1}$ and $\mathbf{\Sigma}_{x_2}$ are the covariance matrices of the normal and abnormal classes of ECG's respectively. The 150 normal and 150 abnormal ECG data is used to compute $\mathbf{\Sigma}_x^*$. Subsequently the transform domain covariance matrix is computed as

$$\tilde{\mathbf{\Sigma}}_x^* = \mathbf{T}\mathbf{\Sigma}_x^*\mathbf{T}' \tag{9.3-2}$$

The variance distributions that result from $\tilde{\mathbf{\Sigma}}_x^*$ are shown in Fig. 9.5. It is apparent that the variance distribution of the DCT compares closely with

[1] The recording was done at the Dykstra Veterinary Hospital of Kansas State University, Manhattan, Kansas, USA and was supervised by Dr. S. G. Harris, Department of Surgery and Medicine.

[2] The digitization and subsequent data processing was done at the Department of Electrical Engineering and the Computing Center of Kansas State University.

[3] We assume that $P_1 = P_2 = 1/2$.

that of the KLT which is obtained by using the eigenvalues of Σ_x^*. We also observe that almost all of the signal energy (i.e., area under the curve) is "packed" into about 45 DCT and KLT components. In contrast, the energy is essentially "spread" over all of the 128 components for the IT. Based on the observation that the signal energy is essentially contained in 45 of the 128 DCT components, a 3:1 data compression is considered.

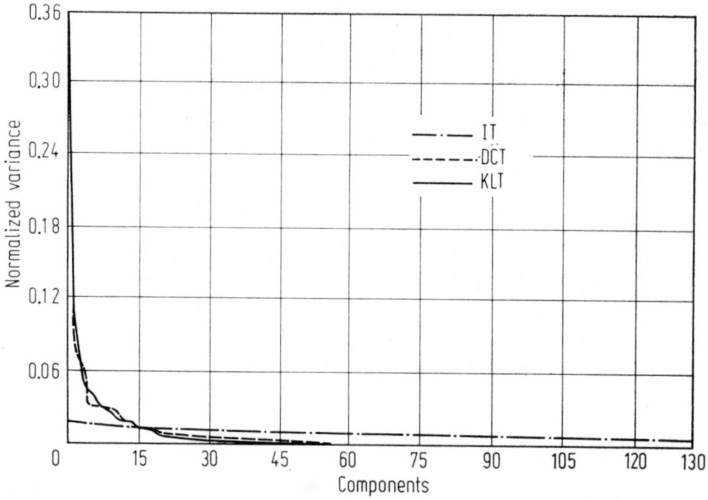

Fig. 9.5 Variance distributions for the IT, DCT, and KLT

For a 3:1 data compression, 43 DCT components corresponding to the largest diagonal terms of $\tilde{\Sigma}_x^*$ are selected. Table 9.3-1 lists the set of DCT components selected. These components are then used to reconstruct the corresponding ECG as summarized in Fig. 9.6.

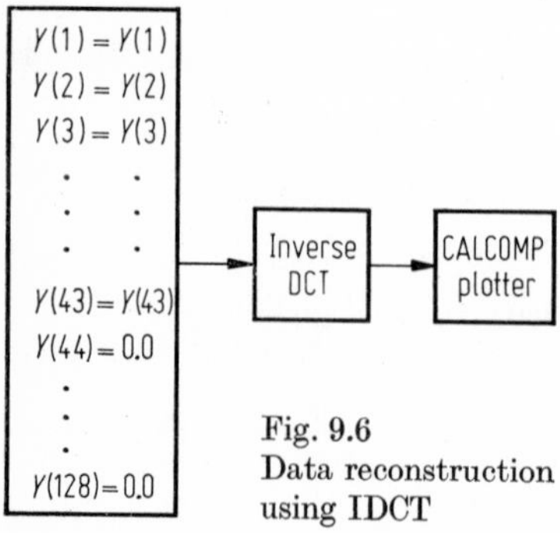

Fig. 9.6
Data reconstruction using IDCT

Table 9.3-1 Dominant 43 DCT Components Used for Data Compression

$Y(1)$	$Y(2)$	$Y(3)$	$Y(4)$	$Y(5)$
$Y(6)$	$Y(7)$	$Y(8)$	$Y(9)$	$Y(10)$
$Y(11)$	$Y(12)$	$Y(13)$	$Y(14)$	$Y(15)$
$Y(16)$	$Y(17)$	$Y(18)$	$Y(19)$	$Y(20)$
$Y(21)$	$Y(22)$	$Y(23)$	$Y(24)$	$Y(25)$
$Y(26)$	$Y(27)$	$Y(28)$	$Y(29)$	$Y(30)$
$Y(31)$	$Y(32)$	$Y(33)$	$Y(34)$	$Y(35)$
$Y(36)$	$Y(37)$	$Y(38)$	$Y(39)$	$Y(40)$
$Y(41)$	$Y(42)$	$Y(43)$		

As can be seen, all the DCT components except the 43 retained to represent the ECG are set equal to zero. The IDCT is then taken and the resulting ECG is plotted. Figures 9.7 and 9.8 show the ECGs reconstructed in this manner along with the ECGs plotted from the original 128 data points. It is apparent that the information lost using the 3:1 DCT data compression is negligible.

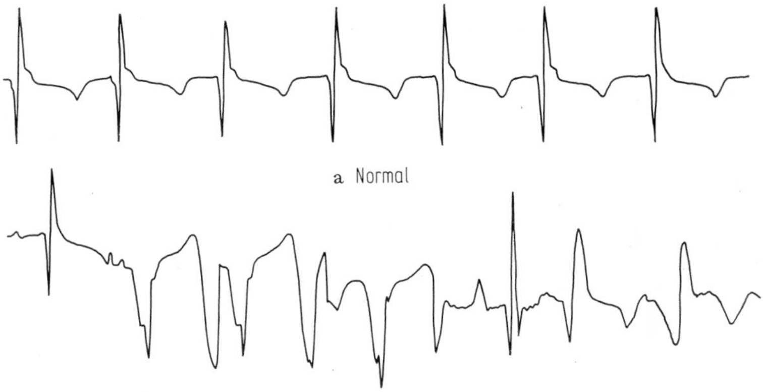

a Normal

b Abnormal

Fig. 9.7 Original ECG's using all the 128 data points

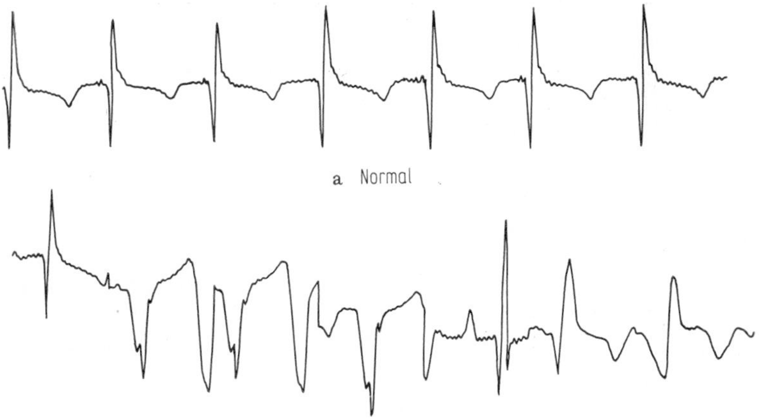

a Normal

b Abnormal

Fig. 9.8 Reconstructed ECG's using the dominant 43 DCT components

Using the above procedure with the HT, we obtain the corresponding 3:1 data compression [see Fig. 9.9]. The dominant 43 components used are listed in Table 9.3-2.

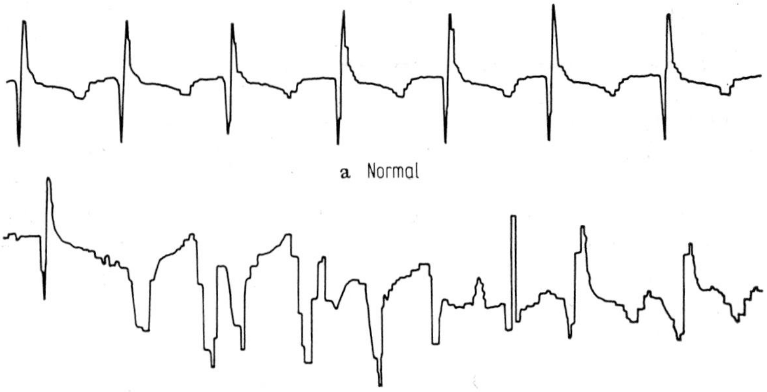

Fig. 9.9 Reconstructed ECG's using the dominant 43 HT components

The above results demonstrate that the variance criterion and variance distribution can be used as effective tools to secure data compression. The general model for this type of data compression is shown in Fig. 9.10.

Table 9.3-2 Dominant 43 HT Components Used for Data Compression

$Y(1)$	$Y(3)$	$Y(2)$	$Y(4)$	$Y(6)$	$Y(12)$	$Y(5)$	$Y(11)$
$Y(8)$	$Y(7)$	$Y(9)$	$Y(24)$	$Y(13)$	$Y(10)$	$Y(25)$	$Y(19)$
$Y(23)$	$Y(21)$	$Y(26)$	$Y(15)$	$Y(16)$	$Y(18)$	$Y(14)$	$Y(20)$
$Y(12)$	$Y(49)$	$Y(46)$	$Y(48)$	$Y(81)$	$Y(73)$	$Y(36)$	$Y(67)$
$Y(41)$	$Y(50)$	$Y(32)$	$Y(45)$	$Y(40)$	$Y(29)$	$Y(51)$	$Y(35)$
$Y(38)$	$Y(27)$	$Y(37)$					

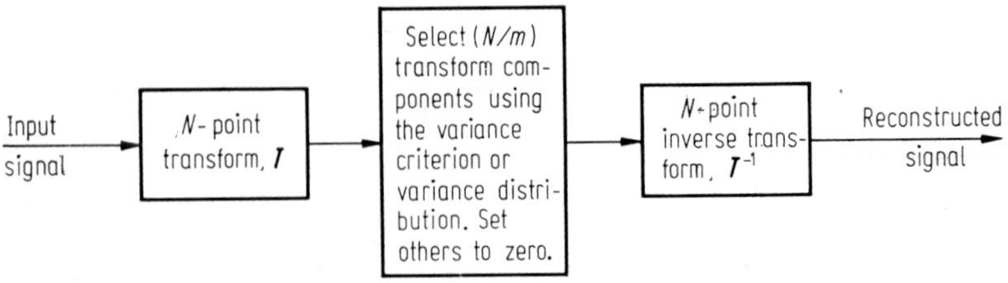

Fig. 9.10 Model for $m:1$ data compression

The above data compression technique can be extended to the 2-dimensional case which is important since it is applicable to image processing. We will illustrate image data compression by means of examples. However prior to doing so, it is worthwhile discussing some fundamentals.

9.4 Image Data Compression Considerations

In digital image processing problems, it is usual to be confronted with large amounts of data. Such data is generally highly correlated along the rows and columns as illustrated in Fig. 9.11. Each picture element is commonly referred to as a *pixel*, and is frequently coded as a 6-bit word. Thus a pixel can be represented by a decimal number between 1 and 64 or 0 and 63 (i.e. 64 levels). Such data is commonly processed in $(N \times N)$ blocks as illustrated in Fig. 9.12.

20	19	21	19	20	20	19	20	21	20	19	20	21	21	20	21
20	17	19	21	17	19	20	20	20	21	19	22	21	21	21	20
19	20	20	21	21	21	20	21	21	21	21	21	20	20	21	21
20	20	20	19	20	19	19	21	24	24	24	24	20	19	18	19
23	23	18	20	21	20	15	16	21	21	20	23	23	21	20	20
21	15	16	21	22	16	15	22	21	16	16	21	23	21	21	21
20	13	16	22	22	15	13	13	13	15	16	20	22	22	21	21
20	15	13	19	16	12	15	17	19	20	21	21	22	25	22	21
21	15	12	12	12	14	14	15	21	21	22	24	22	21	21	19
15	21	19	14	12	13	19	19	15	20	25	26	25	24	25	20
16	23	18	11	11	14	18	20	23	23	21	21	20	23	23	18
15	21	16	12	12	17	22	22	23	21	23	23	19	16	16	19
13	14	14	11	13	19	22	22	20	22	21	21	21	16	15	20
11	12	12	12	15	20	22	20	13	16	17	19	21	20	15	15
12	12	12	11	13	14	13	13	15	17	20	20	14	13	13	15
15	13	12	12	12	12	15	19	21	21	24	24	18	13	13	15

Fig. 9.11 Some 6-bit coded image data

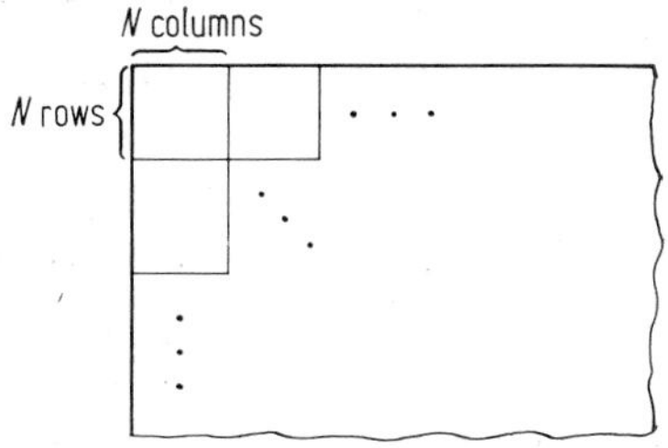

Fig. 9.12 Image processing in $(N \times N)$ blocks

We can represent the pixel in the i-th row and j-th column by the random variable $f(i, j)$, and the matrix of $(N \times N)$ random variables by $[f(i, j)]$. Then the 2-dimensional transform of $[f(i, j)]$ and its inverse are given by

$$[F(u, v)] = \mathbf{T}[f(i, j)]\mathbf{T}'$$

and

$$[f(i, j)] = \mathbf{T}'[F(u, v)]\mathbf{T}$$

where $[F(u, v)]$ denotes the transform matrix, and $\mathbf{T}$ denotes the transformation matrix. For convenience, $\mathbf{T}$ is assumed to be a real-valued matrix such that $\mathbf{T}'\mathbf{T} = \mathbf{I}$.

Let $\sigma^2(i, j)$ and $\tilde{\sigma}^2(u, v)$ denote the variances of $f(i, j)$ and $F(u, v)$ respectively. When the variance function $\tilde{\sigma}^2(u, v)$ is not known for a particular image or a class of images to be transformed, the function can usually be modeled [3]. One approach is to assume that the images to be transformed can be statistically described by a first-order Markov process, and that the rows and columns can be treated independently. We also assume that the variance of each of the row and column random variables equals σ^2. Then the covariance matrices for the rows and the columns are given by [see Eqs. (8.4-1), (A8.1-5), and (A8.1-6)]

$$\mathbf{\Sigma}_k = \sigma^2 \mathbf{R}_k \tag{9.4-1}$$

where

$$\mathbf{R}_k = \begin{bmatrix} 1 & \varrho_k & \varrho_k^2 & \cdots & \varrho_k^{N-1} \\ \varrho_k & 1 & \varrho_k & \cdots & \cdot \\ \varrho_k^2 & \varrho_k & 1 & \cdots & \cdot \\ \cdot & \cdot & \cdot & \cdots & \cdot \\ \varrho_k^{N-1} & \cdot & \cdot & \cdots & 1 \end{bmatrix}, \quad k = 1, 2$$

where ϱ_1 and ϱ_2 are the correlation coefficients associated with the row and column random variables respectively.

The transform domain matrices corresponding to $\mathbf{\Sigma}_k$ in Eq. (1) are

$$\tilde{\mathbf{\Sigma}}_k = \sigma^2[\mathbf{T}\mathbf{R}_k\mathbf{T}'], \qquad k = 1, 2 \tag{9.4-2}$$

The variance function $\tilde{\sigma}^2(u, v)$ is then computed as a function of $\tilde{\Sigma}_1(s, s)$ and $\tilde{\Sigma}_2(s, s)$, which are the diagonal elements of $\tilde{\mathbf{\Sigma}}_1$ and $\tilde{\mathbf{\Sigma}}_2$ respectively. For the purposes of illustration, $\tilde{\Sigma}_k(s, s)/\sigma^2$ for $N = 16$ and $\varrho_1 = \varrho_2 = 0.95$ are plotted in Fig. 9.13 in decreasing order of magnitude. From Fig. 9.13 it is apparent that the DCT compares closely with the KLT. Consequently the corresponding 2-dimensional variances $\tilde{\sigma}^2(u, v)$ must also compare closely. The variance function $\tilde{\sigma}^2(u, v)$ is obtained as

$$\tilde{\sigma}^2(u, v) = \tilde{\Sigma}_1(u, u)\,\tilde{\Sigma}_2(v, v) \tag{9.4-3}$$

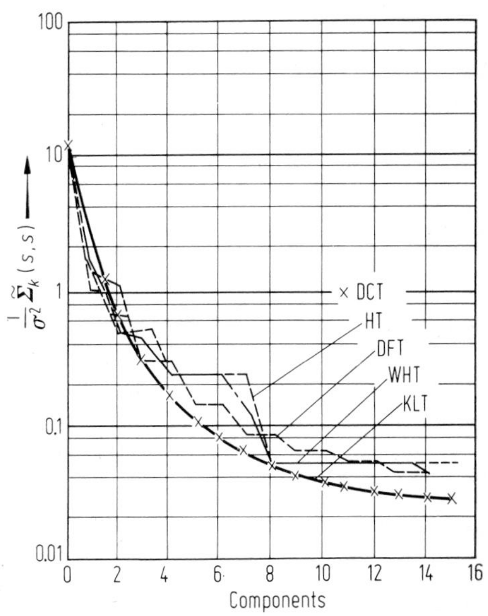

Fig. 9.13
$\tilde{\Sigma}_k(s, s)/\sigma^2$ versus components for various orthogonal transforms

Computing $\tilde{\sigma}^2(u, v)/\sigma^4$ for $N = 16$, and $\varrho_1 = \varrho_2 = 0.9$, we obtain the matrix shown in Fig. 9.14. The elements of this matrix have been computed to two digits beyond the decimal, using truncation; the blanks "–" denote zeros. Again, since the matrix is symmetric, only the upper triangular portion of the matrix is shown. We observe that these 2-dimensional variances for the DCT represent a variance function which is maximum at the origin, circularly symmetric, and decreases in magnitude monotonically toward the higher spatial sequencies.

9.5 Image Data Compression Examples

Suppose a given image, or a class of images, is encoded using k bits per pixel. Then one form of data compression can be expressed in terms of a *bit reduction factor* m, which is defined as [4]

$$m = \frac{\text{number of original image code bits}}{\text{number of transformed image code bits}}$$

We consider a simple method of realizing data compression which is known as *zonal coding*. It involves the following steps.

$$\frac{1}{\sigma^4}\,[\tilde{\sigma}^2(u,v)] = \begin{bmatrix}
97.72 & 28.89 & 11.91 & 5.92 & 3.42 & 2.27 & 1.65 & 1.27 & 1.03 & 0.86 & 0.74 & 0.66 & 0.60 & 0.56 & 0.53 & 0.52 \\
 & 8.6 & 3.55 & 1.7 & 1.02 & 0.67 & 0.49 & 0.37 & 0.30 & 0.25 & 0.22 & 0.19 & 0.17 & 0.16 & 0.16 & 0.15 \\
 & & 1.46 & 0.70 & 0.42 & 0.28 & 0.20 & 0.15 & 0.12 & 0.10 & 0.09 & 0.08 & 0.07 & 0.06 & 0.06 & 0.06 \\
 & & & 0.33 & 0.20 & 0.13 & 0.09 & 0.07 & 0.06 & 0.05 & 0.04 & 0.03 & 0.03 & 0.03 & 0.03 & 0.03 \\
 & & & & 0.12 & 0.08 & 0.05 & 0.04 & 0.03 & 0.03 & 0.02 & 0.02 & 0.02 & 0.02 & 0.01 & 0.01 \\
 & & & & & 0.05 & 0.03 & 0.02 & 0.02 & 0.02 & 0.01 & 0.01 & 0.01 & 0.01 & 0.01 & 0.01 \\
 & & & & & & 0.02 & 0.02 & 0.01 & 0.01 & 0.01 & 0.01 & 0.01 & - & - & - \\
 & & & & & & & 0.01 & 0.01 & 0.01 & - & - & - & - & - & - \\
 & & & & & & & & 0.01 & - & - & - & - & - & - & - \\
 & & & & & & & & & - & - & - & - & - & - & - \\
 & & & & & & & & & & - & - & - & - & - & - \\
 & & & & & & & & & & & - & - & - & - & - \\
 & & & & & & & & & & & & - & - & - & - \\
 & & & & & & & & & & & & & - & - & - \\
 & & & & & & & & & & & & & & - & - \\
 & & & & & & & & & & & & & & & -
\end{bmatrix}$$

Fig. 9.14 2-dimensional variances for the DCT

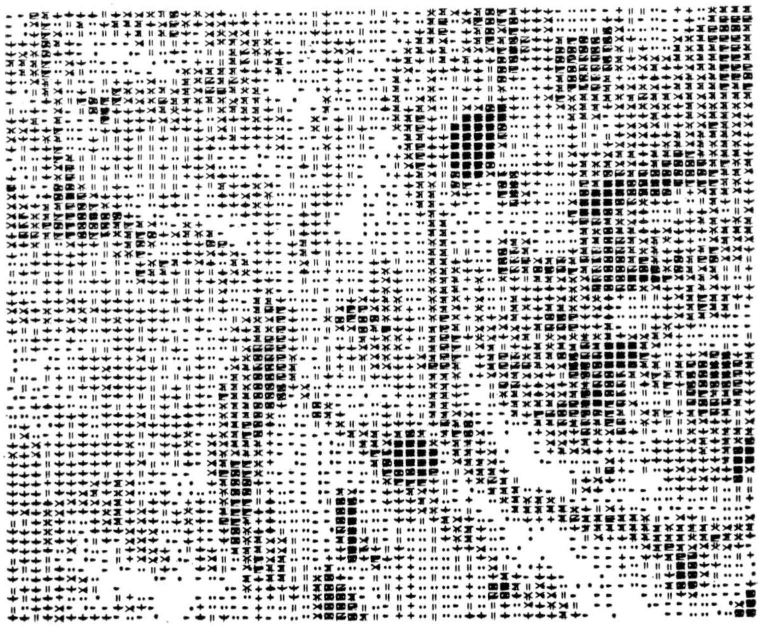

Fig. 9.15(a) Original ERTS image; 6 bits/pixel

Fig. 9.15(b) DCT ERTS image; 1.5 bits/pixel

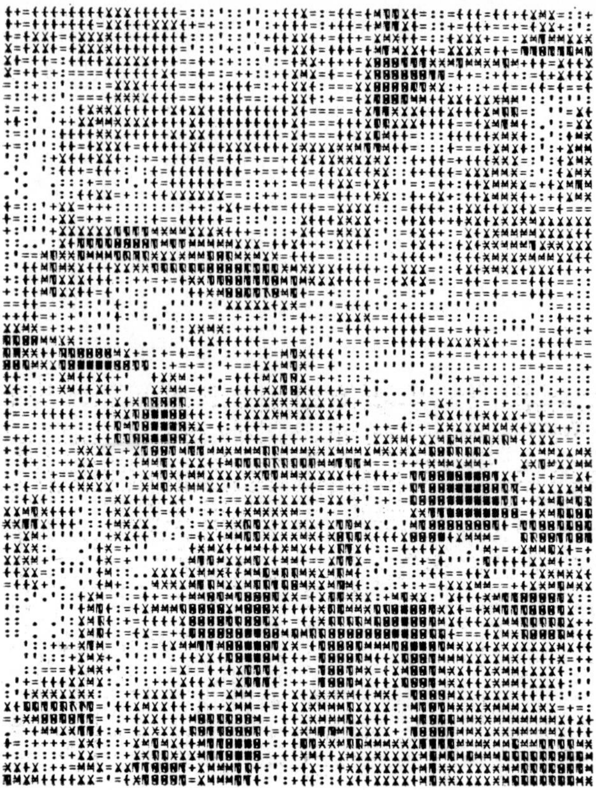

Fig. 9.15(c) $(\mathrm{WHT})_h$ ERTS image; 1.5 bits/pixel

(i) Obtain the 2-dimensional transform of the given image by processing it in $(N \times N)$ blocks.

(ii) Of the N^2 transform components obtained, retain N^2/m of those with the largest 2-dimensional variances $\tilde{\sigma}^2(u, v)$. Set all other components to zero.

(iii) Encode each of the N^2/m retained components using k bits per component. Reconstruct the corresponding $(N \times N)$ block using the inverse transform.

Since we need to code N^2/m transform components rather than N^2 data points for each $(N \times N)$ block, the *average* number of bits per pixel in the reconstructed image is k/m.

The first example concerns a (64×64) block of ERTS[1] image data. The processing of the image was done in (16×16) blocks. Using Eq. (9.4-3),

[1] ERTS abbreviates *E*arth *R*esources *T*echnological *S*atellite.

sixty four transform components with the largest variances were retained. The retained components were then coded using 6 bits, and subsequently the reconstructed images were obtained via the inverse transforms. The original and reconstructed images which consist of 64 levels (i.e. 6 bits) were reduced to 13 equiprobable levels to obtain line printer plots which were then photographed (Fig. 9.15). From these plots it is evident that the DCT reconstructed image compares more closely to the original than that of the $(\mathrm{WHT})_h$. On the other hand, the $(\mathrm{WHT})_h$ can be computed faster and has a simpler implementation. Hence the choice of a transform for a specified m depends upon computation, implementation, and image quality considerations.

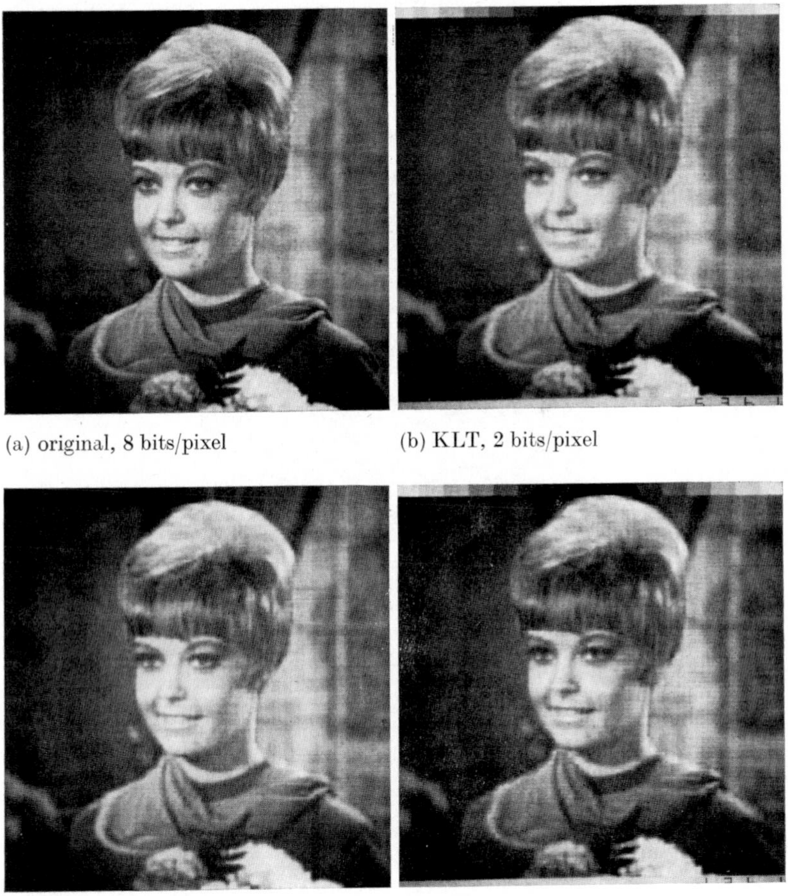

(a) original, 8 bits/pixel

(b) KLT, 2 bits/pixel

(c) DCT, 2 bits/pixel

(d) ST, 2 bits/pixel

Fig. 9.16 Example of image data compression

(e) DFT, 2 bits/pixel

(f) $(WHT)_h$, 2 bits/pixel

(g) HT, 2 bits/pixel

Fig. 9.16 Example of image data compression

The results pertaining to the second example[1] are presented in Fig. 9.16. Each pixel of the original (256×256) image in Fig. 9.16(a) was coded using 8 bits. Figures 9.16(b) through (g) show the reconstructed images for the case $m = 4$. The processing was done in (16×16) blocks. Since $k = 8$ and $m = 4$, the reconstructed images have been coded with an average of 2 bits per pixel.

[1] The inclusion of this example was made possible by Drs. W. K. Pratt, H. C. Andrews, and W. H. Chen, of the Image Processing Institute, University of Southern California, Los Angeles, California.

9.6 Additional Considerations

It has been demonstrated that the variance criterion enables one to predict the relative performances of the various transforms considered for data compression purposes. The variance criterion states that the set of transform components with the largest variances should be retained, and the rest discarded. This approach can also be explained in terms of the rate-distortion function associated with a basis restricted (BR) transformation. The notion of a BR transformation was introduced by Pearl [5, 6] as shown in Fig. 9.17. It consists of a transform **T**, an operation Q (not necessarily linear) on the set of transform components y_i, $i = 1, 2, \ldots, N$, and the inverse transform $\mathbf{T}^{-1}$. This structure is subjected to the constraint that for a given transform, Q operates *separately* on each y_i, $i = 1, 2, \ldots, N$. For example, with $q_i(y_i) = q_i y_i$, the BR transformation yields scalar Wiener filtering which was discussed in Section 8.4. Again, with $q_i(y_i) = y_i$, $i = 1, 2, \ldots, M$ and $q_i(y_i) = 0$, $i = M + 1, \ldots, N$, the BR transformation yields the data compression model in Fig. 9.10.

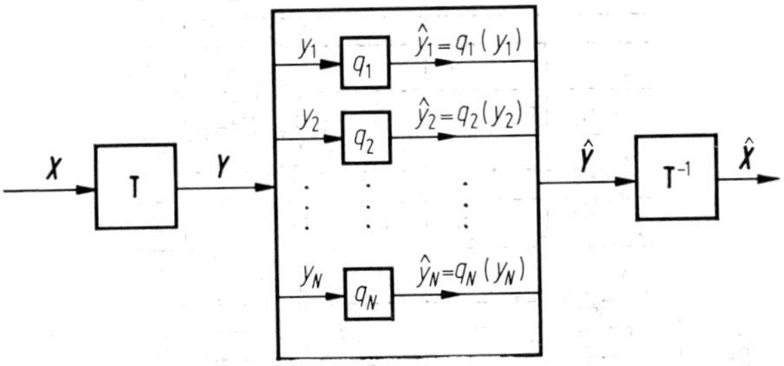

Fig. 9.17 Basis restricted transformation

In essence, the rate-distortion function yields the minimum information rate in bits per transform component needed for coding such that the average distortion is less than or equal to a chosen value D. The designer attempts to distribute the distortion over all the components such that the number of bits per component (i.e., rate) is minimized. We consider the case when the transform components have a Gaussian distribution and distortion is measured using the mean-square error criterion. Then it can be shown [5–7] that the rate-distortion function is defined by the following expression.

$$R(T, D) = \frac{1}{2N} \sum_{i=1}^{N} \max \left\{0, \log \left(\frac{\lambda_i}{\theta}\right)\right\} \tag{9.6-1}$$

where R is the minimum rate in bits per component, λ_i is the i-th eigenvalue if **T** represents the KLT, or is the i-th diagonal element of the transform domain covariance matrix $\tilde{\Sigma}_x$, if it represents any other transform, and θ is a parameter satisfying

$$D(\theta) = \frac{1}{N} \sum_{i=1}^{N} \min(\theta, \lambda_i) \tag{9.6-2}$$

Fig. 9.18 shows a set of rate-distortion[1] curves for a first-order Markov process whose covariance matrix Σ_x is given by Eq. (8.4-1) with $N = 16$, and $\varrho = 0.9$ [8]. For example, if the distortion level allowed is equal to 0.3, then the rates necessary for coding with the transforms considered are such that

$$R(\text{KLT}, 0.3) \simeq R(\text{DCT}, 0.3) < R(\text{DFT}, 0.3) < R(\text{IT}, 0.3).$$

This result can be interpreted qualitatively as follows: the IT leaves all the correlation in the data as is; the DFT tends to decorrelate the data although not totally; the KLT completely decorrelates the data while the DCT comes very close to it. Thus decorrelation via transforms results in much smaller rates than when the data is directly coded as though it consisted of uncorrelated samples.

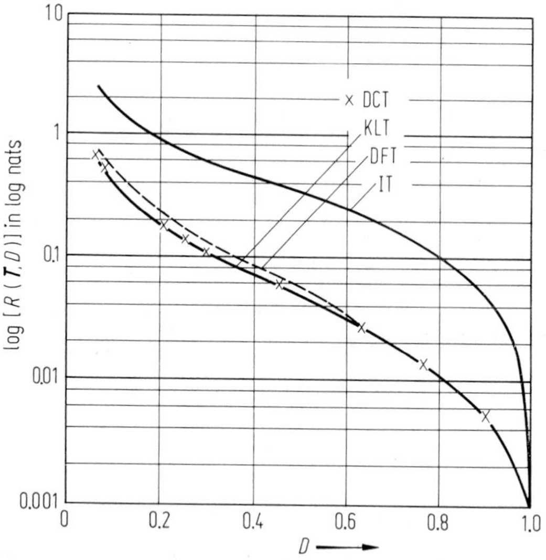

Fig. 9.18 Rate versus distortion for a first-order Markov process, $N = 16$, and $\varrho = 0.9$

9.7 Summary

In this chapter we have demonstrated that orthogonal transforms can be used to achieve data compression. The basic strategy used was to discard transform coefficients with relatively small variances. Examples of 1-dimen-

[1] The relation between nats (see Fig. 9.18) and bits is as follows: 1 nat = 1.44 bits.

sional and 2-dimensional data compression were included. It was shown that the KLT is the optimum transform for data compression, with respect to the mean-square error criterion. The variance criterion was also explained in terms of the rate-distortion function.

In Chapter 10 we will show that the variance criterion introduced in this chapter can also be used for feature selection in pattern recognition.

Appendix 9.1 Lagrange Multipliers [9]

The method of Lagrange multipliers for optimizing a function of n variables subject to some constraints is described below:

The necessary condition that a function $f(x_1, x_2, \ldots, x_n)$ of n independent variables $x_1, x_2, \ldots, x_n$ has a stationary value is

$$df = \sum_{i=1}^{n} \frac{\partial f}{\partial x_i} dx_i = 0 \tag{A9.1-1}$$

which is equivalent to the n conditions

$$\frac{\partial f}{\partial x_1} = \frac{\partial f}{\partial x_2} = \cdots = \frac{\partial f}{\partial x_n} = 0 \tag{A9.1-2}$$

If however, some of the n variables are dependent, then the method of Lagrange multipliers is most useful for optimizing a function. Let N of the n variables $(N < n)$ be dependent and be described by

$$\phi_K(x_1, x_2, \ldots, x_n) = 0, \qquad K = 1, 2, \ldots, N. \tag{A9.1-3}$$

Then the problem of optimizing $f(x_1, x_2, \ldots, x_n)$ subject to the N constraints described in Eq. (A9.1-3) reduces to optimizing the new function (no constraints)

$$f(x_1, x_2, \ldots, x_n) + \sum_{K=1}^{N} \beta_K \phi_K(x_1, x_2, \ldots, x_n) \tag{A9.1-4}$$

where β_K are the Lagrange multipliers. A necessary condition that $f(x_1, x_2, \ldots, x_n)$ subject to the N constraints described earlier has a maximum or minimum is

$$\frac{\partial f}{\partial x_i} + \sum_{K=1}^{N} \beta_K \frac{\partial \phi_K}{\partial x_i} = 0, \qquad i = 1, 2, \ldots, n \tag{A9.1-5}$$

Any N of these n equations can be solved for $\beta_1, \beta_2, \ldots, \beta_N$ which can then be substituted in the remaining $n - N$ equations. This yields the necessary conditions for optimizing $f(x_1, x_2, \ldots, x_n)$ with the prescribed constraints.

References

1. Andrews, H. C.: Multidimensional Rotations in Feature Selection. *IEEE Trans. Computers* C-20 (1971) 1045–1051.
2. Milne, P. J.: Orthogonal Transform Processing of Electrocardiograph Data. Ph. D. dissertation, Kansas State University, Manhattan, Kansas, USA, 1973.
3. Andrews, H. C.: *Computer Techniques in Image Processing.* New York, London: Academic Press, 1970, 135–151.
4. Pratt, W. K.: Walsh Functions in Image Processing and Two-dimensional Filtering. *Proc. 1972 Symp. Applications of Walsh Functions,* 14–22.
5. Pearl, J.: Basis Restricted Transformations and Performance Measures for Spectral Representation. *IEEE Trans. Info. Theory* IT-17 (1971) 751–752.
6. —, Walsh Processing of Random Signals. *IEEE Trans. Electromagnetic Compatability* (special issue) EMC-13 (1971) 137–141.
7. Pearl, J., Andrews, H. C., and Pratt, W. K.: Performance Measures for Transform Data Coding. *IEEE Trans. Communications* COM-20 (1972) 411–415.
8. Ahmed, N., Natarajan, T., and Rao, K. R.: Discrete Cosine Transform. *IEEE Trans. Computers* C-23 (1974) 90–93.
9. Hildebrand, F. B.: *Methods of Applied Mathematics.* Englewood Cliffs, N. J.: Prentice-Hall, 1952, 120–125.

Problems

9-1 If $\mathbf{\Sigma}$ is a symmetric matrix and $\boldsymbol{\phi}$ is an N-vector, show that

$$\nabla_{\phi}[\boldsymbol{\phi}'\mathbf{\Sigma}\boldsymbol{\phi}] = 2\mathbf{\Sigma}\boldsymbol{\phi}$$

and

$$\nabla_{\phi}[\boldsymbol{\phi}'\boldsymbol{\phi}] = 2\boldsymbol{\phi}$$

9-2 Let the covariance matrix of a class of signals represented by 8-vectors be denoted by $\mathbf{\Sigma}_x$. Then the variances of the transform components are given by the diagonal elements of

$$\tilde{\mathbf{\Sigma}}_x = \mathbf{T}\mathbf{\Sigma}_x(\mathbf{T}^*)'$$

where $\mathbf{T}$ denotes the transformation matrix such that $\mathbf{T}(\mathbf{T}^*)' = \mathbf{I}$.

Suppose the diagonal elements of $\tilde{\mathbf{\Sigma}}_x$ are as shown in Table P9-2-1. List a set of transform components y_k, $k = 1, 2, \ldots, 8$ that can be retained for each of these transforms if a 2:1 data compression is desired.

Table P9-2-1 Diagonal elements of $\tilde{\mathbf{\Sigma}}_x$

Diagonal element #	KLT	DCT	DFT	$(\mathrm{WHT})_h$	HT	MWHT
1	6.203	6.186	6.186	6.186	6.186	6.186
2	1.007	1.006	0.585	0.088	0.864	0.088
3	0.330	0.346	0.175	0.246	0.276	0.176
4	0.165	0.166	0.103	0.105	0.276	0.176
5	0.104	0.105	0.088	0.864	0.100	0.344
6	0.076	0.076	0.103	0.103	0.100	0.344
7	0.062	0.062	0.175	0.305	0.100	0.344
8	0.055	0.055	0.585	0.104	0.100	0.344

9-3 Let a (4×4) block of image data be represented by the matrix of random variables

$$\mathbf{X} = \begin{bmatrix} X_{11} & X_{12} & X_{13} & X_{14} \\ X_{21} & X_{22} & X_{23} & X_{24} \\ X_{31} & X_{32} & X_{33} & X_{34} \\ X_{41} & X_{42} & X_{43} & X_{44} \end{bmatrix}$$

The 2-dimensional $(\mathrm{WHT})_h$ of $\mathbf{X}$ is given by

$$\mathbf{Y} = \frac{1}{16}\mathbf{H}_h(2)\,\mathbf{X}\,\mathbf{H}_h(2)$$

where $\mathbf{Y}$ is the (4×4) matrix of transform components, and

$$\mathbf{H}_h(2) = \begin{bmatrix} 1 & 1 & 1 & 1 \\ 1 & -1 & 1 & -1 \\ 1 & 1 & -1 & -1 \\ 1 & -1 & -1 & 1 \end{bmatrix}$$

Using a given set of image data, the variances of the transform components y_{ij} are computed and found to be as follows:

$$\boldsymbol{\sigma}_y = \begin{bmatrix} 10.000 & 0.370 & 2.240 & 0.664 \\ 0.635 & 0.076 & 0.035 & 0.155 \\ 2.55 & 0.183 & 0.925 & 0.315 \\ 1.055 & 0.095 & 0.407 & 0.232 \end{bmatrix}$$

(a) Compute the 2-dimensional $(\mathrm{WHT})_h$ $\mathbf{Y}_1$ of the (4×4) data block

$$\mathbf{X}_1 = \begin{bmatrix} 20 & 19 & 21 & 19 \\ 20 & 17 & 19 & 21 \\ 19 & 20 & 20 & 21 \\ 20 & 20 & 20 & 19 \end{bmatrix}$$

(b) Retain 25 % of the components in $\mathbf{Y}_1$ with the largest variances in accordance with $\boldsymbol{\sigma}_y$. Set the remaining components to zero, and reconstruct the corresponding image using the 2-dimensional $(\mathrm{IWHT})_h$.

Answer:

$$\hat{\mathbf{X}}_1 = \begin{bmatrix} 19.250 & 19.250 & 19.875 & 19.875 \\ 19.125 & 19.125 & 19.750 & 19.750 \\ 19.50 & 19.50 & 20.125 & 20.125 \\ 19.625 & 19.625 & 20.250 & 20.250 \end{bmatrix}$$

Comment. We observe that rather than storing N^2 elements of the original image, one can store $N^2/4$ transform components with the largest variances. Thus a 4:1 data compression is realized. This type of data compression is expressed in terms of a *sample reduction factor* $\hat{m}$, which is defined as [4]

$$\hat{m} = \frac{\text{number of original image samples}}{\text{number of transformed image samples}}$$

9-4 The diagonal elements of $\tilde{\mathbf{\Sigma}}_x = \mathbf{T}\mathbf{\Sigma}_x\mathbf{T}'$ are listed in Table P9-4-1 for the case when Σ_x is given by Eq. (8.4-1), with $N = 16$ and $\varrho = 0.9$.

(a) Compute the corresponding rate-distortion functions $R(T, D)$ using Eq. (9.6-1).

(b) Plot the results obtained in (a) on semi-logarithmic paper [see Fig. 9.18].

Table P9-4-1 Diagonal elements of $\tilde{\mathbf{\Sigma}}_x$

Diagonal element #	IT	$(\text{WHT})_h$	HT	MWHT
1	1.0	9.835	9.835	9.835
2	1.0	0.78	2.537	0.078
3	1.0	0.206	0.864	0.155
4	1.0	0.105	0.864	0.155
5	1.0	0.706	0.276	0.305
6	1.0	0.103	0.276	0.305
7	1.0	0.307	0.276	0.305
8	1.0	0.104	0.276	0.305
9	1.0	2.536	0.100	0.570
10	1.0	0.098	0.100	0.570
11	1.0	0.283	0.100	0.570
12	1.0	0.105	0.100	0.570
13	1.0	1.020	0.100	0.570
14	1.0	0.102	0.100	0.570
15	1.0	0.303	0.100	0.570
16	1.0	0.104	0.100	0.570

Chapter Ten

Feature Selection in Pattern Recognition

10.1 Introduction

The subject of pattern recognition can be divided into two main areas of study: (1) feature selection and (2) classifier design, as summarized in Fig. 10.1. $x(t)$ is a signal that belongs to K classes denoted by $C_1, C_2, \ldots, C_K$.

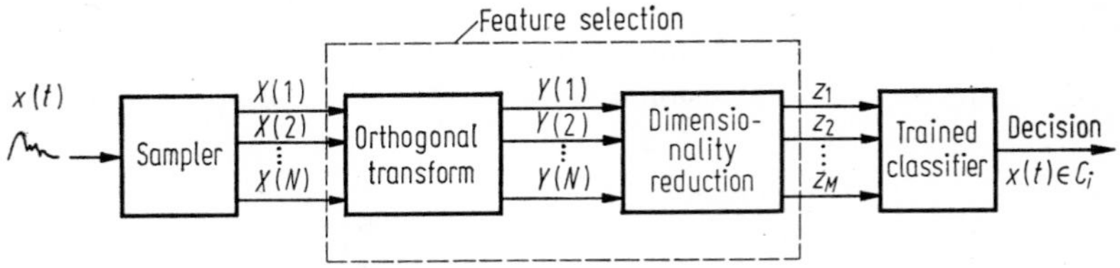

Fig. 10.1 A pattern recognition system

The sampling process shown yields the data sequence

$$\{X(m)\} = \{X(1)\, X(2) \cdots X(N)\}$$

which can be represented by an N-vector. The first stage of the feature selection consists of an orthogonal transformation, material pertaining to which was covered in Chapters 3 through 7. Since such transformations provide a one-to-one mapping, the transform sequence

$$\{Y(m)\} = \{Y(1)\, Y(2) \cdots Y(N)\}$$

also has an N-vector representation. In practice the value of N is usually quite large. Intuitively, as the number of inputs to the classifier is reduced, the design and implementation of the classifier becomes simpler. Thus the second stage in the feature selection process is a dimensionality reduction stage, the output of which is a subset of M features $z_1, z_2, \ldots, z_M$ from $\{Y(m)\}$, such that $M \ll N$. This dimensionality reduction must be achieved in such a way that the corresponding increase in classification error is relatively small. The vector $\boldsymbol{Z}' = [z_1\, z_2 \cdots z_M]$ will be referred to as a *pattern* or *pattern vector*.

The classifier in Fig. 10.1 is a decision making device which is *trained* to classify an incoming signal $x(t)$ as belonging to one of the K classes. The subject of pattern classifiers is indeed a vast one. Several textbooks in this area are available [1–7].

Our main objective in this chapter is to show how the variance criterion (developed in the last chapter) can be used to secure dimensionality reduction, with relatively small increase in classification error. We will do so by means of specific examples. To this end, it is necessary to study some simple classification algorithms and their implemention. Our attention will be restricted to a group of classifiers called *minimum-distance classifiers*.

10.2 The Concept of Training

The training concept is best introduced by means of a simple example. Suppose we wish to train a classifier such that it is capable of automatically classifying a pattern $\boldsymbol{Z}$ as belonging to either C_1 or C_2. Further, suppose the *training set* (i.e., a set whose true classification is known) consists of the following set of 2-dimensional patterns, $\boldsymbol{Z}_{ij}$, where $\boldsymbol{Z}_{ij}$ denotes the j-th pattern belonging to C_i, $i = 1, 2$.

$$C_1:\ \boldsymbol{Z}_{11} = \begin{bmatrix} 5 \\ 5 \end{bmatrix},\quad \boldsymbol{Z}_{12} = \begin{bmatrix} 6 \\ 5 \end{bmatrix},\quad \boldsymbol{Z}_{13} = \begin{bmatrix} 6 \\ 6 \end{bmatrix},\quad \boldsymbol{Z}_{14} = \begin{bmatrix} 6 \\ 7 \end{bmatrix},\quad \boldsymbol{Z}_{15} = \begin{bmatrix} 7 \\ 5 \end{bmatrix}$$

$$C_2:\ \boldsymbol{Z}_{21} = \begin{bmatrix} 0 \\ 3 \end{bmatrix},\quad \boldsymbol{Z}_{22} = \begin{bmatrix} -1 \\ 3 \end{bmatrix},\quad \boldsymbol{Z}_{23} = \begin{bmatrix} -2 \\ 3 \end{bmatrix},\quad \boldsymbol{Z}_{24} = \begin{bmatrix} -3 \\ 3 \end{bmatrix},\quad \boldsymbol{Z}_{25} = \begin{bmatrix} -4 \\ 3 \end{bmatrix} \tag{10.2-1}$$

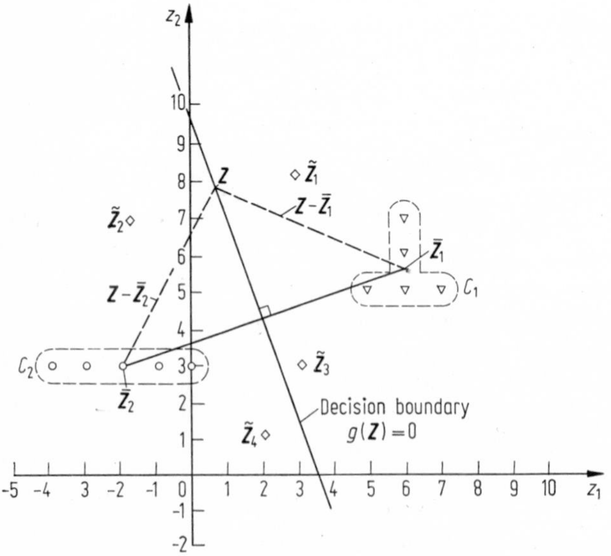

Fig. 10.2
A 2-dimensional feature space

The patterns $\{\boldsymbol{Z}_{ij}\}$ which belong to C_1 and C_2 are located in the 2-dimensional *feature space* shown in Fig. 10.2.

Let $\bar{\boldsymbol{Z}}_1$ and $\bar{\boldsymbol{Z}}_2$ denote the mean pattern vectors associated with C_1 and C_2 respectively. Then we have

$$\bar{\boldsymbol{Z}}_i = \frac{1}{5} \sum_{j=1}^{5} Z_{ij}, \qquad i = 1, 2 \tag{10.2-2}$$

which yields

$$\bar{\boldsymbol{Z}}_1 = \begin{bmatrix} 6 \\ 5.6 \end{bmatrix}, \quad \text{and} \quad \bar{\boldsymbol{Z}}_2 = \begin{bmatrix} -2 \\ 3 \end{bmatrix}$$

Inspection of Fig. 10.2 reveals that a reasonable decision boundary (i.e. a line in two dimensions) which separates C_1 and C_2, is the perpendicular bisector of the line joining $\bar{\boldsymbol{Z}}_1$ and $\bar{\boldsymbol{Z}}_2$. The next step is to describe the proposed decision boundary by means of an equation.

Consider any point $\boldsymbol{Z}$ on the decision boundary as shown in Fig. 10.2. Since the decision boundary is the perpendicular bisector of the line joining $\bar{\boldsymbol{Z}}_1$ and $\bar{\boldsymbol{Z}}_2$, we must have

$$||\boldsymbol{Z} - \bar{\boldsymbol{Z}}_1||^2 = ||\boldsymbol{Z} - \bar{\boldsymbol{Z}}_2||^2$$

which simplifies to

$$(\bar{\boldsymbol{Z}}_1 - \bar{\boldsymbol{Z}}_2)'\boldsymbol{Z} = \frac{1}{2}\{||\bar{\boldsymbol{Z}}_1||^2 - ||\bar{\boldsymbol{Z}}_2||^2\} \tag{10.2-3}$$

Substituting $\boldsymbol{Z}' = [z_1\ z_2]$, and Eq. (2) in Eq. (3), we obtain the equation for the decision boundary as

$$8z_1 + 2.6z_2 = 27.18 \tag{10.2-4}$$

The quantity $\frac{1}{2}(||\bar{\boldsymbol{Z}}_1|||^2 - ||\bar{\boldsymbol{Z}}_2||^2) = 27.18$ is called the *threshold* of the classifier.

Equation (4) provides us with all the information required to design the classifier. The basic ingredient used to represent the classifier is a *discriminant function* $g(\boldsymbol{Z})$ which is defined as

$$g(\boldsymbol{Z}) = 8z_1 + 2.6z_2 - 27.18 \tag{10.2-5}$$

The important question is, "what does $g(\boldsymbol{Z})$ really do?" To answer this, let us consider several *test patterns*[1] which are denoted by $\tilde{\boldsymbol{Z}}_k$, $k = 1, 2, 3, 4$ [see Fig. 10.2]. We now compute the value of $g(\boldsymbol{Z})$ to obtain:

$$\begin{aligned}
\tilde{\boldsymbol{Z}}_1&: g(\boldsymbol{Z}) = 8(3) + 2.6(8) - 27.18 > 0 \\
\tilde{\boldsymbol{Z}}_2&: g(\boldsymbol{Z}) = 8(-2) + 2.6(7) - 27.18 < 0 \\
\tilde{\boldsymbol{Z}}_3&: g(\boldsymbol{Z}) = 8(3) + 2.6(3) - 27.18 > 0 \\
\tilde{\boldsymbol{Z}}_4&: g(\boldsymbol{Z}) = 8(2) + 2.6(1) - 27.18 < 0
\end{aligned} \tag{10.2-6}$$

[1] A test pattern is one whose true classification is not known.

Examination of Eq. (6) reveals that whenever a pattern $\tilde{\boldsymbol{Z}}$ lies to the "right" of the decision boundary, then $g(\boldsymbol{Z}) > 0$. Conversely if $\tilde{\boldsymbol{Z}}$ lies to the "left" of the decision boundary, then $g(\boldsymbol{Z}) < 0$. Further, we note that any point to the right (or positive side) of the boundary is closer to $\bar{\boldsymbol{Z}}_1$, while that to the left (or negative side) of the boundary is closer to $\bar{\boldsymbol{Z}}_2$. Thus by virtue of the discriminant function $g(\boldsymbol{Z})$, we have obtained the following simple *decision rule:*

$$\text{If } g(\boldsymbol{Z}) > 0, \quad \text{then} \quad \boldsymbol{Z} \in C_1$$

and

$$\text{if } g(\boldsymbol{Z}) < 0, \quad \text{then} \quad \boldsymbol{Z} \in C_2$$

The above classifier is called a linear *threshold logic unit* (TLU) and is implemented as shown in Fig. 10.3. The term "linear" implies that $g(\boldsymbol{Z})$ is a linear functional of the features z_1 and z_2.

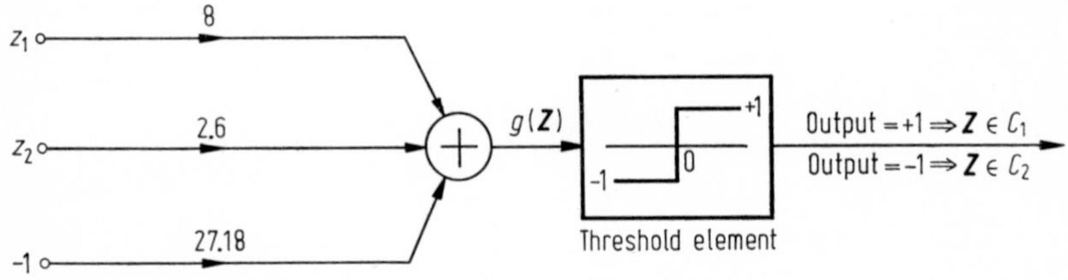

Fig. 10.3 A linear threshold logic unit (TLU)

In summary, a training set of patterns are used to design a classifier via a discriminant function $g(\boldsymbol{Z})$. Once $g(\boldsymbol{Z})$ is obtained, the classifier is said to be trained, in the sense that it is then able to classify patterns using an appropiate decision rule. We observe that the TLU is a *minimum-distance* classifier since its decision rule can be equivalently stated as follows:

$$\text{If } \boldsymbol{Z} \text{ is closer to } \bar{\boldsymbol{Z}}_1, \text{ then } \boldsymbol{Z} \in C_1$$

and

$$\text{if } \boldsymbol{Z} \text{ is closer to } \bar{\boldsymbol{Z}}_2, \text{ then } \boldsymbol{Z} \in C_2$$

10.3 *d*-dimensional Patterns

It is now straight forward to think in terms of a linear TLU when $(d \times 1)$ patterns are involved, where $d > 2$. In such cases $\boldsymbol{Z}$ is of the form $\boldsymbol{Z}' = [z_1\, z_2 \cdots z_d]$. The corresponding discriminant function is given by

$$g(\boldsymbol{Z}) = w_1 z_1 + w_2 z_2 + \cdots + w_d z_d - \theta \tag{10.3-1}$$

where the w_i are known as the *weights* or *parameters* of the classifier, and θ is the threshold. The w_i and θ are obtained from the training set. The boundary in this case is a hyperplane defined by

$$g(\boldsymbol{Z}) = 0 \tag{10.3-2}$$

For the special case $d = 3$, this hyperplane reduces to a plane.

The above classifier is implemented as shown in Fig. 10.4.

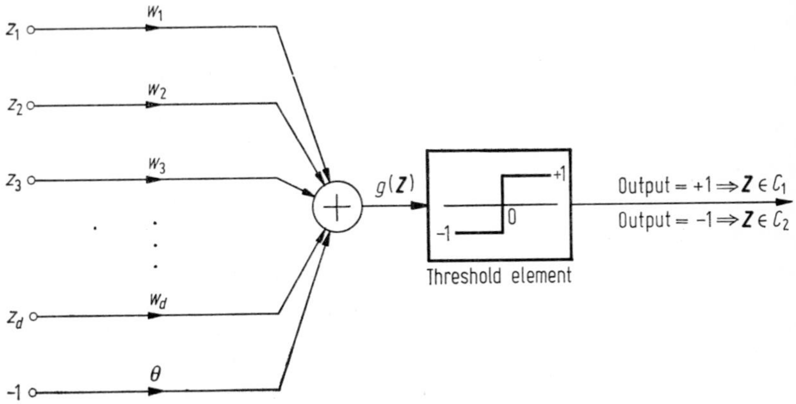

Fig. 10.4 A linear TLU for d-dimensional patterns

10.4 The 3-class Problem

In the previous two sections we have discussed some basic aspects of the so-called 2-class problem. We now extend these concepts to a more general case which involves three classes C_1, C_2, and C_3. Suppose the training set is as follows:

$$\begin{aligned}
&C_1\colon\ \boldsymbol{Z}_{11} = \begin{bmatrix}0\\3\end{bmatrix},\ \boldsymbol{Z}_{12} = \begin{bmatrix}-1\\3\end{bmatrix},\ \boldsymbol{Z}_{13} = \begin{bmatrix}-2\\3\end{bmatrix},\ \boldsymbol{Z}_{14} = \begin{bmatrix}-3\\3\end{bmatrix},\ \boldsymbol{Z}_{15} = \begin{bmatrix}-4\\3\end{bmatrix}\\
&C_2\colon\ \boldsymbol{Z}_{21} = \begin{bmatrix}5\\5\end{bmatrix},\ \boldsymbol{Z}_{22} = \begin{bmatrix}6\\5\end{bmatrix},\ \boldsymbol{Z}_{23} = \begin{bmatrix}6\\6\end{bmatrix},\ \boldsymbol{Z}_{24} = \begin{bmatrix}6\\7\end{bmatrix},\ \boldsymbol{Z}_{25} = \begin{bmatrix}7\\5\end{bmatrix}\\
&C_3\colon\ \boldsymbol{Z}_{31} = \begin{bmatrix}6\\-1\end{bmatrix},\ \boldsymbol{Z}_{32} = \begin{bmatrix}7\\0\end{bmatrix},\ \boldsymbol{Z}_{33} = \begin{bmatrix}8\\1\end{bmatrix},\ \boldsymbol{Z}_{34} = \begin{bmatrix}9\\1\end{bmatrix},\ \boldsymbol{Z}_{35} = \begin{bmatrix}10\\1\end{bmatrix}
\end{aligned} \tag{10.4-1}$$

From Eq. (1) it follows that the mean pattern vectors are given by

$$\bar{\boldsymbol{Z}}_1 = \begin{bmatrix} -2 \\ 3 \end{bmatrix}, \quad \bar{\boldsymbol{Z}}_2 = \begin{bmatrix} 6 \\ 5.6 \end{bmatrix}, \quad \text{and} \quad \bar{\boldsymbol{Z}}_3 = \begin{bmatrix} 8 \\ 0.4 \end{bmatrix}$$

which are plotted in Fig. 10.5.

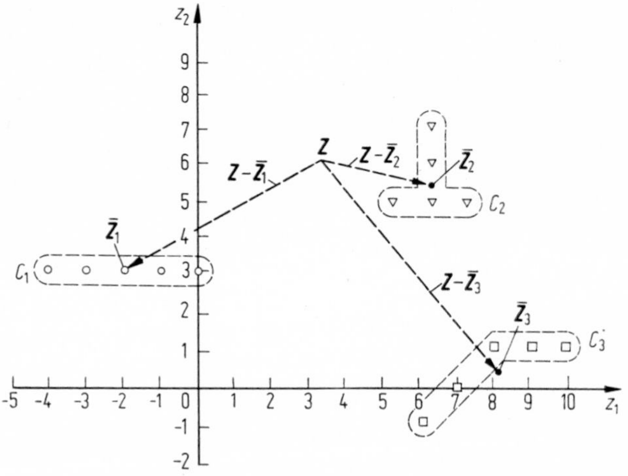

Fig. 10.5 A 2-dimensional feature space associated with C_1, C_2, and C_3

The minimum-distance classifier is designed to operate on the following decision rule:

A given pattern $\boldsymbol{Z}$ belongs to C_i if $\boldsymbol{Z}$ is closest to $\bar{\boldsymbol{Z}}_i$, $i = 1, 2, 3$.

Let D_i denote the distance of $\boldsymbol{Z}$ from $\bar{\boldsymbol{Z}}_i$, $i = 1, 2, 3$. Then we have [see Fig. 10.5]

$$D_i^2 = ||\boldsymbol{Z} - \bar{\boldsymbol{Z}}_i||^2 = (\boldsymbol{Z} - \bar{\boldsymbol{Z}}_i)' (\boldsymbol{Z} - \bar{\boldsymbol{Z}}_i) \tag{10.4-2}$$

Simplification of D_i^2 yields

$$D_i^2 = ||\boldsymbol{Z}||^2 - 2\{\bar{\boldsymbol{Z}}_i'\boldsymbol{Z} - \frac{1}{2} ||\bar{\boldsymbol{Z}}_i||^2\}, \qquad i = 1, 2, 3 \tag{10.4-3}$$

Clearly, D_i^2 is a *minimum*, when the quantity $\left\{\bar{\boldsymbol{Z}}_i'\boldsymbol{Z} - \frac{1}{2} ||\bar{\boldsymbol{Z}}_i||^2\right\}$ is a *maximum*. Thus, rather than having the classifier compute D_i^2 in Eq. (3), it is simpler to require it to compute the quantity $\left\{\bar{\boldsymbol{Z}}_i'\boldsymbol{Z} - \frac{1}{2} ||\bar{\boldsymbol{Z}}_i||^2\right\}$. The classifier is then described by the discriminant functions

$$g_i(\boldsymbol{Z}) = \bar{\boldsymbol{Z}}_i'\boldsymbol{Z} - \frac{1}{2} ||\bar{\boldsymbol{Z}}_i||^2, \qquad i = 1, 2, 3 \tag{10.4-4}$$

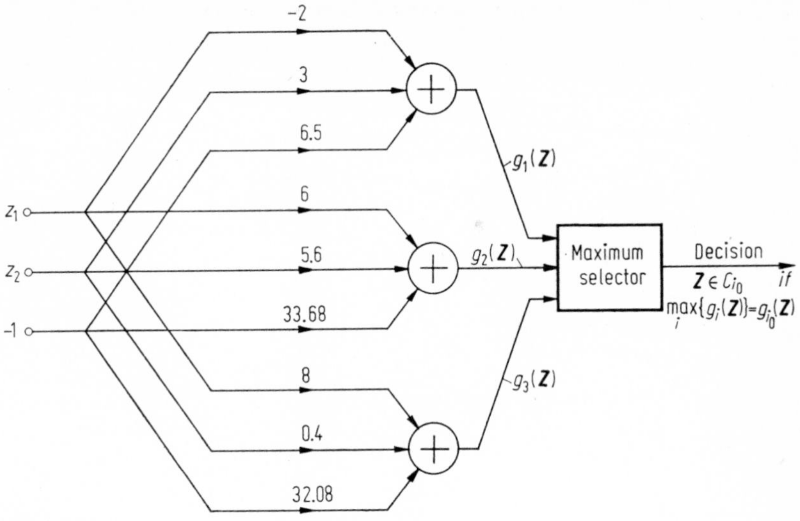

Fig. 10.6 3-class minimum-distance classifier

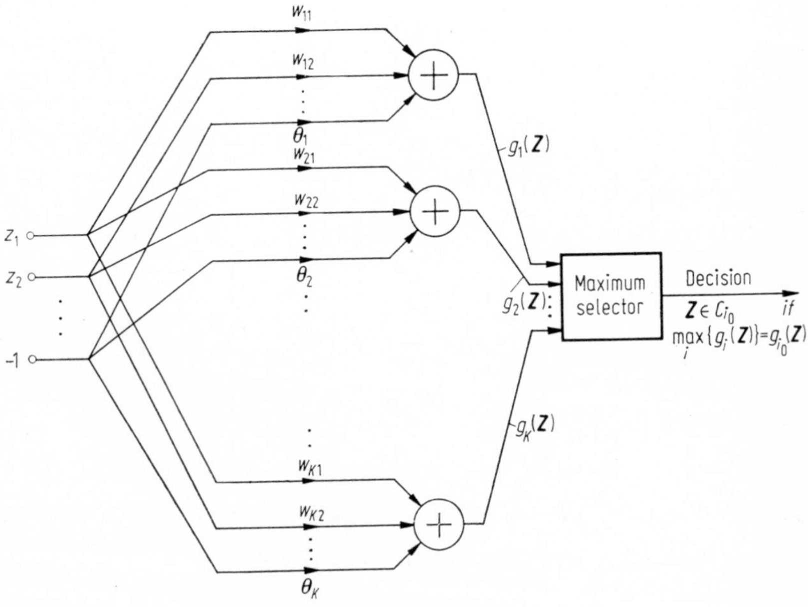

Fig. 10.7 K-class minimum-distance classifier

Substitution of numerical values for $\bar{\boldsymbol{Z}}_i$ and $||\bar{\boldsymbol{Z}}_i||^2$ results in

$$g_1(\boldsymbol{Z}) = -2z_1 + 3z_2 - 6.5$$

$$g_2(\boldsymbol{Z}) = 6z_1 + 5.6z_2 - 33.68$$

and

$$g_3(\boldsymbol{Z}) = 8z_1 + 0.4z_2 - 32.08$$

The classifier thus computes three *numbers* $g_1(\boldsymbol{Z})$, $g_2(\boldsymbol{Z})$, and $g_3(\boldsymbol{Z})$ as shown in Fig. 10.6, and then compares them. It assigns $\boldsymbol{Z}$ to C_1 if $g_1(\boldsymbol{Z})$ is maximum, to C_2 if $g_2(\boldsymbol{Z})$ is maximum, or to C_3 if $g_3(\boldsymbol{Z})$ is maximum. The implementation associated with the general K-class linear minimum-distance classifier is shown in Fig. 10.7.

10.5 Image Classification Experiment

An interesting experiment conducted by Andrews [7, 8] consisted of numeral character recognition where the number of classes was ten ($K = 10$), corresponding to the numerals 0 through 9. The characters were handwritten

2	4									2	1
2	4									5	5
	6									4	5
1	5									2	5
1	5									3	6
	5	1								3	7
	5									3	6
	5	1								3	7
	6	5	6	5	5	5	4	4	5	5	6
	4	5	4	3	4	4	3	5	5	5	5
										3	6
										3	5
										4	5
										2	5
										2	6

Fig. 10.8
Eight level representation of numeral 4 [8]

and digitized according to a (16×12) raster scan of eight levels of gray. An example in computer printout form is illustrated in Fig. 10.8. This information was represented as a 192-vector. For computational purposes, each of these vectors was appended with zeros to obtain a 256-vector denoted by [see Fig. 10.1]

$$\boldsymbol{X}' = [X(1)\ X(2) \cdots X(256)]$$

Feature selection process. The overall covariance matrix was computed as [see Eq. (9.2-1)]

$$\Sigma_x^* = P_1\Sigma_{x_1} + P_2\Sigma_{x_2} + \cdots + P_{10}\Sigma_{x_{10}} \tag{10.5-1}$$

where Σ_{xk} is the covariance matrix of the k-th class; P_k denotes the a priori probability of class C_k which was assumed to equal 1/10. The variance distributions[1] computed using Σ_x^* are shown in Fig. 10.9. If a transform vector $\boldsymbol{Y} = \mathbf{T}\boldsymbol{X}$ is denoted by

$$\boldsymbol{Y}' = [Y(1)\, Y(2) \cdots Y(256)]$$

then it is evident that the variance energy (indicated by the area under each curve) for the IT is spread over many more components, relative to the other transforms. In accordance with the variance criterion, the components with the M largest variances are selected as features. This process results in a set of patterns each of which is denoted by

$$\boldsymbol{Z}' = [z_1\ z_2 \cdots z_M]$$

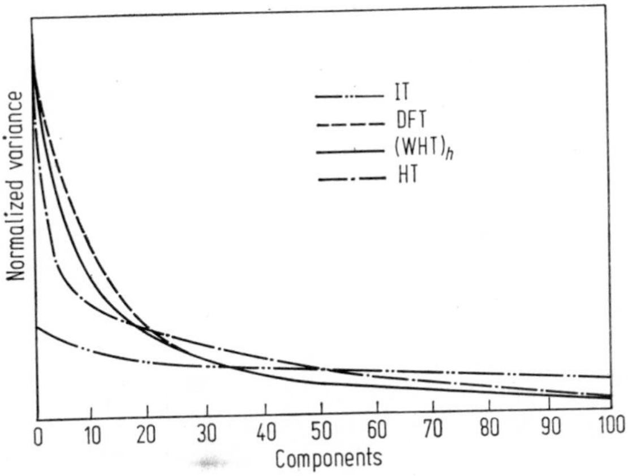

Fig. 10.9 Variance distributions associated with various discrete transforms [8]

Classification considerations. A 10-class minimum distance classifier was used, whose discriminant functions are given by [see Eq. (10.4-4)]

$$g_k(\boldsymbol{Z}) = \bar{\boldsymbol{Z}}_k'\boldsymbol{Z} - \frac{1}{2}\,||\bar{\boldsymbol{Z}}_k||^2, \qquad k = 1, 2, \ldots, 10 \tag{10.5-2}$$

where $\bar{\boldsymbol{Z}}_k$ is the mean pattern vector for the k-th class. Since $\boldsymbol{Z}' = [z_1\ z_2 \cdots z_M]$,

[1] Note that only 100 components are plotted.

the k-th discriminant function is of the form

$$g_k(\mathbf{Z}) = w_{k1}z_1 + w_{k2}z_2 + \cdots + w_{kM}z_M - \theta_k, \qquad k = 1, 2, \ldots, 10 \quad (10.5\text{-}3)$$

where $\mathbf{Z}_k' = [w_{k1}\ w_{k2} \cdots w_{kM}]$ and $\theta_k = \frac{1}{2}||\bar{\mathbf{Z}}_k||^2$. The implementation of this algorithm is as shown in Fig. 10.7.

The above classifier was trained using 500 patterns for each class. Figure 10.10 displays the results of correct classification versus features retained. These results were obtained by classifying the training set only. It is evident that the DFT feature space provides considerable improvement over the IT feature space.

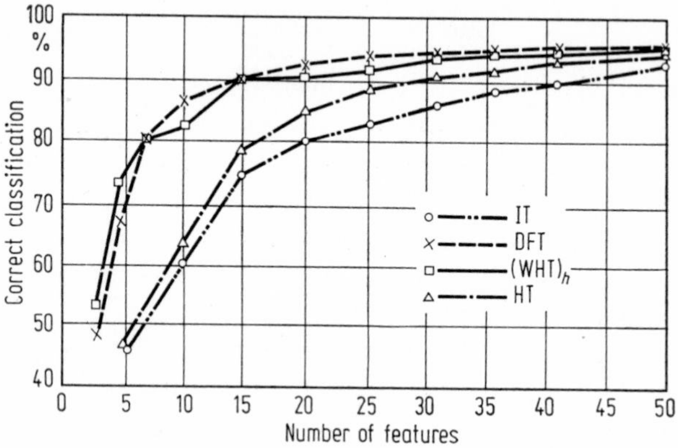

Fig. 10.10 Correct classification versus features retained – training set [8]

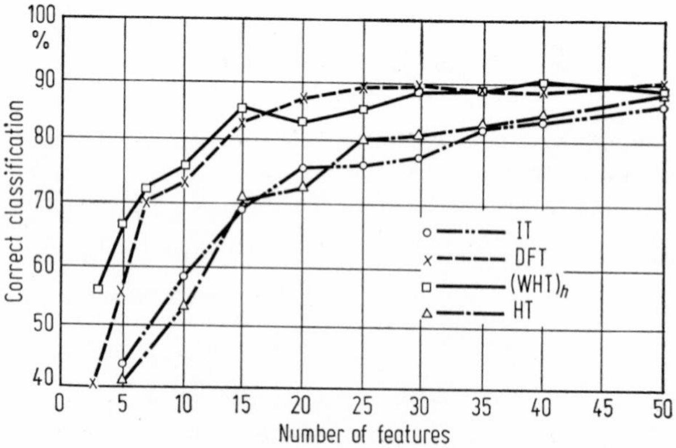

Fig. 10.11 Correct classification versus features retained – test set [8]

The above experiment was repeated with a test set of 500 patterns, in place of the training set which defined the $\bar{\boldsymbol{Z}}_k$ in Eq. (2). The classification results were typically 2% to 7% worse than those obtained with the training set [see Fig. 10.11], which is as would be expected considering that the means were not defined by the test set. Again, the increase in misclassification error can also be attributed to the classification algorithm which is rather simple. However, the point to be made is of feature selection and not of classification. From Fig. 10.11 it can be seen that to obtain an 88% correct classification, the number of features required for the DFT, $(\mathrm{WHT})_h$ and HT are 25, 35, and 50 respectively. In contrast, it is reported that 120 IT features are necessary to obtain an 88% correct classification.

10.6 Least-Squares Mapping Technique

In discussing the minimum-distance classifiers, we have inherently assumed that the pattern classes in the feature space cluster around their respective means $\bar{\boldsymbol{Z}}_i$, $i = 1, 2, \ldots, K$. This however may not be a reasonable assumption in many applications. In such cases, the classifier must first map the patterns into a *decision space* wherein the patterns belonging to C_i are made to cluster around a preselected point $\boldsymbol{V}_i$, $i = 1, 2, \ldots, K$. The transformation $\mathbf{A}$ which enables this mapping from the feature space to the decision space is generally chosen such that the overall mean-square mapping error is minimized. To classify a given pattern, it is first mapped into the decision space and then classified as belonging to C_{i_0} if it is mapped closest to $\boldsymbol{V}_{i_0}$. Thus this type of a classifier will be referred to as a *least-squares minimum distance* classifier, the development of which will be considered in Section 10.8. To this end, we introduce a least-squares mapping technique which will form the core of the least-squares minimum-distance classifier.

For the purposes of discussion, consider a set of M-dimensional patterns $\boldsymbol{Z}_{ij}$, $j = 1, 2, \ldots, N_i$ which are to be mapped into a specified point in a K-dimensional space, denoted by $\boldsymbol{V}_i' = [v_1\, v_2 \cdots v_K]$. We seek the transformation $\mathbf{A}$ which maps $\{\boldsymbol{Z}_{ij}\}$ into $\boldsymbol{V}_i$ such that the overall mean-square error incurred in the mapping is minimized. Suppose the mapping of a pattern $\boldsymbol{Z}_{ij}$ is denoted by $\boldsymbol{L}_{ij}$. Then the corresponding error vector is

$$\boldsymbol{\varepsilon}_j = \boldsymbol{L}_{ij} - \boldsymbol{V}_i = \mathbf{A}\boldsymbol{Z}_{ij} - \boldsymbol{V}_i \qquad (10.6\text{-}1)$$

From Eq. (1), the total mean-square error in mapping $\{\boldsymbol{Z}_{ij}\}$ into $\boldsymbol{V}_i$ is given by

$$\varepsilon = \frac{1}{N_i} \sum_{j=1}^{N_i} ||\boldsymbol{\varepsilon}_j||^2 \qquad (10.6\text{-}2)$$

Substitution of Eq. (1) in Eq. (2) leads to

$$\varepsilon = \frac{1}{N_i} \sum_{j=1}^{N_i} \{\boldsymbol{Z}'_{ij}\mathbf{A}'\mathbf{A}\boldsymbol{Z}_{ij} - 2\boldsymbol{Z}'_{ij}\mathbf{A}'\boldsymbol{V}_i + ||\boldsymbol{V}_i||^2\} \tag{10.6-3}$$

Since $\mathbf{A}$ must be chosen such that ε is a minimum, it is obtained by solving the equation

$$\nabla_{\mathbf{A}}\, \varepsilon = 0$$

which yields

$$\frac{1}{N_i} \sum_{j=1}^{N_i} \nabla_{\mathbf{A}} \{\boldsymbol{Z}'_{ij}\mathbf{A}'\mathbf{A}\boldsymbol{Z}_{ij}\} - \frac{2}{N_i} \sum_{j=1}^{N_i} \nabla_{\mathbf{A}} \{\boldsymbol{Z}'_{ij}\mathbf{A}'\boldsymbol{V}_i\} + \nabla_{\mathbf{A}} \{||\boldsymbol{V}_i||^2\} = 0 \tag{10.6-4}$$

From Eqs. (8.1-3), (8.1-7), and (8.1-8), it follows that

$$\nabla_{\mathbf{A}} \{\boldsymbol{Z}'_{ij}\mathbf{A}'\mathbf{A}\boldsymbol{Z}_{ij}\} = 2\mathbf{A}(\boldsymbol{Z}_{ij}\boldsymbol{Z}'_{ij})$$

$$\nabla_{\mathbf{A}} \{\boldsymbol{Z}'_{ij}\mathbf{A}'\boldsymbol{V}_i\} = \boldsymbol{V}_i\boldsymbol{Z}'_{ij}$$

and

$$\nabla_{\mathbf{A}} \{||\boldsymbol{V}_i||^2\} = 0$$

Application of the above identities to Eq. (4) results in

$$\mathbf{A}\left[\frac{2}{N_i} \sum_{j=1}^{N_i} (\boldsymbol{Z}_{ij}\boldsymbol{Z}'_{ij})\right] = \frac{2}{N_i} \sum_{j=1}^{N_i} \boldsymbol{V}_i\boldsymbol{Z}'_{ij} \tag{10.6-5}$$

which yields $\mathbf{A}$ as

$$\mathbf{A} = \left[\sum_{j=1}^{N_i} \boldsymbol{V}_i\boldsymbol{Z}'_{ij}\right]\left[\sum_{j=1}^{N_i} \boldsymbol{Z}_{ij}\boldsymbol{Z}'_{ij}\right]^{-1} \tag{10.6-6}$$

To illustrate, let $\{\boldsymbol{Z}_{ij}\}$ be as follows:

$$\boldsymbol{Z}_{i1} = \begin{bmatrix} 5 \\ 5 \end{bmatrix}, \quad \boldsymbol{Z}_{i2} = \begin{bmatrix} 5 \\ 6 \end{bmatrix}, \quad \boldsymbol{Z}_{i3} = \begin{bmatrix} 5 \\ 7 \end{bmatrix}, \quad \boldsymbol{Z}_{i4} = \begin{bmatrix} 6 \\ 6 \end{bmatrix}, \quad \boldsymbol{Z}_{i5} = \begin{bmatrix} 6 \\ 7 \end{bmatrix},$$

which implies that $N_i = 5$. Again, let $\boldsymbol{V}_i = \begin{bmatrix} 1 \\ 1 \end{bmatrix}$. Then it follows that

$$\sum_{j=1}^{5} \boldsymbol{V}_i\boldsymbol{Z}'_{ij} = \begin{bmatrix} 27 & 31 \\ 27 & 31 \end{bmatrix}, \quad \text{and} \quad \sum_{j=1}^{5} \boldsymbol{Z}_{ij}\boldsymbol{Z}'_{ij} = \begin{bmatrix} 147 & 168 \\ 168 & 195 \end{bmatrix} \tag{10.6-7}$$

Substitution of Eq. (7) in Eq. (6) results in

$$\mathbf{A} = \begin{bmatrix} 0.129 & 0.0475 \\ 0.129 & 0.0475 \end{bmatrix} \tag{10.6-8}$$

Evaluating $\boldsymbol{L}_{ij} = \mathbf{A}\boldsymbol{Z}_{ij}$ we obtain

$$\boldsymbol{L}_{i1} = \begin{bmatrix} 0.895 \\ 0.895 \end{bmatrix}, \quad \boldsymbol{L}_{i2} = \begin{bmatrix} 0.932 \\ 0.932 \end{bmatrix}, \quad \boldsymbol{L}_{i3} = \begin{bmatrix} 0.982 \\ 0.982 \end{bmatrix}, \quad \boldsymbol{L}_{i4} = \begin{bmatrix} 1.06 \\ 1.06 \end{bmatrix}, \quad \boldsymbol{L}_{i5} = \begin{bmatrix} 1.11 \\ 1.11 \end{bmatrix}$$

The sets of patterns $\{\boldsymbol{Z}_{ij}\}$ and $\{\boldsymbol{L}_{ij}\}$ are displayed in Fig. 10.12.

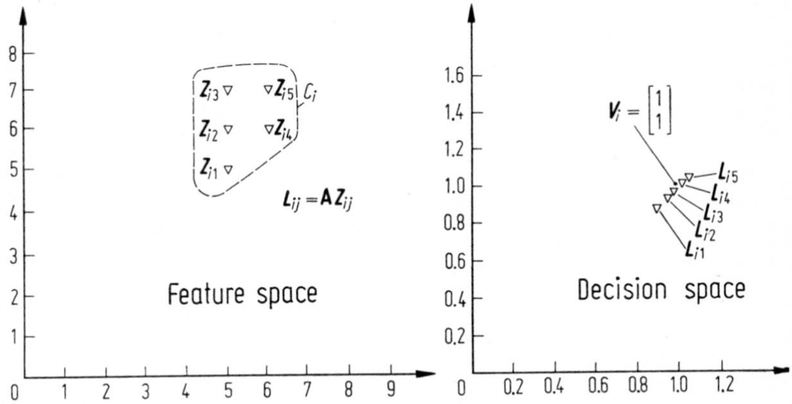

Fig. 10.12 Pertaining to a numerical example illustrating the least-squares mapping technique

10.7 Augmented Feature Space

The notion of an augmented feature space is convenient for the development of the least-squares minimum distance classifier. It follows naturally from the definition of a discriminant function, which is

$$g(\boldsymbol{Z}) = w_1 z_1 + w_2 z_2 + \cdots + w_d z_d - \theta$$

We rewrite $g(\boldsymbol{Z})$ in the form

$$g(\boldsymbol{Z}) = \boldsymbol{W}'\hat{\boldsymbol{Z}} \tag{10.7-1}$$

where

$$\boldsymbol{W}' = [w_1\ w_2 \cdots w_d\ \theta]$$

and

$$\hat{\boldsymbol{Z}}' = [z_1\ z_2 \cdots z_d\ -1] = [\boldsymbol{Z}'\ -1]$$

From Eq. (1) it follows that $\hat{\boldsymbol{Z}}$ is easily obtained from a given pattern $\boldsymbol{Z}$ by merely augmenting it with an additional component, equal to -1. The space which consists of the $(d+1)$-dimensional patterns $\hat{\boldsymbol{Z}}$ is called an augmented feature space.

10.8 3-class Least-Squares Minimum Distance Classifier [9–11]

The development of a least-squares minimum distance classifier is best illustrated by the case when three classes are present. Corresponding to a d-dimensional feature space, we obtain a $(d + 1)$-dimensional augmented feature space by augmenting each pattern $\boldsymbol{Z}$, by a component equal to -1 to obtain $\hat{\boldsymbol{Z}}' = [\boldsymbol{Z}' \;\; -1]$. This is illustrated for the case $d = 2$ in Figs. 10.13 (a) and (b) respectively.

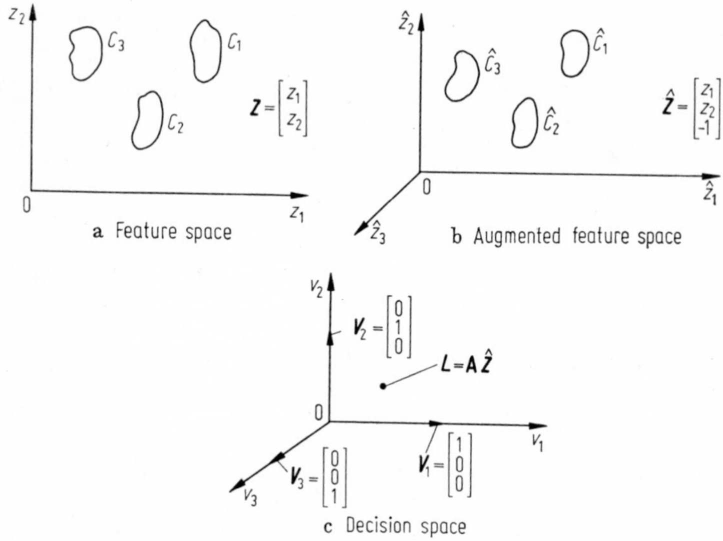

Fig. 10.13 Pertaining to the 3-class least-squares minimum-distance classifier (a) feature space; (b) augmented feature space; (c) decision space

We seek a transformation $\mathbf{A}$ which maps $\hat{\boldsymbol{Z}} \in C_k$, $k = 1, 2, 3$ into V_k, $k = 1, 2, 3$, in the least-squares sense. Now, we choose the V_i to be the vertices of the three *unit vectors* as illustrated in Fig. 10.13 (c). Again, let the a priori probabilities of C_k be denoted by P_k, $k = 1, 2, 3$. The transformation $\mathbf{A}$ follows readily from the discussion in Section 10.6. In particular, corresponding to Eq. (10.6-5) we have

$$\mathbf{A}\left[\frac{P_1}{N_1}\sum_{j=1}^{N_1}\hat{\boldsymbol{Z}}_{1j}\hat{\boldsymbol{Z}}'_{1j} + \frac{P_2}{N_2}\sum_{j=1}^{N_2}\hat{\boldsymbol{Z}}_{2j}\hat{\boldsymbol{Z}}'_{2j} + \frac{P_3}{N_3}\sum_{j=1}^{N_3}\hat{\boldsymbol{Z}}_{3j}\hat{\boldsymbol{Z}}'_{3j}\right]$$

$$= \frac{P_1}{N_1}\sum_{j=1}^{N_1}\boldsymbol{V}_1\hat{\boldsymbol{Z}}'_{1j} + \frac{P_2}{N_2}\sum_{j=1}^{N_2}\boldsymbol{V}_2\hat{\boldsymbol{Z}}'_{2j} + \frac{P_3}{N_3}\sum_{j=1}^{N_3}\boldsymbol{V}_3\hat{\boldsymbol{Z}}'_{3j} \qquad (10.8\text{-}1)$$

Solving for $\mathbf{A}$ we obtain

$$\mathbf{A} = \mathbf{S}_{v\hat{z}}\mathbf{S}_{\hat{z}\hat{z}}^{-1} \tag{10.8-2}$$

where

$$\mathbf{S}_{v\hat{z}} = \sum_{i=1}^{3}\sum_{j=1}^{N_i}\frac{P_i}{N_i}(\boldsymbol{V}_i\hat{\boldsymbol{Z}}'_{ij}) = \mathrm{E}\{\boldsymbol{V}\hat{\boldsymbol{Z}}'\}$$

and

$$\mathbf{S}_{\hat{z}\hat{z}} = \sum_{i=1}^{3}\sum_{j=1}^{N_i}\frac{P_i}{N_i}(\hat{\boldsymbol{Z}}_{ij}\hat{\boldsymbol{Z}}'_{ij}) = \mathrm{E}\{\hat{\boldsymbol{Z}}\hat{\boldsymbol{Z}}'\}$$

We observe that $\mathbf{S}_{v\hat{z}}$ and $\mathbf{S}_{\hat{z}\hat{z}}$ are crosscorrelation and autocorrelation matrices respectively [see Eq. (A8.1-4)]. From Eq. (2) it is evident that $\mathbf{A}$ is a $[3\times(d+1)]$ matrix.

To classify a pattern $\hat{\boldsymbol{Z}}' = [z_1\ z_2 \cdots z_d\ -1]$, the classifier first computes

$$\boldsymbol{L} = \mathbf{A}\hat{\boldsymbol{Z}}$$

and thereby maps $\hat{\boldsymbol{Z}}$ in the decision space [see Fig. 10.13]. Next, it uses the following minimum-distance decision rule:

If $\boldsymbol{L}$ is closest to $\boldsymbol{V}_{i_0}$, then $\boldsymbol{Z}$ is classified as belonging to C_{i_0}. Now, the distances the classifier must compute are given by

$$D_i^2 = ||\boldsymbol{L} - \boldsymbol{V}_i||^2, \qquad i = 1, 2, 3$$

That is

$$D_i^2 = ||\boldsymbol{L}||^2 - 2\boldsymbol{V}_i'\boldsymbol{L} + ||\boldsymbol{V}_i||^2, \qquad i = 1, 2, 3 \tag{10.8-3}$$

In Eq. (3), $||\boldsymbol{V}_i||^2 = 1$, $i = 1, 2, 3$ and hence it follows that D_i^2 is a *minimum* when $\boldsymbol{V}_i'\boldsymbol{L}$ is a *maximum*. Thus rather than computing D_i^2 in Eq. (3), it is sufficient for the classifier to compute

$$d_i = \boldsymbol{V}_i'\boldsymbol{L}, \qquad i = 1, 2, 3 \tag{10.8-4}$$

Substituting $\boldsymbol{V}_1' = [1\ 0\ 0]$, $\boldsymbol{V}_2' = [0\ 1\ 0]$, $\boldsymbol{V}_3' = [0\ 0\ 1]$, and $\boldsymbol{L} = \mathbf{A}\hat{\boldsymbol{Z}}$ in Eq. (4), we obtain

$$\begin{bmatrix} d_1 \\ d_2 \\ d_3 \end{bmatrix} = \begin{bmatrix} 1 & 0 & 0 \\ 0 & 1 & 0 \\ 0 & 0 & 1 \end{bmatrix} \mathbf{A}\hat{\boldsymbol{Z}} = \mathbf{A}\hat{\boldsymbol{Z}} \tag{10.8-5}$$

If the transformation matrix is denoted as

$$\mathbf{A} = \begin{bmatrix} a_{11} & a_{12} & \cdots & a_{1d} & \theta_1 \\ a_{21} & a_{22} & \cdots & a_{2d} & \theta_2 \\ a_{31} & a_{32} & \cdots & a_{3d} & \theta_3 \end{bmatrix}$$

then Eq. (5) implies that the d_i are the following discriminant functions which specify the classifier.

$$g_i(\boldsymbol{Z}) = d_i = a_{i1}z_1 + a_{i2}z_2 + \cdots + a_{id}z_d - \theta_i, \qquad i = 1, 2, 3 \tag{10.8-6}$$

From Eq. (6) it is clear that the classifier is completely defined by the transformation matrix **A**, which is obtained from the training set patterns. To classify a given pattern, the classifier computes three numbers $g_1(\boldsymbol{Z})$, $g_2(\boldsymbol{Z})$ and $g_3(\boldsymbol{Z})$ using Eq. (6). If $\max_i \{g_i(\boldsymbol{Z})\} = g_{i_0}(\boldsymbol{Z})$, then the pattern is assigned to C_{i_0}. This classifier is implemented as shown in Fig. 10.14. In what follows, we consider two numerical examples.

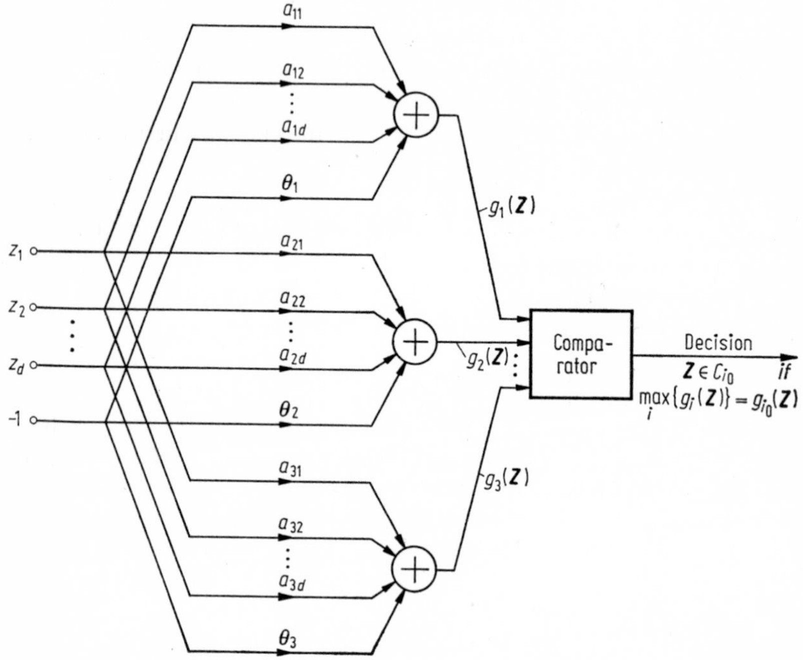

Fig. 10.14 Implementation of the 3-class least-squares minimum distance classifier

Example 10.8-1

Consider a 2-dimensional (i.e. $d = 2$) training set to be as follows:

$$C_1: \begin{bmatrix} 0 \\ 3 \end{bmatrix}, \begin{bmatrix} -1 \\ 3 \end{bmatrix}, \begin{bmatrix} -2 \\ 3 \end{bmatrix}, \begin{bmatrix} -3 \\ 3 \end{bmatrix}, \begin{bmatrix} -4 \\ 3 \end{bmatrix}$$

$$C_2: \begin{bmatrix} 5 \\ 5 \end{bmatrix}, \begin{bmatrix} 6 \\ 5 \end{bmatrix}, \begin{bmatrix} 6 \\ 6 \end{bmatrix}, \begin{bmatrix} 6 \\ 7 \end{bmatrix}, \begin{bmatrix} 7 \\ 5 \end{bmatrix}$$

$$C_3: \begin{bmatrix} 6 \\ -1 \end{bmatrix}, \begin{bmatrix} 7 \\ 0 \end{bmatrix}, \begin{bmatrix} 8 \\ 1 \end{bmatrix}, \begin{bmatrix} 9 \\ 1 \end{bmatrix}, \begin{bmatrix} 10 \\ 1 \end{bmatrix}$$

(a) Assuming that $P_1 = P_2 = P_3 = 1/3$, find the discriminant functions that specify the classifier.

(b) Using $g_i(Z)$, $i = 1, 2, 3$, find the equations to the decision boundaries and locate them in the feature space.

Solution: (a) **A** is computed using Eq. (2) as follows.

$$\mathbf{S}_{v\hat{z}} = \frac{1}{15}\left\{\begin{bmatrix}1\\0\\0\end{bmatrix}[0\ 3\ -1] + \begin{bmatrix}1\\0\\0\end{bmatrix}[-1\ 3\ -1] + \cdots + \begin{bmatrix}1\\0\\0\end{bmatrix}[-4\ 3\ -1]\right.$$
$$+ \begin{bmatrix}0\\1\\0\end{bmatrix}[5\ 5\ -1] + \begin{bmatrix}0\\1\\0\end{bmatrix}[6\ 5\ -1] + \cdots + \begin{bmatrix}0\\1\\0\end{bmatrix}[7\ 5\ -1]$$
$$\left. + \begin{bmatrix}0\\0\\1\end{bmatrix}[6\ -1\ -1] + \begin{bmatrix}0\\0\\1\end{bmatrix}[7\ 0\ -1] + \cdots + \begin{bmatrix}0\\0\\1\end{bmatrix}[10\ 1\ -1]\right\}$$

which yields

$$\mathbf{S}_{v\hat{z}} = \begin{bmatrix}-0.667 & 1.000 & -0.333\\ 2.000 & 1.866 & -0.333\\ 2.666 & 0.133 & -0.333\end{bmatrix} \tag{10.8-7}$$

$$\mathbf{S}_{\hat{z}\hat{z}} = \frac{1}{15}\left\{\begin{bmatrix}0\\3\\-1\end{bmatrix}[0\ 3\ -1] + \begin{bmatrix}-1\\3\\-1\end{bmatrix}[-1\ 3\ -1] + \cdots + \begin{bmatrix}-4\\3\\-1\end{bmatrix}[-4\ 3\ -1]\right.$$
$$+ \begin{bmatrix}5\\5\\-1\end{bmatrix}[5\ 5\ -1] + \begin{bmatrix}6\\5\\-1\end{bmatrix}[6\ 5\ -1] + \cdots + \begin{bmatrix}7\\5\\-1\end{bmatrix}[7\ 5\ -1]$$
$$\left. + \begin{bmatrix}6\\-1\\-1\end{bmatrix}[6\ -1\ -1] + \begin{bmatrix}7\\0\\-1\end{bmatrix}[7\ 0\ -1] + \cdots + \begin{bmatrix}10\\1\\-1\end{bmatrix}[10\ 1\ -1]\right\} \tag{10.8-8}$$

which yields

$$\mathbf{S}_{\hat{z}\hat{z}} = \begin{bmatrix}36.130 & 10.599 & -4.000\\ 10.599 & 13.932 & -3.000\\ -4.000 & -3.000 & 1.000\end{bmatrix}$$

Evaluation of the inverse of $\mathbf{S}_{\hat{z}\hat{z}}$ yields

$$\mathbf{S}_{\hat{z}\hat{z}}^{-1} = \begin{bmatrix} 0.051 & 0.014 & 0.246 \\ 0.014 & 0.207 & 0.678 \\ 0.246 & 0.678 & 4.017 \end{bmatrix} \tag{10.8-9}$$

Substituting Eqs. (7) and (9) in Eq. (2) we obtain

$$\mathbf{A} = \begin{bmatrix} -0.101 & -0.209 & -0.825 \\ 0.046 & 0.189 & 0.418 \\ 0.055 & -0.160 & -0.593 \end{bmatrix} \tag{10.8-10}$$

Thus the discriminant functions which specify the classifier are given by [see Eq. (6)]

$$g_1(\boldsymbol{Z}) = -0.101 z_1 - 0.029 z_2 + 0.825$$

$$g_2(\boldsymbol{Z}) = 0.046 z_1 + 0.189 z_2 - 0.418$$

and

$$g_3(\boldsymbol{Z}) = 0.055 z_1 - 0.160 z_2 + 0.593 \tag{10.8-11}$$

Figure 10.15 shows the related implementation.

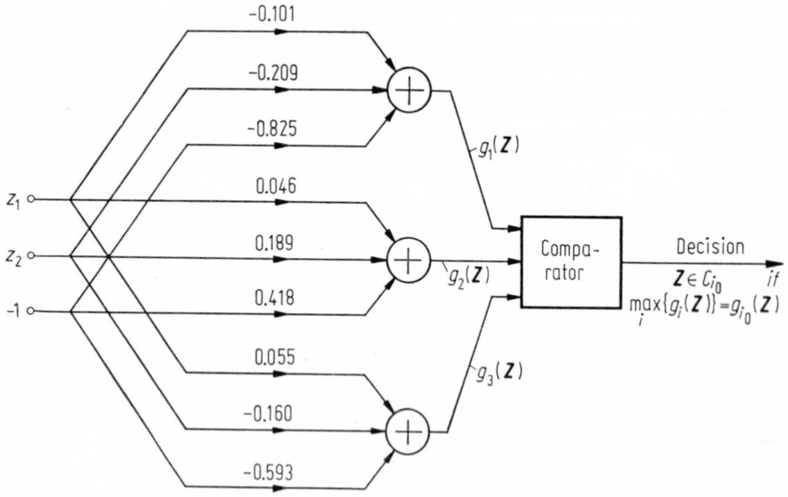

Fig. 10.15 Implementation of classifier pertaining to Example 10.8-1

(b) The equations to the boundaries b_{12}, b_{23}, and b_{31} which respectively separate (C_1, C_2), (C_2, C_3), and (C_3, C_1) are obtained from the $g_i(\boldsymbol{Z})$ as fellows:

$$b_{12}:\ g_1(\mathbf{Z}) = g_2(\mathbf{Z})$$
$$b_{23}:\ g_2(\mathbf{Z}) = g_3(\mathbf{Z})$$
$$b_{31}:\ g_3(\mathbf{Z}) = g_1(\mathbf{Z}) \quad (10.8\text{-}12)$$

Evaluation of Eq. (12) yields the decision boundaries shown in Fig. 10.16.

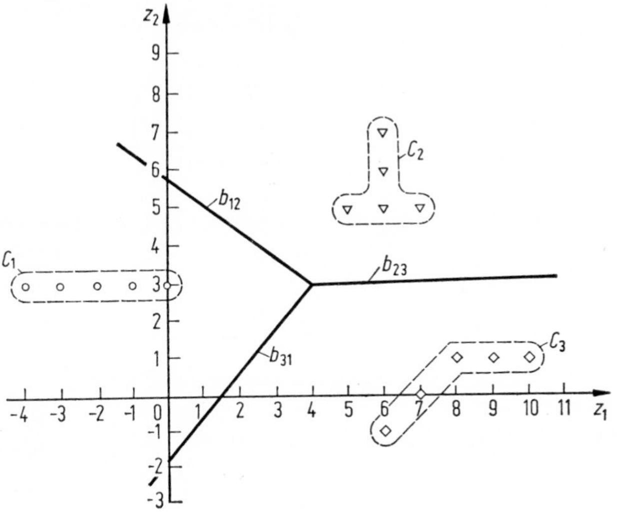

Fig. 10.16 Decision boundaries for the classifier in Example 10.8-1

Example 10.8-2

Consider the 2-dimensional training set shown in Fig. 10.17[1].

(a) Find $g_i(Z)$, $i = 1, 2, 3$ assuming that $P_1 = P_2 = P_3 = 1/3$.

(b) Use the classifier to classify the patterns in the training set.

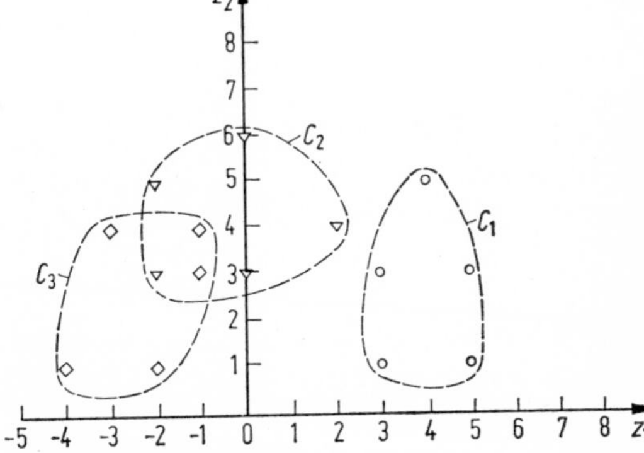

Fig. 10.17
Training set pertaining to Example 10.8-2

[1] The symbols ○, ▽ and ◇ denote patterns belonging to C_1, C_2 and C_3 respectively.

Solution: (a) Details pertaining to the various computations are the same as those presented in Example 10.8-1. It can be shown that

$$\mathbf{A} = \begin{bmatrix} 0.141 & -0.063 & -0.465 \\ -0.032 & 0.147 & 0.111 \\ -0.110 & -0.084 & -0.647 \end{bmatrix}$$

Hence the discriminant functions are given by

$$g_1(\boldsymbol{Z}) = \quad 0.141z_1 - 0.063z_2 + 0.465$$

$$g_2(\boldsymbol{Z}) = -0.032z_1 + 0.147z_2 - 0.111$$

and

$$g_3(\boldsymbol{Z}) = -0.110z_1 - 0.084z_2 + 0.647 \qquad (10.8\text{-}13)$$

(b) We classify the training set patterns by computing $g_i(\boldsymbol{Z})$, $i = 1, 2, 3$ for each $\boldsymbol{Z}$ belonging to the set. Then $\boldsymbol{Z}$ is assigned to C_{i_0} if $\max\{g_i(\boldsymbol{Z})\} = g_{i_0}(\boldsymbol{Z})$. These classification results are generally summarized in the form of the following *confusion matrix* **F**:

$$\mathbf{F} = \begin{bmatrix} 5 & 0 & 0 \\ 1 & 2 & 2 \\ 0 & 1 & 4 \end{bmatrix} \qquad (10.8\text{-}14)$$

In Eq. (14), non-zero off diagonal terms reflect errors incurred by the classifier; i.e. if the element in the k-th row and l-th column of **F** is denoted by f_{kl}, $k \neq l$, and $f_{kl} = s$, then s patterns belonging to C_k have been erroneously classified as belonging to C_l.

Comments: It is instructive to justify the errors reflected by **F** in Eq. (14). We do so by plotting the decision boundaries implemented by the classifier. The procedure is omitted since it parallels that in Example 10.8-1. Figure 10.18 shows the decision boundaries b_{12}, b_{23}, and b_{31}. Examining Fig. 10.18, we observe that two triangles are erroneously classified as belonging to C_3, while one triangle is erroneously classified as belonging to C_1. This accounts for the second row of **F**. Again, we observe that one square is incorrectly classified as belonging to C_2, which explains the third row of **F**. Finally all patterns of C_1 have been classified correctly, which agrees with the first row of **F**.

In conclusion we note that any classifier that implements *linear* boundaries is bound to make errors as far as the training set in Fig. 10.18 is concerned, since C_2 and C_3 are *not linearly separable*.

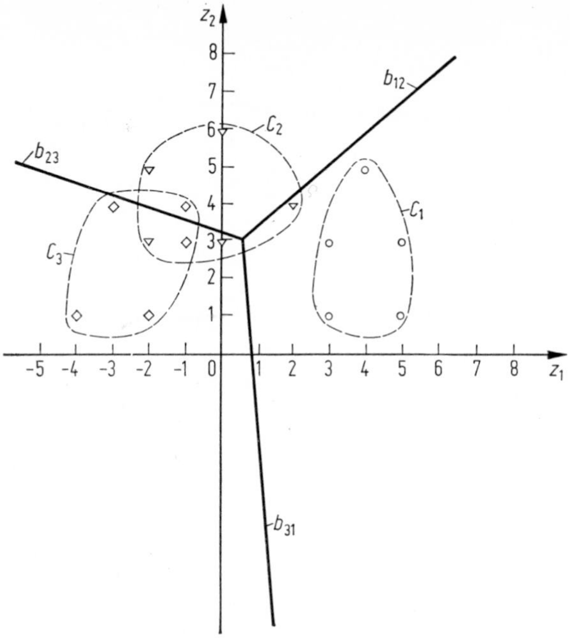

Fig. 10.18 Decision boundaries associated with the classifier in Example 10.8-2

10.9 *K*-class Least-Squares Minimum Distance Classifier [9–11]

An extension of the 3-class development to the K-class case is straight forward. The augmented patterns of C_i are mapped into the vertices of the unit K-vector $\boldsymbol{V}_i$,

$$\boldsymbol{V}_i' = [0 \ \cdots \ 0 \ 1 \ 0 \ \cdots \ 0] \tag{10.9-1}$$

where the nonzero element "1" is in the i-th row of $\boldsymbol{V}_i$, $i = 1, 2, \ldots, K$. In place of the $(3 \times (d+1))$ transformation matrix $\mathbf{A}$ in Eq. (10.8-2), we obtain the $(K \times (d+1))$ matrix

$$\mathbf{A} = \mathbf{S}_{v\hat{z}} \mathbf{S}_{\hat{z}\hat{z}}^{-1} \tag{10.9-2}$$

where

$$\mathbf{S}_{v\hat{z}} = \sum_{i=1}^{K} \sum_{j=1}^{N_i} \frac{P_i}{N_i} (\boldsymbol{V}_i \hat{\boldsymbol{Z}}_{ij}') = \mathrm{E}\{\boldsymbol{V} \hat{\boldsymbol{Z}}'\}$$

and

$$\mathbf{S}_{\hat{z}\hat{z}} = \sum_{i=1}^{K} \sum_{j=1}^{N_i} \frac{P_i}{N_i} (\hat{\boldsymbol{Z}}_{ij} \hat{\boldsymbol{Z}}_{ij}') = \mathrm{E}\{\hat{\boldsymbol{Z}} \hat{\boldsymbol{Z}}'\}$$

Thus the discriminant functions that specify the classifier are

$$g_i(\mathbf{Z}) = a_{i1}z_1 + a_{i2}z_2 + \cdots + a_{id}z_d - \theta_i, \qquad i = 1, 2, \ldots, K \quad (10.9\text{-}3)$$

The implementation of the above classifier is shown in Fig. 10.19.

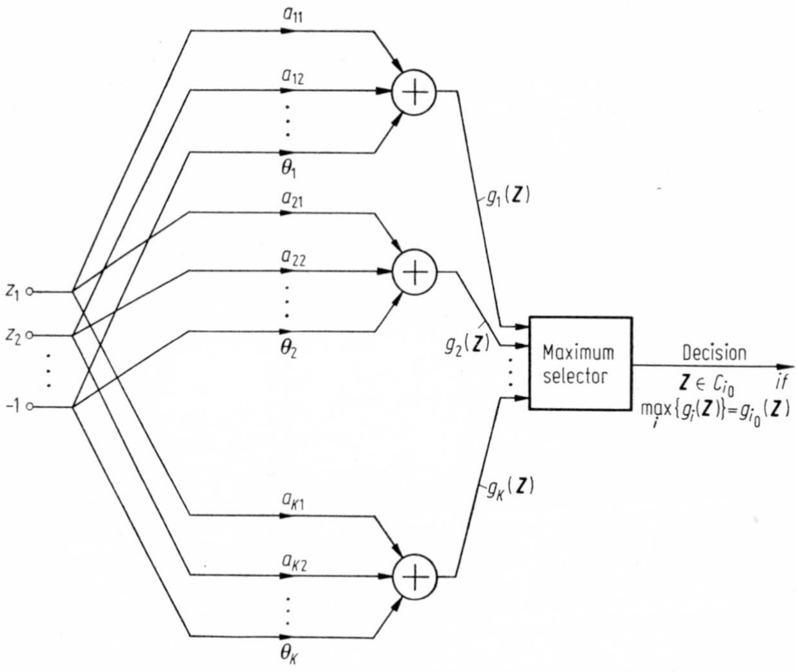

Fig. 10.19 Implementation of a K-class least-squares minimum distance classifier

Example 10.9-1

Given the following training set patterns:

$$C_1: \begin{bmatrix}2.5\\5.5\end{bmatrix}, \begin{bmatrix}2\\6.5\end{bmatrix}, \begin{bmatrix}3.5\\5\end{bmatrix}, \begin{bmatrix}3\\6.3\end{bmatrix}, \begin{bmatrix}2.5\\7.5\end{bmatrix}, \begin{bmatrix}3.5\\7\end{bmatrix}, \begin{bmatrix}4\\8\end{bmatrix}, \begin{bmatrix}4.5\\7\end{bmatrix}, \begin{bmatrix}5\\6\end{bmatrix}, \begin{bmatrix}6\\7\end{bmatrix}$$

$$C_2: \begin{bmatrix}0.5\\1\end{bmatrix}, \begin{bmatrix}0.5\\0.5\end{bmatrix}, \begin{bmatrix}0.5\\2\end{bmatrix}, \begin{bmatrix}0.5\\3\end{bmatrix}, \begin{bmatrix}1\\1\end{bmatrix}, \begin{bmatrix}1.5\\0.5\end{bmatrix}, \begin{bmatrix}1.5\\1.5\end{bmatrix}, \begin{bmatrix}1.5\\2.5\end{bmatrix}, \begin{bmatrix}2.5\\1\end{bmatrix}, \begin{bmatrix}2.5\\2\end{bmatrix}$$

(a) Find the discriminant functions $g_1(\mathbf{Z})$ and $g_2(\mathbf{Z})$ and hence show how the classifier is implemented. Use $P_1 = P_2 = 1/2$.

(b) Plot the above training set and locate the decision boundary in the atfeure space.

(c) Plot the positions of the training set patterns in the decision space.

Solution: (a) Since $K = 2$, $V_1' = [1\ 0]$ and $V_2' = [0\ 1]$. Using Eq. (2), $\mathbf{A}$ is evaluated. The computations are as follows:

$$\mathbf{S}_{\hat{z}\hat{z}} = \frac{1}{20}\left\{\begin{bmatrix}2.5\\5.5\\-1\end{bmatrix}[2.5\ 5.5\ -1] + \begin{bmatrix}2\\6.5\\-1\end{bmatrix}[2\ 6.5\ -1] + \cdots + \begin{bmatrix}6\\7\\-1\end{bmatrix}[6\ 7\ -1]\right.$$

$$\left. + \begin{bmatrix}0.5\\1\\-1\end{bmatrix}[0.5\ 1\ -1] + \begin{bmatrix}0.5\\0.5\\-1\end{bmatrix}[0.5\ 0.5\ -1] + \cdots + \begin{bmatrix}2.5\\2\\-1\end{bmatrix}[2.5\ 2\ -1]\right\}$$

That is

$$\mathbf{S}_{\hat{z}\hat{z}} = \begin{bmatrix} 8.425 & 13.02 & -2.450 \\ 13.020 & 23.472 & -4.040 \\ -2.450 & -4.040 & 1.000 \end{bmatrix}$$

$$\mathbf{S}_{v\hat{z}} = \frac{1}{20}\left\{\begin{bmatrix}1\\0\end{bmatrix}[2.5\ 5.5\ -1] + \begin{bmatrix}1\\0\end{bmatrix}[2\ 6.5\ -1] + \cdots + \begin{bmatrix}1\\0\end{bmatrix}[6\ 7\ -1]\right.$$

$$\left. + \begin{bmatrix}0\\1\end{bmatrix}[0.5\ 1\ -1] + \begin{bmatrix}0\\1\end{bmatrix}[0.5\ 0.5\ -1] + \cdots + \begin{bmatrix}0\\1\end{bmatrix}[2.5\ 2\ -1]\right\}$$

which yields

$$\mathbf{S}_{v\hat{z}} = \begin{bmatrix} 1.825 & 3.290 & -0.500 \\ 0.625 & 0.750 & -0.500 \end{bmatrix}$$

Evaluating $\mathbf{A} = \mathbf{S}_{v\hat{z}}\mathbf{S}_{\hat{z}\hat{z}}^{-1}$ we obtain

$$\mathbf{A} = \begin{bmatrix} 0.043 & 0.159 & 0.247 \\ -0.043 & -0.159 & -1.247 \end{bmatrix} \tag{10.9-4}$$

From Eq. (4) we have

$$g_1(\boldsymbol{Z}) = 0.043z_1 + 0.159z_2 - 0.247$$

$$g_2(\boldsymbol{Z}) = -0.043z_1 - 0.159z_2 + 1.247$$

(b) The equation to the decision boundary b_{12} is given by

$$g_1(\boldsymbol{Z}) = g_2(\boldsymbol{Z})$$

which yields

$$0.086z_1 + 0.318z_2 = 1.494$$

This boundary is located in the feature space shown in Fig. 10.20(a).

(c) The locations of the training set patterns in the decision space are obtained as

$$C_1:\ \mathbf{L}_{1j} = \mathbf{A}\hat{\mathbf{Z}}_{1j}$$
$$C_2:\ \mathbf{L}_{2j} = \mathbf{A}\hat{\mathbf{Z}}_{2j}, \qquad j = 1, 2, \ldots, 10 \qquad (10.9\text{-}5)$$

The results of the above computations are plotted in Fig. 10.20(b).

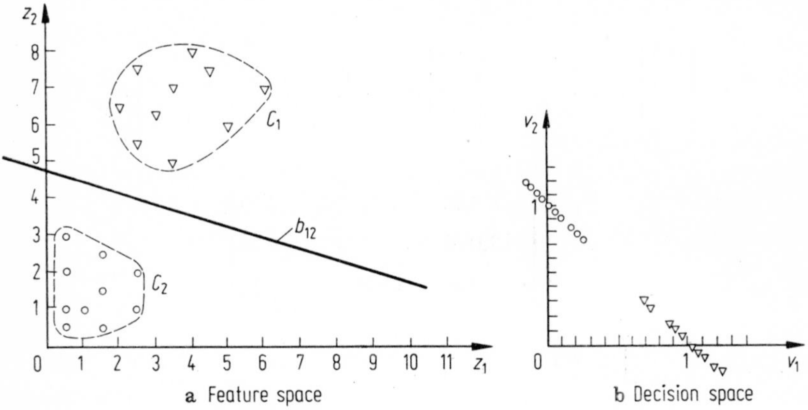

Fig. 10.20 Pertaining to Example 10.9-1
(a) Feature space; (b) Decision space

10.10 Quadratic Classifiers [2]

In the earlier discussions related to minimum distance classifiers, we restricted our attention to linear classifiers; i.e., classifiers which implement linear decision boundaries in the feature space. The corresponding training procedure can easily be extended to implement quadratic boundaries. The price one pays is that the implementation of the classifier is more involved. A quadratic classifier is specified by the discriminant function

$$g_i(\mathbf{Z}) = \underbrace{\sum_{j=1}^{d} w_{jj} z_j^2 + \sum_{j=1}^{d-1} \sum_{k=j+1}^{d} w_{jk} z_j z_k}_{\text{quadratic terms}} + \underbrace{\sum_{j=1}^{d} w_j z_j}_{\text{linear terms}} - \theta_i, \qquad i = 1, 2, \ldots, K \quad (10.10\text{-}1)$$

From Eq. (1) it follows that a quadratic discriminant function $g_i(\mathbf{Z})$ has $[(d+1)(d+2)]/2$ weights or parameters consisting of

d parameters as coefficients of z_j^2 terms	w_{jj}
d parameters as coefficients of z_j terms	w_j
$d(d-1)/2$ parameters as coefficients of $z_j z_k$ terms, $j \neq k$	w_{jk}
a threshold which is not a coefficient	θ_i.

The parameters listed above are obtained from the training process.

To explain the implementation associated with Eq. (1) we first define the Q-dimensional vector $\boldsymbol{G}$, whose components $f_1, f_2, \ldots, f_Q$ are functions of the z_i, $i = 1, 2, \ldots, d$. The first d components of $\boldsymbol{G}$ are $z_1^2, z_2^2, \ldots, z_d^2$; the next $[d(d-1)]/2$ components are all pairs $z_1z_2, z_1z_3, \ldots, z_{d-1}z_d$; the last d components are $z_1, z_2, \ldots, z_d$. The total number of these components are $Q = [d(d+3)]/2$. We denote this correspondence as

$$\boldsymbol{G} = \boldsymbol{G}(\boldsymbol{Z})$$

where $\boldsymbol{G}(\boldsymbol{Z})$ is a one-to-one transformation. Thus, for every pattern $\boldsymbol{Z}' = [z_1\, z_2 \cdots z_d]$ in a d-dimensional space, there is a unique vector $\boldsymbol{G}' = [f_1\, f_2 \cdots f_Q]$ in an Q-dimensional space. This one-to-one correspondence enables us to write $g_i(\boldsymbol{Z})$ as a linear function of the components of $\boldsymbol{G}$ with the result that for every *quadratic* discriminant function of $\boldsymbol{Z}$, there is a corresponding *linear* discriminant function of $\boldsymbol{G}$. Thus Eq. (1) can be written as

$$g_i(\boldsymbol{Z}) = \tilde{w}_1 f_1 + \tilde{w}_2 f_2 + \cdots + \tilde{w}_Q f_Q - \theta_i \tag{10.10-2}$$

The implementation of the quadratic classifier corresponding to $g_i(\boldsymbol{Z})$ in Eq. (2) is shown in Fig. 10.21.

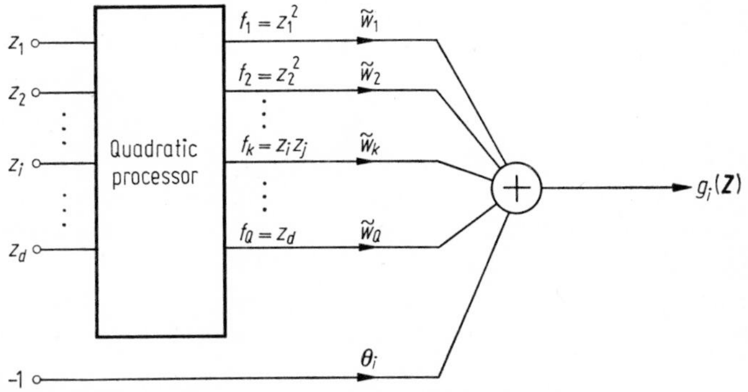

Fig. 10.21 Implementation of a quadratic classifier

10.11 An ECG Classification Experiment

As a second example for illustrating the use of the variance criterion/distribution to achieve feature selection, an ECG classification experiment is considered [12]. This problem concerns the design of a pattern recognition system that is capable of determining whether a given ECG is normal or abnormal. Our study is restricted to ECG data of the canine. Details pertaining to the canine ECG data acquisition were given in Section 9.3.

Feature selection process. The selection process is based on the variance criterion. A 2-lead ECG was used, and 150 normal and 150 abnormal signals were studied. After some preliminary investigations it was decided that the most effective classification results are obtained when information from both leads is used. For each digital ECG, 64 samples from lead 1 were concatenated with the corresponding 64 samples from lead 2, to give a 128 sample representation. The overall data domain covariance matrix was calculated as

$$\boldsymbol{\Sigma}_x^* = \frac{1}{2}\,[\boldsymbol{\Sigma}_{x_1} + \boldsymbol{\Sigma}_{x_2}] \tag{10.11-1}$$

where $\boldsymbol{\Sigma}_{x_1}$ and $\boldsymbol{\Sigma}_{x_2}$ are the covariance matrices of the normal and abnormal classes of ECG's respectively. The variance distributions computed using $\boldsymbol{\Sigma}_x^*$ for the KLT, DCT, $(\mathrm{WHT})_h$, and IT are shown in Fig. 10.22. From these distributions it is apparent that the ranking of the most effective to the least effective, for dimensionality reduction in the feature selection process is as follows:

$$\mathrm{KLT} > \mathrm{DCT} > (\mathrm{WHT})_h > \mathrm{IT}$$

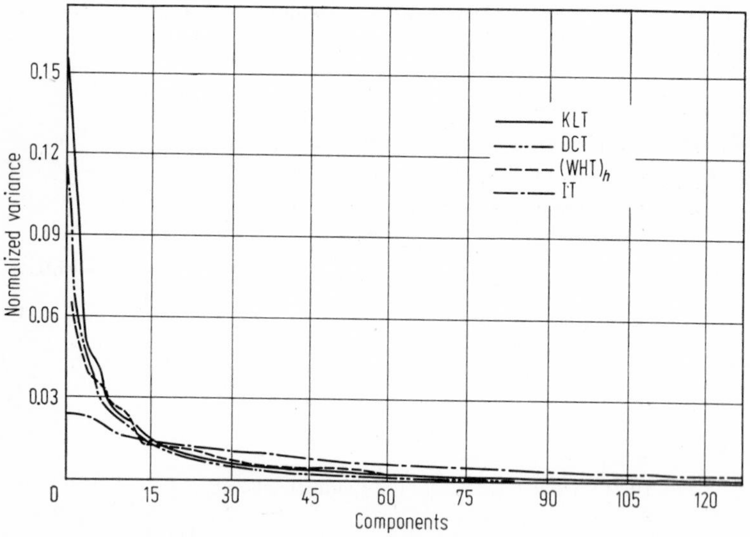

Fig. 10.22 Variance distribution for the IT, KLT, DCT, and $(\mathrm{WHT})_h$

Again, it is observed that beyond 30 components, the variance distributions of the KLT, DCT and $(\mathrm{WHT})_h$ compare very closely. Since there is no fast computational algorithm for the KLT, the DCT, $(\mathrm{WHT})_h$, and IT were chosen for feature selection. In accordance with the variance criterion, the components with the M largest variances were selected as features for classification.

Classification considerations. A 2-class least-squares minimum-distance classifier was used. The classifier was trained using 150 normal and 150 abnormal ECGs. The various steps involved in the feature selection and training processes are summarized in Fig. 10.23.

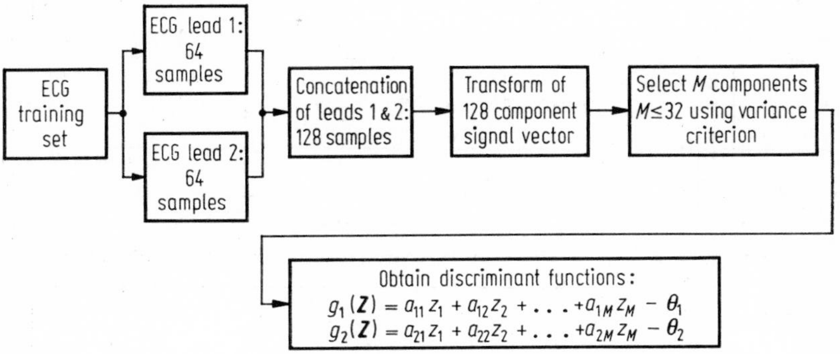

Fig. 10.23 Summary of feature selection and training processes

The performances of the various transforms were compared on the basis of their success in classifying the 300 training set patterns. Figure 10.24 shows the results. The DCT performs most effectively at 23 components with 87 % correct classification, followed by the $(\mathrm{WHT})_h$ and IT.

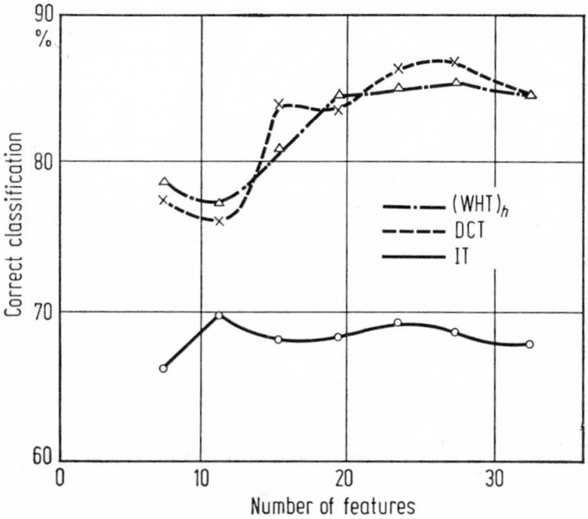

Fig. 10.24 Correct classification versus features retained for the IT, DCT, and $(\mathrm{WHT})_h$

From Fig. **10.24** it is evident that the DCT and $(\mathrm{WHT})_h$ perform appreciably better than the IT. A block diagram of the above ECG pattern recognition system is shown in Fig. **10.25**.

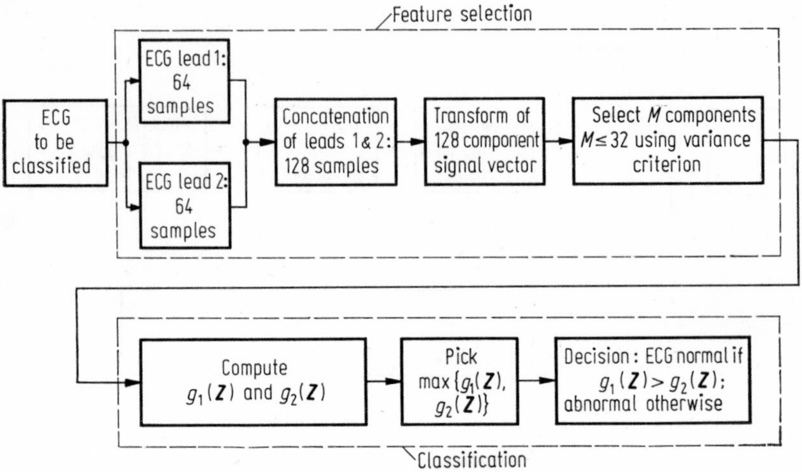

Fig. 10.25 Block diagram for ECG pattern recognition system

Comment: In screening ECG's, the more serious type of error is that of classifying an abnormal ECG as normal. Figure **10.26** shows a plot of this type of error as a function of the number of features used. This type of error can be effectively minimized by using a *piecewise linear* classifier[1],

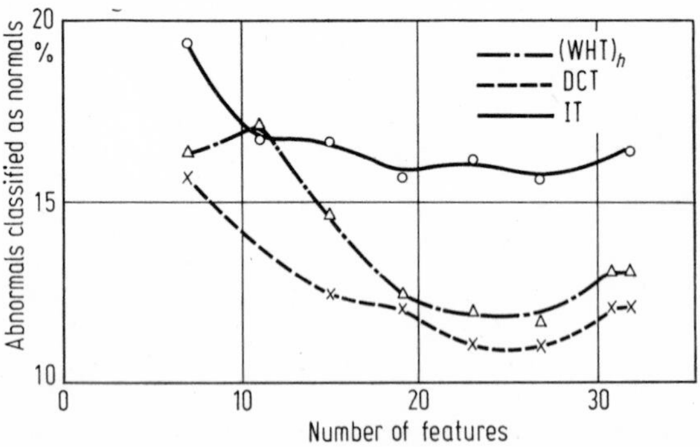

Fig. 10.26 Percentage of abnormals classified as normals versus number of features retained for IT, DCT, and $(\mathrm{WHT})_h$

[1] A piecewise linear classifier is one which implements piecewise linear boundaries in the feature space.

rather than a linear classifier. The feature selection process however, remains unchanged. Details pertaining to this aspect of the ECG classification experiment are discussed in [12].

10.12 Summary

In this chapter, some simple classification algorithms and their implementations were studied. The motivation for doing so was to use these algorithms to demonstrate that effective dimensionality reduction in feature selection can be achieved via the variance criterion/distribution. Two pattern recognition experiments were discussed. The results showed that orthogonal transform processing enables substantial dimensionality reduction with only a small decrease in percent correct classification.

References

1. Sebestyen, G. S.: *Decision-Making Processes in Pattern Recognition.* New York, London: Macmillan, 1962.
2. Nilsson, N. J.: *Learning Machines.* New York, London: McGraw-Hill, 1965.
3. Fu, K. S.: *Sequential Methods in Pattern Recognition and Machine Learning.* New York, London: Academic Press, 1968.
4. Watanabe, S.: *Methodologies in Pattern Recognition.* New York, London: Academic Press, 1969.
5. Mendel, J. M., and Fu, K. S. (editors): *Adaptive, Learning, and Pattern Recognition Systems.* New York, London: Academic Press, 1970.
6. Fukunaga, K.: *Introduction to Statistical Pattern Recognition.* New York, London: Academic Press, 1972.
7. Andrews, H. C.: *Introduction to Mathematical Techniques in Pattern Recognition.* New York, London: Wiley-Interscience, 1972.
8. Andrews, H. C.: Multidimensional Rotations in Feature Selection. *IEEE Trans. Computers* C-20 (1971) 1045–1051.
9. Chaplin, W. G., and Levadi, V. S.: A Generalization of the Linear Threshold Decision Algorithm to N Dimensions. Second Symposium on Computers and Information Sciences, Columbus, Ohio, August 1968.
10. Zagalsky, N. R.: A New Formulation of a Classification Procedure. Thesis for the Degree of Master of Science, University of Minnesota, Minneapolis, Minn., 1968.
11. Wee, W. G.: Generalized Inverse Approach to Adaptive Multiclass Pattern Classification. *IEEE Trans. Computers* C-17 (1968) 1157–1164.
12. Milne, P. J.: Orthogonal Transform Processing of Electrocardiograph Data. Ph.D. Dissertation, Kansas State University, Manhattan, Ks., 1973.

Problems

10-1 (a) The mean patterns corresponding to a set of 3-dimensional training patterns are

$$\bar{Z}_1' = [2 \ 3 \ 4], \quad \text{and} \quad \bar{Z}_2' = [1 \ 2 \ 3]$$

Find the equation to the plane which is the perpendicular bisector of the line joining $\bar{Z}_1$ and $\bar{Z}_2$.

Answer: $z_1 + z_2 + z_3 = 7.5$.

(b) A certain 3-class problem yields the following discriminant functions:

$$g_1(\boldsymbol{Z}) = 2z_1 + 3z_2 + 3$$

$$g_2(\boldsymbol{Z}) = z_1 + 2z_2 + 5$$

$$g_3(\boldsymbol{Z}) = -z_1 + 7z_2 + 2$$

Find the equations to the decision boundaries b_{12}, b_{23}, b_{31}, and their point of intersection.

Hint: The point of intersection is obtained by solving the equantions of any two of the three decision boundaries.

10-2 Let $\boldsymbol{X}_1$ and $\boldsymbol{X}_2$ be two points in a d-dimensional space. Fig. P10-2-1 shows these points when $d = 2$.

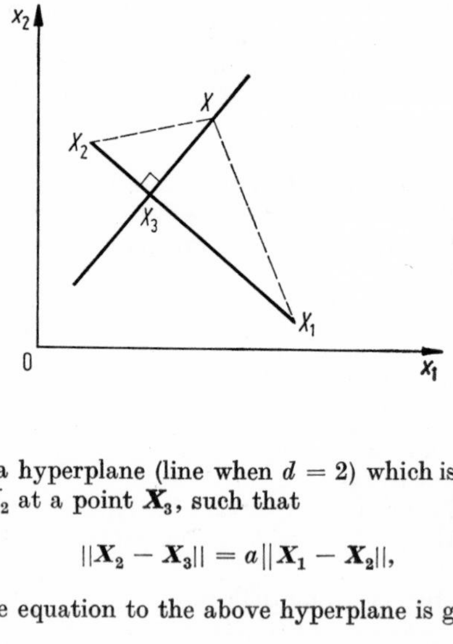

Fig. P10-2-1

Consider a hyperplane (line when $d = 2$) which is perpendicular to the line joining $\boldsymbol{X}_1$ and $\boldsymbol{X}_2$ at a point $\boldsymbol{X}_3$, such that

$$||\boldsymbol{X}_2 - \boldsymbol{X}_3|| = a||\boldsymbol{X}_1 - \boldsymbol{X}_2||, \qquad 0 < a < 1$$

Show that the equation to the above hyperplane is given by

$$\boldsymbol{X}'(\boldsymbol{X}_2 - \boldsymbol{X}_1) + a||\boldsymbol{X}_1||^2 + (a - 1)||\boldsymbol{X}_2||^2 = (2a - 1)\,\boldsymbol{X}_1'\boldsymbol{X}_2$$

Hint: Choose a point $\boldsymbol{X}$ on the hyperplane (see Fig. P10-2-1) and note that

$$||\boldsymbol{X} - \boldsymbol{X}_2||^2 = ||\boldsymbol{X} - \boldsymbol{X}_3||^2 + ||\boldsymbol{X}_2 - \boldsymbol{X}_3||^2$$

and

$$||\boldsymbol{X} - \boldsymbol{X}_1||^2 = ||\boldsymbol{X} - \boldsymbol{X}_3||^2 + ||\boldsymbol{X}_1 - \boldsymbol{X}_3||^2$$

10-3 Consider $\mathbf{A}\boldsymbol{X} = \boldsymbol{Y}$, where $\mathbf{A}$ is a $(m \times n)$ matrix, $\boldsymbol{X}$ is a $(n \times 1)$ vector, and $\boldsymbol{Y}$ is a $(m \times 1)$ vector. Show that

$$\nabla_{\boldsymbol{X}} \|\mathbf{A}\boldsymbol{X} - \boldsymbol{Y}\|^2 = 0$$

yields

$$\boldsymbol{X} = [\mathbf{A}'\mathbf{A}]^{-1}\mathbf{A}'\boldsymbol{Y} \qquad \text{(P10-3-1)}$$

where $[\mathbf{A}'\mathbf{A}]$ is assumed to be non-singular. In matrix theory, $[\mathbf{A}'\mathbf{A}]^{-1}$ is referred to as the "generalized" or "pseudo" inverse of the matrix $\mathbf{A}$.

Hint: Use the following identities.

$$\nabla_{\boldsymbol{X}}[\boldsymbol{X}'\mathbf{A}'\mathbf{A}\boldsymbol{X}] = 2(\mathbf{A}'\mathbf{A})\boldsymbol{X}$$

$$\nabla_{\boldsymbol{X}}[\boldsymbol{X}'\mathbf{A}'\boldsymbol{Y}] = \nabla_{\boldsymbol{X}}[\boldsymbol{Y}'\mathbf{A}\boldsymbol{X}] = \mathbf{A}'\boldsymbol{Y}$$

$$\nabla_{\boldsymbol{X}}[\|\boldsymbol{Y}\|^2] = 0$$

10-4 Consider the matrix equation

$$\begin{bmatrix} 1 & 1 \\ 1 & 0 \\ 1 & 2 \end{bmatrix} \begin{bmatrix} x_1 \\ x_2 \end{bmatrix} = \begin{bmatrix} 2 \\ 1 \\ 2 \end{bmatrix} \qquad \text{(P10-4-1)}$$

The three linear simultaneous equations corresponding to the above matrix equation are used to locate the lines l_1, l_2, and l_3 shown in Fig. P10-4-1.

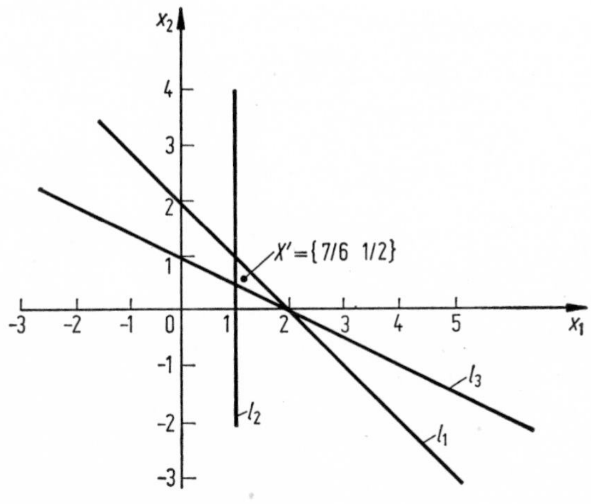

Fig. P10-4-1

In Fig. P10-4-1 we note that the three lines l_i, $i = 1, 2, 3$ *do not* intersect at a point, which implies that no unique solution exists. In other words, Eq. (P10-4-1) constitutes a set of *inconsistent* equations.

Use Eq. (P10-3-1) to show that the least-squares solution to the set of inconsistent equations represented by Eq. (P10-4-1) is given by $\boldsymbol{X}' = [7/6 \;\; 1/2]$.

10-5 Let $\mathscr{J}$ represent a digital device whose response to an excitation x_i is denoted by y_i as shown in Fig. P10-5-1.

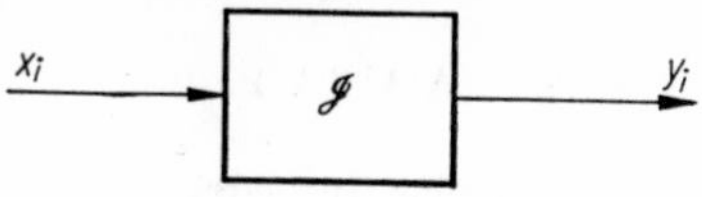

Fig. P10-5-1

Consider four pairs of excitation-response points (x_i, y_i), $i = 1, 2, 3, 4$ as shown in Fig. P10-5-2.

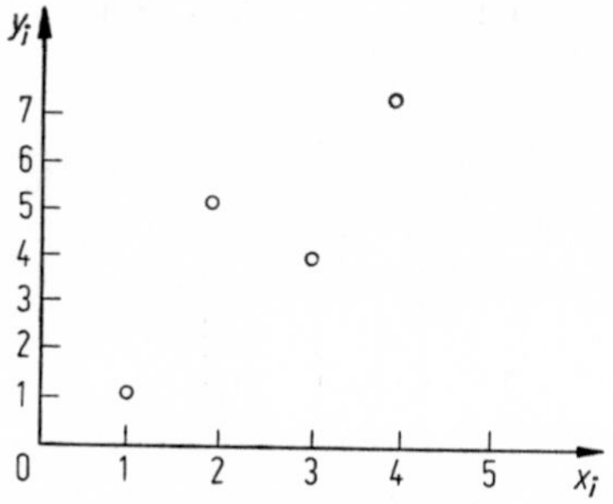

Fig. P10-5-2

Compute the values of a and b which define the straight line $y = a + bx$ such that the mean-square error of the approximation is minimized. Locate this straight line approximation in Fig. P10-5-2.

Hint: The mean-square error of the straight line approximation is given by

$$\overline{\varepsilon^2} = \frac{1}{4} \sum_{i=1}^{4} [y_i - (ax_i + b)]$$

where a and b are evaluated by solving the linear simultaneous equations which result from

$$\frac{\partial \overline{\varepsilon^2}}{\partial a} = 0 \quad \text{and} \quad \frac{\partial \overline{\varepsilon^2}}{\partial b} = 0$$

Answer: $y = 1.7x$.

10-6 Consider the 2-dimensional patterns given in Fig. P10-6-1. Write a computer program which implements a 3-class least-squares minimum distance classifier, and hence show that:

(a) The confusion matrix obtained by classifying the training set is

$$\mathbf{F} = \begin{bmatrix} 16 & 0 & 0 \\ 0 & 15 & 1 \\ 0 & 2 & 14 \end{bmatrix}$$

(b) The decision boundaries implemented by the classifier are as shown in Fig. P10-6-1.

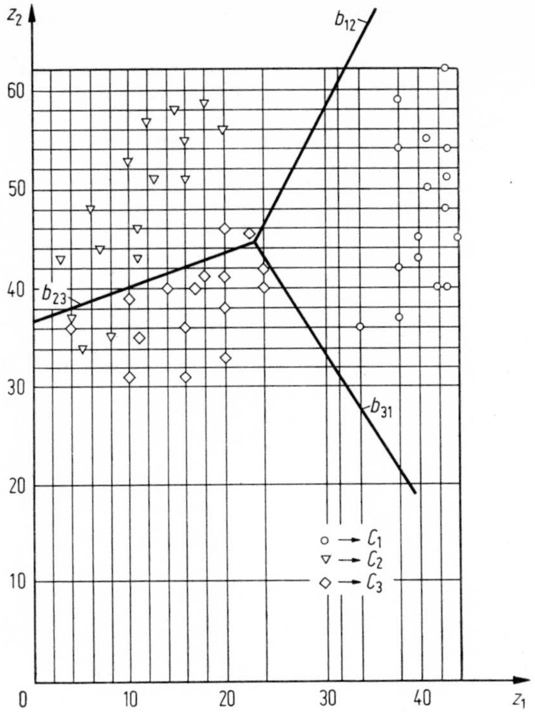

Fig. P10-6-1

Table P10-7-1

Class #	z_1	z_2	Class #	z_1	z_2	Class #	z_1	z_2
1	44	61	2	19	59	3	20	46
1	40	59	2	13	58	3	22	45
1	41	55	2	11	56	3	18	42
1	39	53	2	20	56	3	24	42
1	44	53	2	16	55	3	14	41
1	43	50	2	10	53	3	20	41
1	41	50	2	15	50	3	17	40
1	43	46	2	12	50	3	20	38
1	40	45	2	6	48	3	24	38
1	44	44	2	11	46	3	9	39
1	40	42	2	8	44	3	4	36
1	39	40	2	11	43	3	15	36
1	42	40	2	3	41	3	10	35
1	36	36	2	5	37	3	19	34
1	43	36	2	8	36	3	10	31
1	39	33	2	5	34	3	15	31

10-7 Consider the 2-dimensional patterns listed in Table P10-7-1. For each pattern $\boldsymbol{Z}' = [z_1 \; z_2 \; -1]$, evaluate the corresponding quadratic pattern $\boldsymbol{G}' = [z_1^2, z_2^2, z_1 z_2, z_1, z_2, -1]$. Make appropiate changes in the computer program written for Prob. 10-6 and then use it to show that the confusion matrix associated with a least-squares minimum distance quadratic classifier for classifying the training set is

$$\mathbf{F} = \begin{bmatrix} 16 & 0 & 0 \\ 0 & 15 & 1 \\ 0 & 1 & 15 \end{bmatrix}$$

Author Index

Subject Index

721/9/75